THE LATER HUSSERL AND THE IDEA OF PHENOMENOLOGY

ANALECTA HUSSERLIANA

THE YEARBOOK OF PHENOMENOLOGICAL RESEARCH

VOLUME II

Editor:

ANNA-TERESA TYMIENIECKA

THE LATER HUSSERL AND THE IDEA OF PHENOMENOLOGY

Idealism-Realism, Historicity and Nature

PAPERS AND DEBATE OF THE INTERNATIONAL
PHENOMENOLOGICAL CONFERENCE
HELD AT THE UNIVERSITY OF WATERLOO, CANADA,
APRIL 9–14, 1969

Edited by

ANNA-TERESA TYMIENIECKA

D. REIDEL PUBLISHING COMPANY / DORDRECHT-HOLLAND

Library of Congress Catalog Card Number 78–25369

ISBN 90 277 0223 3

Printed in The Netherlands by D. Reidel, Dordrecht.

TABLE OF CONTENTS

PART II / PHENOMENOLOGY AND HERMENEUTICS

PART III / PHENOMENOLOGY AND NATURAL SCIENCE

COMPLEMENTARY ESSAYS

ANNEX

ACKNOWLEDGEMENTS

Thanks are due to Mr. John Levonik, Instructor in Philosophy, for his expert and dedicated help in transcribing the tapes of the proceedings of the conference and to the Polish Academy of Science, Warszawa for their kind permission to publish the original letter of Roman Ingarden to Husserl.

Thanks are also due to the Canada Council and to the University of Waterloo for a grant of funds in support of the conference.

Most of all, however, this volume owes much to Prof. Lawrence Haworth, Chairman of the Philosophy Department of the University of Waterloo. Prof. Haworth took an active part in the organization, conduct and publication of the conference, which would not have succeeded without him.

INAUGURAL LECTURE

ANNA-TERESA TYMIENIECKA

PHENOMENOLOGY REFLECTS UPON ITSELF*, II

The ideal of the universal science: the original project of Husserl reinterpreted with reference to the acquisitions of phenomenology and the progress of contemporary science.

Opening this conference I will ask a paradoxical question the answer to which determines the scope, the significance and the aim of our gathering: *Is phenomenology still alive?*

It is quite clear that the great wave of philosophical inspiration which has radiated from phenomenology for the last three decades, exercising a sort of hegemony all over the European continent and beyond, and penetrating all the fields of human endeavour, has recently lost in intensity and vitality, while the new wave of interest in phenomenology which is now visible in the Western Hemisphere has not yet produced mature fruits. There is certainly a considerable number of scholars in philosophy as well as in other fields still working along phenomenological lines, but the question which I have put at the outset has a deeper meaning.

What is at stake is not a restricted historical work in relation to the past of the phenomenological movement, or some limited pursuit to elucidate phenomenologically formulated problems, both of which are certainly important endeavours but would be either what Austin called 'polishing the tombs' or would remain restricted to individual pursuits, whereas phenomenology as inaugurated by Edmund Husserl and undertaken by his immediate disciples has raised a vaster claim. As we well know, it was meant not only as a *philosophia prima* but also as a *universal science*.

It is as a vast project of universal, fundamental inquiry to be carried on by a group of scholars, each in his own sector, that phenomenology has been conceived, has received the basic formulation of its task, field of research, and methods and has been taken up by philosophers and scholars. However, as it is well known, the very enthusiasm which made it radiate led to such a diversification of tasks, differentiation of approaches, ways of procedure, points of view and conclusions and terminated in such disparate types of discourse, formulation of problems and

Tymieniecka (ed.), Analecta Husserliana, Vol. II, 3–17.

linguistic expression that we may ask ourselves whether this actual, concrete accomplishment of the Husserlian inspiration corresponds in any sense to his original project of phenomenology as a universal science which has been initiated and carried on at least at the time of the Göttingen School. Husserl himself, having in his later development switched to a historical, that is genetico-transcendental foundation of phenomenology, seems to have despaired of his initial aspiration, which then appeared to him as an unfulfilled dream.

Even if we could admit that there are criteria by which the phenomenological work being presently carried on still falls into this initial framework, we have to ask whether the present weakening of its vitality is not a sign that the task proposed by Husserl did not contain within itself the germ to fulfil its promise.

The question then which I propose for your attention is: 'How does phenomenology stand today with respect to its original proposal of a fundamental, universal science?'

At this basic level of consideration I challenge the contention that phenomenology has come to the end of its essential development. I will try, on the contrary, to establish that precisely now, when the two great lines of phenomenological inquiry outlined by Husserl, the *eidetic* and the *transcendental*, have been fully developed by himself and his followers, we enter into a new, *self-reflective* and *self-critical* phase of research. Comparable to that of the progress in science, which points out further perspectives into ever deeper dimensions of man's transaction with the universe, this new phase establishes phenomenology as an open field of continuous philosophical scholarship.

It has still to be seen whether this work at the level of foundations can correspond to the aspiration and claim of a *universal science;* for can philosophy ever satisfy its postulates?

Raising the question of the validity of Husserl's initial project regarding the task of phenomenology, we have to immediately point out that the contemporary development of scientific inquiry opens up a different and larger perspective of its requirements than the one at the end of the nineteenth century within which it was conceived. With science having evolved and changed its own criteria for validity, adequacy and truth, requirements demanded of philosophy to be at the same time its foundation and a 'science' itself take on a transformed form.

I. HUSSERL'S PROJECT AND THE NINETEENTH CENTURY STANDARDS

Would present day phenomenology comply with these requirements? As is well known, Husserl's idea that philosophy should be the foundation of science stems from his teacher Franz Brentano. To lay down a foundation for a *philosophia prima* at the end of the nineteenth century called for a *mathesis universalis,* which, as Leibniz had seen, should comprise not only the foundation for the basic inventory of cognition but also an *ars inveniendi,* not only the analytic of the givenness but also its hidden modes of operations. It could only really be the last ultimate explanation of human intercourse with the universe if it clarifies the given as it gives itself and accounts also for its origin and becoming. But in the actual formulation and elaboration it had to meet specifically the criteria and standards both for validity and excellence as set by the positive natural science of the period. Indeed, the enormous advance of natural science has given it an unquestionable prestige.

These postulates and standards were, first of all: *cognitive immediacy* (as opposed to speculation), *intersubjective verifiability, precision and clarity of concepts used for formulation,* and a *common language.* This is to provide an experiential ground of cognition together with the objective rationality of the discourse. Furthermore, to satisfy the foundational demands of science its subject matter should be determined with precision, its methods elucidated and its objectives clearly outlined. So much for general requirements. But since the specific task of philosophy as a universal science is to give an ultimate grasp of the empirical, contingent reality caught in a continuous progress and flux, it had as its last point of reference to establish data of a universal nature and validity, to surpass the fluidity of the senses by an apodictically certain principle.

The early Husserl, inspired by Brentano, who saw the "immediately given" and the self-evident in the autonomous contents of psychic acts, has seen and never disclaimed the first basis of the *philosophia prima* in the ontological conditions of all cognition, that is in the *formal structures.* Their self-evident givenness served to secure the apodictic certainty of cognition of objects in the world as well as the regulative principle of their constitution. Coupling it with the conception of the eidetic insight that is with the clarification of its cognitive mode, Husserl inaugurated a specific field of research.

This research field has as its subject matter the ontological *conditions of human cognition,* conceived not only as an inventory of the formal structures of the givenness, but as revealing also the formally possible ways of coming to be and passing away, of action and passion, of generation and corruption of the various types of the given. This clearly defined domain of research initiated by Husserl, taken up by the Gottingen School, has found its monumental accomplishment in the work of Roman Ingarden[1]. We can see, indeed, in Ingarden's extensive ontological inquiry not only the analytic disclosure of the ultimate structures of objectivity but also by the same stroke the establishment for the first time in the history of Western philosophy of a conceptual apparatus required to treat with clear and subtle distinctions major philosophical questions according to the postulate of rational intersubjectivity of the universal science.

But Husserl moves further in his quest for autonomous, self-justifying contents of cognition. Again he follows Brentano and yet again goes beyond Brentano's intentions. The central role within the *complete scheme* of human knowledge was attributed by Brentano to conscious processes understood as an autonomous, irreducible ground for investigation. The realm of conscious processes and acts in their specific intentional nature was also for Husserl the basic level of inquiry, but, going beyond Brentano, he assumed it to be the ultimate ground for the *final explicitation of cognition.* At the same time, the foundation of cognition became for Husserl the foundation and source of the human universe itself.

The ground of the *philosophia prima* is thus shifted to a two-fold field of investigation: the constitutive function of human consciousness and the explication through its genesis of the human world. Husserl has not only prepared and outlined this project but has also worked it out to quite a considerable degree. It has also captivated the attention of a large number of philosophers who, albeit in various and divergent ways, have, like Sartre, Merleau-Ponty, and Levinas attempted to pursue further Husserl's inspiration.

To transpose the ground of the universal science into the investigation of the *life-world* as co-natural with the genetic progress of human consciousness meant, however firstly, to open it, to make it relative also to experience no longer distilled through the postulates of certainty but expanding in all the directions of life into manifold nuances, shades, and intensities with a misty and blurred horizon of associative links. Did this switch to a directly intangible and nearly undefinable historical genesis of man and his lived world, itself an ever-advancing flux, self-generating, never at rest, itself in need of explanatory principles or rules to account for its specific nature and origin, not mean an abandonment of the positive ideal of ultimate, intersubjective validity? The expansion of phenomenology into the investigation of the life-world meant at least the abandonment of modes of expression consisting of clear, distinct and strictly definable rational concepts.

Secondly, Husserl's radicalism in his search for self-justifying ultimate ground, which led him to see it in the constitutive consciousness (on account of the assumed 'absolute existential status' of 'pure' transcendental consciousness) resulted in the implicit decision to deprive the contingent reality of its autonomous existential rights. Is it not the crucial experiential acuteness of the contingent reality, of empirical nature, which makes it a focal dimension among all other dimensions of man's cognition that the quest for an apodictic foundation of knowledge stems?

We may well attribute qualitative autonomy to psychic acts and to human consciousness specific and irreducible *modi operandi,* but to claim that they have a transcendental character entirely different from nature and originate in their own right means separating the transcendental consciousness from the empirical one. The empirical world and nature were thereby transposed into the realm of absolute constitutive consciousness, and not only their qualitative content but even their empirical presence was supposed to be made explicit by the constitutive operations of transcendental consciousness.

The 'transcendental turn' of Husserl meant in the first place a radical devaluation of reality. It resulted also in a radical separation between intentional acts and processes and the empirical ones still responsible for their singular, contingent occurrence; intentional consciousness, in spite of all ulterior efforts of Husserl became separated from its concatenations with the empirical world, from the workings of nature. Finally, assuming the transcendental consciousness to be a unique, complete, self-contained and self-explicit source of the human universe meant making nature and cosmos either radically relative to consciousness or reduced it to the status of brute, unaccountable matter.

I have already indicated that Husserl's transposition of fundamental inquiry from the ontology of rational structures into the field of transcendental consciousness raised the difficulty of intersubjectivity and of univocal rationality. The separation of the transcendental and the empirical consciousness compromised the relation of the phenomenological explication of cognition to natural science. In our subsequent scanning of the acquisitions of past phenomenological research we have to keep these two points well in mind.

II. THE ACQUISITIONS OF PHENOMENOLOGY AND THE OPENING OF NEW AVENUES OF RESEARCH

At the present moment we see fully developed in front of us the two major lines of inquiry outlined by Husserl; the attention of probing phenomenological scholars focuses upon their results but also critically scrutinizes their assumptions and presuppositions[2].

In fact, the accomplishment of the original Husserlian project may be envisaged at various levels.

First of all we have a far-reaching elaboration of the *ultimate rational foundation of the human discourse* in the regional ontologies on the one hand and in the formal ontology of Ingarden on the other. The latter presents all the structural types of objectivity that can possibly be constituted, including the form of the region of the world and of consciousness.

Concurrently, we have the two-fold project of the transcendental explication of the given. That is, through Husserl's own work but also through that of a whole cluster of phenomenologically-oriented philosophers and psychologists an extensive and profound knowledge is gained of the cognizing-constitutive functions of human consciousness together with the explication of the genesis and meaning of the lived-in world.

The specific clarification of concepts, elaboration of basic notions like essence, idea, etc., the role of transcendental consciousness in human relations in the social realm of man's functioning, and the extensive study of man's artistic, aesthetic, and social modalities may be considered as constituting a *second level* of the acquisitions of the phenomenological inquiry[3].

The third level which emerges in the present research is that of self-criticism. In point of fact, the extensive and nuanced investigation of the transcendental constitution on the one hand, seems to have been extended to its very limits but the difficulty of the relation between the transcendental and the natural consciousness posed at the outset remains postponed rather than resolved. The meaning of transcendental idealism now appears to be far from being definitely settled; the status of the natural world surges forward to become the focus of attention.

Furthermore, the manifold results obtained by phenomenologically-inspired psychiatric research do not contribute any longer to the legitimacy of basic phenomenological claims, but on the contrary call for their critical re-evaluation. Interpreted with reference to some of the main phenomenological principles they seem to indicate that, for example, the idea of transcendental consciousness, which through its horizon was meant to account for *all* actual, past and possible structures of experience appearing in the field of consciousness needs to be submitted to critical examination. Indeed, it falls short of accounting for a whole range of phenomenologically inspectable structures, which appear in instances of various 'abnormal mental conditions' (e.g., in psychosis).[4] On the other hand, structural ontology also appears as non-final. It may well account for order, for the possibility of a relatively stable world within a flux; it may account for motion, for the mechanisms of coming to be and passing away of things; but it remains on a static level of approach and cannot account for the *initial spontaneity* of the contingent nature or for that of the life of intentional consciousness. The scrutiny of the constitutive consciousness is also not capable of explaining its *initial spontaneity.*

Finally if we approach the given from this two-fold, *eidetic* and *transcendental* perspective we may explain it by *what it is,* but if we ask the final questions: 'Why is the givenness such and no other?' or 'Why should such a selection of forms and their constitution and no other be given?' we cannot seek their answers along these lines.

And yet phenomenology has presented us so completely and exhaustively with the examination of givenness in its forms, laws and genesis that now we naturally will seek its last reasons.

Progressive accomplishments in current critical research seem to raise two kinds of questions.

The first range of questions seem to indicate that in spite of the considerable nucleus of valid results obtained in phenomenological research, we have reached a borderline, a limit which at least puts in doubt, if it does not contradict outright, Husserl's claim to *absolutism of the trans-*

cendental consciousness. As I have pointed out elsewhere, we are bound to go deeper in those directions where cracks appear [5]: that is in the direction of the puzzling borderline between the transcendental and natural consciousness of empirical nature itself. That such attempts have been made is well-known. It is enough to mention the work of the early Heidegger, Sartre and Merleau-Ponty. Their most instructive efforts have emphasized the issue without however solving it. The more specific problems of the embodiment of the conscious functions and of the legitimacy of the subjective referential framework are more than ever under critical scrutiny.

But it is ultimately the status of the constitutive function with its structurizing, constitutive power that is at stake. The crucial questions to be treated arise 'Would constitutive consciousness in its given virtualities be really the *last* factor determining man and his world?' 'Is the constitutive consciousness positing its own final rules and regulations?' 'Or, as I have tried to establish it in my concise treatise on cosmology, can we discover its final structural models and system of reference beyond it, through *a priori* transcendent principles?' [7]

Thus, we are engaged on the other hand, in the direction of a new scrutiny of the relation between the intentional, the conscious and the natural realm. On the other hand, inquiry has begun into the ultimate conditions of consciousness and its functions other than constitutive.[8]

The second range of critical questions culminates in the unavoidable, and at the present stage no more to be postponed quest for the *first principles or rules of the givenness itself,* which would simultaneously guarantee the differentiating of man from the rest of nature and the establishing of the transcendental consciousness within the realm of natural existence, which it postulates and from which it must be converted. Thus the question of *natural, real factual existence* comes into view. Existence in its singular, concrete, fleeting factuality irreducible to its modes of constitution. It lurks from behind the universal structures as the marrow of their bones.

I submit again that the *concrete actual existence of the real individual* is not only the thorny point and the Achilles heel of the constitutive universalism both as regards the universal schemes of the consciousness and of the formal ontological structures as models for their operations but also as regards all inquiry into man's transactions with the universe. [9]. Unless we secure for it an adequate treatment in its own right not by singling out the universal modes of its concretion in things and beings, processes and events, but inversely by considering their distinctiveness as a result of the *context of actual existence,* we shall be always skimming the surface of appearance without reaching down to its conditions.[10]

Phenomenology thus seems to have completed a circle: actual existence devaluated to start with through an ever-more nuanced apparatus to think it away eventually revindicates its inalienable rights. Around this surges a cluster of questions to be answered concerning the transcendent reality of the *Cosmos.* From within the transcendental dimension of man itself we must ask what are the last laws and regulations of the ALL such that they could account for the temporality of consciousness, for its embodied condition and for the specific selection of forms it uses.

Neither the Husserlian transcendental nor the Ingardenian structural eidetic investigations allow for the formulation of these questions; their accomplishments, however, as well as their shortcomings, have prepared for a break-through.

This quick perusal illuminates a striking point. The wholly developed Husserlian outline neither justifies itself completely nor denies its own ultimate validity but through its reflective criticism points towards a further self-discriminatory quest. This line of progress is comparable to the progress in positive science, which exhibits a specific continuity of historical progress. It advanced via a criticism that partly retained and partly rejected the new evidence introduced by current research, in such a way that even the rejection of some of its initial basic assumptions still meant an advance in knowledge.

With this we are brought back to phenomenology's initial aspiration to be a universal science. Can phenomenological research extending over such a vast realm, so diversified in its methods and results, still aspire to this ideal? What is its relation to the positive science of today?

III. THE FIELD OF PHENOMENOLOGICAL RESEARCH WITH RESPECT TO THE NEW STANDARDS OF SCIENCE

There can be no doubt that Husserl's work itself is responsible for the confusion and the serious misunderstandings concerning the relation of phenomenology to science, especially natural science. It is time to re-examine, be it in a most sketchy way, this crucial issue.

If the philosophical quest aims at getting to the "bottom of things' without prejudice and arbitrary restrictions, it is impossible to deny that natural science is one of the most important sources of insight and evidence about the universe and man, about the given and givenness as such. Husserl's battle-cry of "back to the things themselves" ratifies "the validity of all the sources of cognition to the exclusion of none". Phenomenological reduction certainly does not mean what Merleau-Ponty seems to believe that it does, namely the rejection of science. How could philosophy discard or ignore one of the major sources, the source *par excellence,* of reflection upon the contingent reality, its greatest puzzle?

That, in fact, phenomenology did not remain alien to science we see, on the one hand, from the fact that its fundamental guidelines and insights have inspired in different ways the majority of the sciences of man, revolutionizing them as their new theoretical foundations and method. Their results in turn have provided phenomenology with the most precious material to evolve towards further progress.[11] On the other hand, a considerable amount of philosophical reflection and inquiry as performed by phenomenologists, suffice to mention Ingarden, Conrad-Martius, Sartre, Merleau-Ponty, S. Strasser, Gurvitch and Van Peursen, draws directly or indirectly upon the results of science.

Finally yet above all, as we have seen, the state of progress of phenomenological inquiry and its self-evaluation brings into focus the enigmatic links between the strictly 'reduced' realm of consciousness and the 'pre-reduced' realm of nature as investigated by natural science.

To approach the realm of nature through the investigations of natural science in a significant way undoubtedly demands a revision of the reductive procedure of phenomenology as well as the elaboration of methods for the evaluation of the results of natural science, which would at the same time satisfy phenomenological radicalism and the elusive nature of the subject matter. The direction towards nature thus opened appears to be crucial for the progress of phenomenological inquiry into man and the universe. Yet the question arises whether the inquiry can possibly maintain its basic radicalism in the clarification of foundations and at the same time break through the constructivism of science to do justice to its findings.

There is also another channel into which the progress of phenomenology has directed itself. In point of fact, the conceptual apparatus, the linguistic matrix used in phenomenological investigations, has itself become subject of critical query. Indeed, already the analysis of the originary, primordial forms of experience uses present-day language, which is the product of a long genesis in which different stages in the development of mankind progressively passed on to each other their acquisitions in the form of tradition. This language in which we attempt to grasp the initial, the virginal elements of this genesis is already the fruit of the long development of man's expression in the historical process. Furthermore, for these analyses that aim at the disclosure of the constitutive operations of the 'originary', primary experience, we are bound to use a linguistic apparatus which assumes naively, pre-critically a basic grammar and syntax, which themselves are forms established by Western man in the historical process of organising and articulating the given. What then about the root-matrix of language itself, about its origin? The need for a genetic elucidation of the meanings, as moulds in which our experience constitutes itself requires the elucidation of primordial, spontaneous, autochthonous attitudes man assumes towards the elements, towards the vital impulses. Here again we are led into scientific studies of anthropology, psychology, and psychiatry dealing with the emotional expression of myth, symbol, and archetype.

Drilling under the rationality of the originary constitution in order to satisfy one of the postulates of phenomenological radicalism which opens a subterranean perspective calling for *metaphysics of the primeval logos,* we encounter the same dilemma of the indispensable modifications of the approach and method with respect to the subject matter and sciences that we might have to call for evidence.

Finally, does the constitutive analysis grasp satisfactorily the *life* of consciousness in its essential aspect, that of the 'inward life' of man? On the contrary, it is obvious that the constitutive analysis of Husserl and those of the various types of 'experience vécue' made by Sartre, Merleau-Ponty, Levinas operate either exclusively with the rational elements of the human psyche or with the universal schemes, rules and types of the human 'inward life'. The great tragical conflicts, struggles of emotions and aspirations, ideals, successes and failures, which play an essential role in the progress of the conscious life as its *dynamic factors,* remain together with the 'empirical soul' reduced by Husserl and neglected by others. How can we then scrutinize this dimension of the human existence since what phenomenology proposes as object of inquiry turns out to be universal structures?

And yet without taking this dimension into serious consideration our picture of man remains flat, abstract, and dead.

In view of all these directions which open, a fundamental doubt arises: 'Is phenomenology standing on moving sands, unable to stop anywhere to pitch a tent? 'Would the ever further stretching avenues into which we are drawn mean that each time we put our foot down a new shifting of ground occurs which could efface the validity of the previously accomplished effort?'

In view of the demand which these directions postulate to diversify even more the approaches and methods with which the newly proposed objects have to be treated in order to be grasped adequately, we must ask ourselves whether, first of all, they could satisfy the criteria of science, or whether the dream of a "scientific philosophy" must definitely vanish for ever. Secondly, could a unity of the phenomenological work be accomplished? Should the unity of a science not repose in its method?

Are we then bound as individual scholars to close ourselves each in the tiny compartment of this query sticking to it rigidly, because the ever advancing and shifting level of questions differentiates the approaches and starting points and calls for an ever new medium of expression? Would this be a requirement to comply with in order to be meaningful for the scientific inquiry as its foundation? Does science remain within unchanging and rigid limits?

In brief, can we still accept the plausibility of the initial aspiration of phenomenology to be a 'universal science' an idea of which even Husserl himself seemed to have despaired?

Yet since the time of Husserl our science has changed enormously. We are now completely disabused of the dogmatism to which science held rigidly in the nineteenth century and the beginning of our century. In view of the flexibility, expansion and openness of the scientific spirit of today, which does not pose any pre-established criteria for the advance of the search into reality, but on the contrary is ready to experiment with everything available, the postulates to be satisfied by a foundational philosophy themselves also change. At certain levels of formulation and expression rigorous precision and clarity, that is rationality *par excellence,* would be required, but at different levels procedures which by trial and error use all possible means of scrutiny would be a perfectly legitimate approach to the final reasons of the real.

The postulate of immediacy of cognition, derived from the empiricism of the nineteenth century natural science does not hold any more even for that science, which, on the contrary, became highly speculative, discovering the phenomenal character of sensory givenness. However it remains still the basic point of reference, the immediate evidence of science as well as of the phenomenology of the life-world.

Philosophia prima from which concrete scientific endeavour would draw its ultimate explanation and clarification will not lose its efficiency on the grounds of a more or less intuitive or more or less rigorous establishing of its data. On the contrary it has itself to be spread through several levels according to the various levels of reality and corresponding levels of scientific pursuit. Its specific orchestration along essential lines alone to be discovered by philosophy, would by its specific import give to the scientific narrowly conceived compartments the basic unity mirroring that of the universe.[12]

Eidetic analysis which remained throughout Husserl's work as the only valid tool to guarantee apodictic certainty itself falls short of providing more than the groundwork for the philosophic exploration of the universe and man; restricted to itself it would only amount to tautology.[13] It certainly remains the referential framework for the rational skeleton of reality and its expression. However, in order to reach the multiple dimensions of the world-order in its *constitutive* principles and reasons we will have to resort to speculative means and pass from analysis to *conjecture,* from *eidos* to *principles*.[14]

It appears then that both contemporary science and the phenomenological investigation have taken a strikingly congenial line by engaging each on its own the cognitive probing deeper and deeper into the elusive and always remote nature of man and the universe.

It seems that the progress of phenomenology has contributed to the classic ideal of the *mathesis universalis* this open horizon of man's own progress.

From the above-expressed point of view, we may consider the diversified and ramified field of the present-day's phenomenological research, which we have in front of us as the possible fulfillment of the conditions for a *philosophia prima* albeit in a form more elaborated and refined than originally proposed by Husserl. In fact, the basic ground from which all cognition either emerges or partakes of stretches from the ontological foundation of objective rationality to the most intimately sensed and felt regions of man's irrepeatably subjective life in which it is rooted as a part of its mode of being. This spread embraces several dynamic dimensions in a historical genesis. In hermeneutics we have the tool for adequate interpretation of its experiential nuances as they fluctuate in the historical progress. Thus a field of research is outlined. This field, comprising the eidetic conditions of rationality through the constitutive functions of human consciousness and the life-world as well as the hidden rules and laws of the elemental nature in its concatenations within the cosmos and its final conditions, suggests an outline in which the great universal schemes of *Philosophia perennis* is revindicated, albeit in an infinitely more nuanced and refined form, as has emerged spontaneously from the research itself.

What else has always been and will be the true and highest vocation of great philosophy if not to answer man's quest for ultimate ordering, such that it would guarantee the continuity of the past, present and future, explaining man's integral being in terms of spirit as well as of nature, his place in the cosmos as well as his relation to absolute transcendence and could give meaning to his existence that he would always interpret anew?

NOTES

* First part of the 'Phenomenological Selfcriticism' has appeared in *Analecta Husserliana*, Vol. I, 1971.

[1] Cf. the present writer's: *Essence et Existence, Étude a propos de la philosophie de Roman Ingarden et Nicolai Hartmann*, Editions Montaigne, Aubier, 1955. Also *The Pattern of Roman Ingarden's Philosophy*, Studia Filozoficzne, Warszawa, 1971.

[2] Cf. the present writer's 'Die phänomenologische Selbstbesinnung. I: Der Leib und die Transzendentalität in der gegenwärtigen phänomenologischen und psychiatrischen Forschung', in *Analecta Husserliana*, Vol. I, 1971.

[3] Cf. the present writer: *Phenomenology and Science*, Noonday Press, N.Y., 1960.

[4] Cf. the previously quoted 'Die phänomenologische Selbstbesinnung, I'. The whole first volume of the *Analecta Husserliana* is directly and implicitly devoted to a critical revision of the relation between the transcendental consciousness and transcendent reality.

I have also made a direct confrontation between the transcendental conscious function and nature in: 'Cosmos, Nature and Man and the Foundations of Psychiatry', in *Heidegger and the Path of Thinking*, Duquesne Univ. Press, 1970.

[5] Cf. 'Cosmos, Nature and Man and the Foundations of Psychiatry'.

[6] Cf. *Analecta Husserliana*, Vol. I, 1971.

[7] Cf. by the present writer, *Why is There Something Rather than Nothing Prolegomena to the Phenomenology of Cosmic Creation*, Van Gorcum, Assen, 1965.
[8] Cf. my outline of the *creative function* of man analysed in opposition to the constitutive function in: *Eros et Logos, esquisse de phénoménologie de l'intériorité créatrice*, Nauwelearts, Belgium, 1971.
[9] Cf. 'Existence Vindicated', *The Personalist*, 1964 and the clearly quoted *Why is There Something Rather than Nothing* where I have directly approached the question of the actual existence proposing a new formulation and treatment for it within the framework of a cosmological conception that would go beyond the respective pitfalls of both the transcendental and structural universalism and yet take into consideration and re-evaluate their essential contribution.
[10] *Ibid.*
[11] Cf. by the prequoted: *Phenomenology and Science in Contemporary European Thought*.
[12] *Ibid.*, pp. 176–84.
[13] Cf. present writers, 'Phenomenological Analysis and Conjectural Inference', in *Logic, Methodology and Philosophy of Science, Festschrift for I. M. Bochenski* (ed. by A. T. Tymieniecka), North-Holland, Amsterdam, 1966.
[14] Cf. the prequoted, *Why is There Something Rather than Nothing*?

ADDRESS

(Professor Klibansky on April 10, 1969)

"Ladies and Gentlemen,

May I first express our sense of gratitude to the organizers of this Congress who have taken considerable trouble in making this assembly possible. I couple this with an expression of pleasure that this first International Conference on *Husserl and the Idea of Phenomenology* takes place on Canadian soil. Canada, like the United States, has been comparatively slow in recognizing the importance of the phenomenological approach. In 1950, an historian of American philosophy could still write that phenomenology is in its infancy on this Continent. Since that time, much has been done to interest those not acquainted with the problems and the methods of phenomenology. A good beginning has been made in rendering the main writings of phenomenology accessible. No doubt, much more remains to be done.

The links between phenomenology and Canada, though very tenuous, are nonetheless not quite nonexistent. It is sometimes forgotten that one of Husserl's early pupils at Göttingen was a young Canadian philosopher, Winthrop Bell, of Chester, Nova Scotia, whom my lamented friends Jean Hering and Alexandre Koyré held in high esteem; it was through Bell that Husserl acquired a closer knowledge of the thought of Josiah Royce.

Yet another of Husserl's early students links him with this country: Gustav Hübener, who appears in that well known picture of the Göttingen 'Philosophische Gesellschaft' of 1912. Later on Professor of English Literature at Bonn, he was one of the few German scholars who refused to teach under the National Socialist regime and, though not falling under any of the racial laws, freely chose to leave his homeland. He was welcomed in Canada at the University of Mount Allison, Sackville; before his premature death, he taught his pupils there how to apply phenomenological method to the study of literature.

In viewing this distinguished gathering, my grief is all the greater to have to state that the main speaker of the day, Dr. Roman Ingarden, has

Tymieniecka (ed.), Analecta Husserliana, Vol. II, 18–20.

not been able to come. Hard phenomena, such as shipping schedules and the non-arrival of the United States transit visa, prevented him from crossing the Atlantic. He has asked the Chairman of the Waterloo Conference to convey to the Congress his greetings, his regrets that he cannot attend, and his hope that the discussions will be a success.

Roman Ingarden does not need an introduction. However, on this Continent his works are less well known than they deserve to be, mainly owing to the fact that none of his major writings have been translated into English. The very titles of his earliest articles are characteristic of the problems which engaged his attention throughout his life. The first of his German essays known to me, published in 1921, was entitled *About the Danger of a petitio principii in the Theory of Knowledge*; this was followed a few years later by his first book, *Über die Stellung der Erkenntnistheorie im System der Philosophie,* in which he strongly criticized the emphasis placed on the theory of knowledge and the central place it occupied in the neo-Kantian systems in the German universities of that period. At the same time, his interest focussed on what he called in the title of one of his earlier essays, *Essentiale Fragen. Ein Beitrag zum Problem des Wesens.* His concern with the problem of essence and, more generally, with ontology thus goes back to his youth. In his approach both regarding the questions he asked and the way he set about answering them he strongly differed from his fellow philosophers in his native Poland, several of whom have acquired world fame for their work in logic, epistemology and semantics. He stood out alone among them, and throughout his life he pursued the problems which had acted him since the time of his first acquaintance with Husserl.

With apprehension he viewed the new direction which his master's thought had taken since the publication of *Ideen*, and he made it his particular task to clarify and to settle the controversy between idealism and realism. This is the subject of his contribution to the Husserl-Festschrift in 1925, and of his two volumes published after the war by the Polish Academy, *Spór o istnieniu świata* (Controversy about the Existence of the World), now available also in a German translation. He is careful to distinguish various meanings of idealism and to show which meaning he has in mind. I hope that in the discussion account will be taken of his criticism of Husserl's transcendental idealism and of the *Cartesian Meditations.*

During the last decade, Ingarden has become more widely known, especially in Europe, by his application of phenomenological method to aesthetics. In his numerous writings on the subject – I mention only *Das literarische Kunstwerk, Untersuchungen zur Ontologie der Kunst, Vom Erkennen des literarischen Kunstwerks,* and *Erlebnis, Kunstwerk und Wert,* and articles such as 'Sur différentes conceptions de la vérité dans l'art' – he examines in which sense works of art are said to exist and what must be taken to be their general form. In his ontological analysis he shows that, while receiving their objective foundation in physical objects shaped by the artist, they are constituted by purely intentional acts.

The paper he proposed to present at this conference bears the title 'What is New in Husserl's Crisis'. He wrote it in German; at very short notice it was translated by Professor George, of the University of Waterloo. I am now calling on Professor George to read this paper to us."

PART I

THE LATER HUSSERL

ROMAN INGARDEN†

WHAT IS NEW IN HUSSERL'S 'CRISIS'

It is now almost common-place to speak of the 'later' Husserl, as if this 'later' Husserl had assumed an entirely new position. In this connection one usually thinks of his *Crisis*. What are the facts of the matter?

It is much too early to attempt an outline of the development of Husserl's philosophy. Many of his manuscripts have not yet been published, and it will take a long time yet to assess even the published *Husserliana* under the aspect of the historical development of his research. Only the members of the Husserl Archive can have a certain insight into these matters. But if we merely consider the succession of works which were published by Husserl himself, the order in which they became historically effective, then the *Crisis* – at least its first two parts – is his last work; even the rest of the text which now forms the *Crisis* seems to have been prepared for the printer by Husserl himself. It followed immediately upon the *Méditations Cartésiennes* and the *Formale und transzendentale Logik*, and its historical effect was determined by its position in this order. Moreover, it is the last work which Husserl composed specifically for the philosophical public. It was produced after many years of futile attempts at creating a great systematic groundwork of phenomenology, which was to be its ultimate foundation. In addition, it was written in a European intellectual atmosphere in which Husserl sought in phenomenology deliverance for himself and for European culture. It does not appear that the book was intended to convey new analytic inquiries; it is more of a synopsis, perhaps in improved form, of the main results achieved earlier; it was meant to convince the reader that the course toward transcendental phenomenology taken in the *Ideas* is correct, even though the *Ideas* remained a fragment, and in spite of the opposition which it encountered. The *Crisis* was to demonstrate as succinctly as possible, that philosophy must take this course, and it was to open a vista upon the 'promised land' of a new philosophy which would save and enhance the intellectual culture. For this reason it is more directed toward the intellectual public than any of his other writings. Thus if the *Crisis* really were the expression

Tymieniecka (ed.), Analecta Husserliana, Vol. II, 23–47.

of something new, it would run counter to the role its author intended for it, and to Husserl's own expectations.

The *Crisis* grew out of lectures which Husserl held in Vienna and elsewhere during the thirties. Their intention obviously was to propagate, if not to propagandize, his views, and it was probably connected with the political and intellectual situation in Germany. This is the reason why in the *Crisis* phenomenology takes on a role and function which was not previously associated with it. Earlier, before 1914, it was to be the manner in which philosophy would be realized as an exact science. Now it was to influence men, it was to make them conscious of the threatening catastrophe and move them to defend themselves. It was to enrich mankind, it was to rebuild men inwardly and move them onto a new course of inquiry by showing them the error of 'objectivism'. The true philosopher – for Husserl he must be a transcendental phenomenologist – acquires in this way a "calling" within the life-world which Husserl had not thought of before 1930, and which was perhaps first mentioned in his letters to the Prague Congress of 1934. This may be very important from the purely human point of view and for the history of ideas, but it does not add anything *factually* new to the content of phenomenology; indeed, if one wants to protect philosophy from various external influences and usurpations, such a role of the philosopher must be viewed as rather disadvantageous and undesirable.

During his entire life, Husserl was active in his philosophical work. Hence the *Crisis*, also, contains a number of new contributions to particular problems. But the book's main line of thought is merely a new manifestation of a basic idea that was formed more than 25 years earlier. Nonetheless, it is best, first of all, to point out certain innovations, some of which have led to noticeable modifications of earlier conceptions.

Thus a striking feature of the *Crisis* and of the connected *Essays* is that a perspective is given, within a phenomenological context, of problems in the philosophy of history, and of the historicity of pure transcendental consciousness itself. There is a certain similarity here with a historico-philosophical world-view which Husserl had firmly rejected in discussions with Dilthey (1911). This is obviously connected with the strengthening of the influence of Dilthey in Germany, even in Husserl's immediate surroundings, toward the end of the twenties and the beginning of the thirties.[1] But in the *Crisis* it is no more than an overview over a range of problems. Significant analyses of the relevant problems are not given. But if this line of interpretation had been pursued, it would have noticeably modified the character of phenomenology.

In the *Crisis* there is a tendency which was never expressed so clearly before, to view phenomenology as it were as the end state for which European philosophy was yearning, to see phenomenology as the mature state of European philosophy. Husserl's personal situation makes this understandable. But the interpretations of various schools of modern philosophy which he actually carries out are not persuasive, mainly because there is almost a complete lack of precise textual analysis of the works under discussion. Only the discussion of the manner in which Galileo placed modern natural science upon a mathematical foundation, and the related tendency of a rationalizing nature, forms an important enlargement of our stock of knowledge. Together with the Galileo researches of Alexandre Koyré it is among the true achievements of phenomenology.

In the *Crisis* these individual considerations merely contribute to the main line of thought, which leads via universal phenomenological *epoché* and transcendental reduction to the founding and unfolding of the idea of transcendental phenomenology, and which ends by postulating that philosophy should be conducted exclusively on the basis of 'transcendental subjectivity'. More precise formulations of individual thoughts make the realization of this old idea more unambiguous, but this does not lead to an essential alteration of the old viewpoint. On the other hand, several concrete analyses, which were conducted in the *Ideas* are not repeated in *Crisis*, and the new analyses, which could be useful, are not given. Thus, compared with the position in the *Ideas*, the clarification of the essence of 'epoché' is not further advanced. Only the old forms of expression are used, though some of them, for example 'bracketing' (*Einklammern*), or 'to suspend' (*Ausschalten*) are no longer used, and the sense of other expressions such as 'abstention' (*Enthaltung*) from 'acceptances' (*Geltungen*) or 'to put out of action' (*Ausser Aktion Setzen*) is not clarified at all. He launches a vigorous attack upon the so-called 'objectivism', which he sees as the main error of modern natural science, and which he stigmatizes as the source of the 'crisis' of European sciences. Significantly, the otherwise unclarified concept of objectivity goes together with the concept of 'being in itself' (*an sich sein*) and of 'objective truth', so that the struggle against 'objectivism' turns into a denial of the 'being in itself' of the world or the life-world; in this way a standpoint is reached which has been called, on the basis of *Ideas I*, Husserl's 'transcendental idealism'. The 'relativity' of the knowledge of nature is more strongly emphasized than before. There is no attempt to clarify positively the mode of being (*Seinsweise*) of the world (after denying that it has being in itself); but he emphasizes that there can be no doubt about the *being* of the world, and this view is called 'the extreme realism of phenomenology'. At the same time he says clearly that the world or life-world, or the things in the life-world are nothing but correlates of the corresponding sensebestowing (*sinngebend*) acts of pure consciousness, i.e., purely intentional objects of certain combinations of acts of consciousness. He usually avoids the expression 'pure consciousness' and replaces it by the phrase 'absolute subjectivity'. Compared with Husserl's earlier writings, there is an increased dogmatism in the tone with which he discusses all questions regarding the world, or the life-world; this may have to do with the fact that hardly any concrete analyses are given. The most important points seem to have been decided in advance, and the solution of the question concerning the existence of the world and the constitution of the things in the life-world is not presented as the final analytical result of a phenomenological analysis which follows upon *epoché* and reduction. In the last section on 'pure psychology', which he holds to be identical with transcendental phenomenology, Husserl treats several questions, e.g., the analysis of 'empathy' and of the intersubjective constitution of things in the world, the world, and the life-world, questions which he had only opened up in the *Meditations*, as if they were answered and fully clarified, as if there could be no further scruples or doubts concerning the efficacy of 'empathy'. This is undoubtedly due to the brevity of the discussion as a whole and to the fact that the *Crisis* was meant for a broad public, who would not have been capable of following the kind of individual analysis to be found in Husserl's earlier writings. A conceivable psychological explanation is that for several decades Husserl had carried out such a mass of analyses that in the *Crisis* he only gives the bare results. Perhaps there are manuscripts in the Husserl Archives which contain these missing analyses and which would support the assertions made in the *Crisis* (as it is done, in part, by the Annexe). The present form, however, will cause a certain resistence and reservations in the reader, since many basic claims are simply asserted without further commentary. They give the book a certain character of torpidity and dogmatism which is alien to Husserl's other works.

Certain particulars in the main line of development, i.e., pre-given (*vorgegebene*) life-world – *epoché* – transcendental subjectivity – differ from the version which is found, for example, in *Ideas I*. It is particularly noticeable that Husserl now makes a clear conceptual, as well as terminological distinction between universal *epoché* and phenomenological reduction; in *Ideas I* he did indeed use these two expressions, but gave the impression that they referred to one and the same thing. In *Crisis* the genuine transcendental *epoché* is sharply contrasted with reduction as something which makes the latter possible (heading in §41); at the same time it is distinguished from the 'first' *epoché* in the objective sciences. Both rest upon the 'abstention' (*Enthaltung*) from acceptances (*Geltungen*) which are already present. While the 'first' *epoché* is the abstention from 'any participation in the cognitions of objective science, and from any critical evaluation which aims at truth or falsity (p. 138); it makes possible the grasping of the pre-given (*vorgegeben*) life-world in its full intuitiveness (*Anschaulichkeit*). By contrast, *transcendental epoché* has to do with the 'universal validity of the natural life-within-the-world' (*Weltgeltung des natürlichen Weltlebens*) (in *Ideas* Husserl called this the general thesis of the natural world); it puts 'out of action' the execution of all valuations. It leads to a reorientation from the pre-given natural attitude of man to a new habitual attitude, in which genuine subjectivity arises for us. It makes possible the exploration of the transcendental correlation between world and consciousness. The transcendental reduction consists in inquiring back to the naively pre-given life-world and things in the life-world, to 'phenomena'; they are correlates of subjectivity, and it gives them their sense of being (*Seinssinn*); from its acceptance (*Geltung*) derives the very being of the *life-world.* Hence it is a reduction to the "absolutely final grounds" (p. 157). By virtue of the *epoché* I stand "above the world, which in a very special sense has now become a phenomenon for me" (p. 155). In this reduction becomes obvious the change of the pre-given world into the correlate of ego-centred (*ichlich*) modes of acceptance (*Geltungsmodi*). Their systematic uncovering "makes comprehensible the world as being-for-us, comprehensible as a meaningful structure (*Sinngebilde*) of elementary intentionalities" (p. 171). In the reduction are disclosed "all layers and levels through which the syntheses are interwoven which intentionally reach from subject to subject" and which form a "universal unity of synthesis". Through this synthesis "*comes about* the objective *universe*, the *world* as it is given – concrete and alive, (and as it is pre-given for all possible praxis)" (170/171). "In the reorientation of the *epoché* nothing is lost, none of the interests and purposes of life in the world, and none of the purposes of knowledge. Except that for everything we point out its essential (*wesens mässigen*) subjective correlates, which makes explicit the full and *true* sense of being (*Seinssinn*) of objective being and hence of all objective truth" (p. 179). It is only through the reduction that things which have being in themselves are changed into object-poles, into purely intentional objects. Hence (reduction) is responsible for the formation of the view commonly called 'transcendental idealism'. Where it is carried through to the end, the reduction "to the absolute ego as ultimately the only functional centre of all constitutions", "to the absolutely unique (*einzig*) ultimately functioning ego" is carried out (pp. 190 ff.). This unique ego is apodictically given "with the employment of the *epoché*" (192). As "central position for all constitution" this ego creates the original sense of being (*Seinsinn*) and bestows it upon all there is. We must point out that in the *Crisis* Husserl strongly emphasized that as soon as one "arrives at the apodictically required unique ego" one becomes aware that one is standing within a sphere of evidence of such a nature that any attempt to inquire behind it would be nonsensical" (p. 192). However, this formulation is not clear and unambiguous enough to make obvious what it is that is here stigmatized as nonsensical. But other, subsequent remarks lead to the assumption that Husserl condemns here a whole group of fundamentally important problems. Thus we read "We are not concerned with securing objectivity, but with understanding it". (p. 192) This, too, could be put more clearly. But it apparently was Husserl's opinion that after the discovery of the correlativity between the existing (*seiend*) world and the transcendental subjectivity which lends it its sense of being (*seinssinn*), and after a comprehending investigation of the constitutive processes, in which all possible existential situations (*Seinsbestände*) are 'constituted' and whence they derive their sense of being (*Seinssinn*) or being and particular character (*So-sein*), that after all this no further questions can be asked or answered. It seems as if here the "Questions concerning the justification of reason," which are developed in the last section of *Ideas*, are considered "nonsensical" and hence ruled out. But their point was not merely to investigate the course of the constitution, but in the end also to give a justification of constitution and its result; hence they were to demonstrate not only the possible relativity of the resultant knowledge which relates to the world and its elements, but also of the likewise possible ultimate decision concerning the existence (in whatever sense) of the world. It seems that the above passage in the *Crisis* simply and programmatically declares this whole complex of profound problems to be nonexistent. Accordingly, the ultimate point that could be reached in transcendental phenomenology would be to understand how all being which is not itself pure consciousness issues from the processes of the finally constituted subjectivity.

This then is the upshot of the passages in the *Crisis* which are concerned with *epoché* and reduction. The being of the objective world from the natural attitude and this world itself "are not lost because they are understood from within the sphere of absolute being (of transcendental subjectivity) where they ultimately and truly have their being".

This giving up of questions of justification is indeed an important change in the function of transcendental phenomenology compared with the standpoint of *Ideas*, Vol. I. It removes problems whose solution would give an importance to phenomenology which it cannot attain if it merely satisfies the desire to understand the course of constitutions and to uncover all correlativity. In the last section of the *Crisis*, Husserl seeks to prove that 'pure phenomenological psychology' is identical with transcendental phenomenological philosophy. In view of the fact that Husserl has given up questions of justification, one should rather say that pure psychology is the only thing left over from transcendental philosophy. *Hence the above assertions from the Crisis should be clarified anew, and then we should consider if we can accept them.* This requires not only a special consideration of the various groups of problems of constitution, which are not treated at all in the *Crisis*, but also a discussion of the *question what is it that forms the guiding thread of the constitutional inquiry from the deepest layer of the originally constituted consciousness of time up to constituted units of sense through which things of the world and other objects are given.*

It has been asserted several times that the *Crisis* opens a view upon a sphere of being (*Seinsgebiet*) which Husserl had not investigated earlier, namely the 'life-world'. Allegedly, this is the grand discovery of the *Crisis*. I cannot agree with this. Basically, the life-world is nothing but the 'world from the natural attitude' in *Ideas*, Vol. I. Certainly, the *Crisis* contains several valuable special discussions and remarks concerning the 'life-world' and 'things in the life-world' which did not occur in *Ideas*, Vol. I. But this does not suffice to show that we should consider it to be something other than the world from the natural attitude. On the contrary, the *Crisis* contains several passages which clearly indicate that it (the life-world) is given and always pre-given in the natural attitude which man takes prior to the *epoché*. This 'being pre-given' means nothing but that this presence is original and forms the basis of all acts of cognition and operations which refer to individual occurrences (*Tatbestände*) in the world, and that it *precedes* all contemplation and critical calculation which aim at knowing and knowledge (*Erkennen und die Erkenntnisse*). In particular, it must precede all attempts at conducting a transcendental *epoché*; for the 'total' or universal *epoché* can only be based upon this 'pregivenness' of the world, and must be performed upon it. But then this 'pregivenness' is nothing but the 'general thesis' of the world which is always present from the natural attitude. Hence, that which is given in both cases must be identically the same. It is of course true that mankind lives in this world of the natural attitude, and that it may therefore be called the 'life-world'. But if we wanted to make the fact of living in the world into an essential characteristic (*Bestimmungsgrund*) of the life-world, we would run into certain difficulties in view of the fact that Husserl recommends at the same time the so-called 'first' *epoché* from positive (objective) science, especially natural science, as a method for a pure grasping of the life-world with its full intuitive content. For this *epoché* excludes from the world of the natural attitude all facts discovered by natural science, which go far beyond the intuitive form of the world, e.g. the whole range of microscopic facts or chemical processes which are explored through theoretical considerations. Strictly speaking, (and Husserl justly emphasizes this) objects and processes which are idealized through mathematization also go beyond the simply observable. But all these objects lie within the world in which we lead our daily life; we reckon with them in our actions, and we must do it more and more, for the advancing technology is rooted in science and has repercussions upon the development of natural science itself. Although we do not perceive them directly, we constantly encounter facts (*Tatsachen*) which make them appear, and they are always present and condition our actions and our daily life. With the development of natural science in all its branches and the advancing mechanization of life, the limits of the life-world are pushed back, since the life-world is materially determined by the historical development of mankind. On the other hand, the world of the natural attitude is sharply and constantly fixed by definition as the world which is given *before* the universal *epoché*. It is true that we neither live nor (what comes to the same), act within that which is uncovered by universal *epoché*; it has made the world inaccessible, and we can have no practical interest in the residue.

This then is a brief summary of the most important new developments in the *Crisis*. One can hardly speak of a new phase of Husserl's phenomenology. As before, the emphasis is upon opening a way to transcendental phenomenology by means of the *epoché*. One can also not claim seriously that in this book Husserl uncovered the basis for the crisis of European science, and that it lies in the fallacy of objectivism. It is clear that we must not forget that the work is unfinished. But it seems highly improbable that Husserl could succeed in showing that the source of this crisis lies in certain purely theoretical decisions. The true greatness of this book lies in something quite different, namely in the fact that Husserl did not despair in the final and most difficult years of his life, but that he defended the legitimacy of his philosophy quietly and through scholarly work.

No matter how highly we value and honour this final writing as the deed of his spirit, we must consider, in an unbiased way, the reservations which will be occasioned by some of the fundamental assertions of the *Crisis*. Did this book overcome the various difficulties which remained after *Ideas* and the *Cartesian Meditations*? Although the methodological procedure in which transcendental phenomenology is founded is more correct, it must be said that at decisive junctures the text shows hardly any progress. Difficulties in some important points could perhaps have been overcome by some very far-ranging and innovative analyses. But the role which Husserl had intended for the *Crisis* made it impossible for him to put before the reader the concrete results of his studies, which had lasted for decades. The attempt would have taken several years of work, and the *Crisis* would have reached a size of several volumes, and would thus have failed in its purpose. If we today compare the

Crisis with the other volumes of the *Husserliana*, in which concrete individual analyses are collected, we see that the *Crisis* gives no indication of how far Husserl was advanced in many individual problems, and that his truly creative period does not lie toward the end of the thirties, but much earlier. If it becomes possible to publish Husserl's most important writings concerning the phenomenology of time, we shall see before us the supremely masterful uncoverings of matters that are extremely difficult to grasp; by comparison, all his efforts to invent a method for the correct founding of transcendental phenomenology will seem unimportant. Nonetheless, we must here call attention to some points of the *Crisis* that could stimulate further discussions.

Above all, what is actually meant by the sentence:

> Once one reaches the ego, one becomes aware that one is standing within a sphere of evidence of such a nature that any attempt to inquire behind it would be nonsensical,

and by the second sentence:

> The task is not to secure objectivity, but to understand it. (pp. 192/3).

This "sphere of evidence" of the individual ego is undoubtedly the same thing that was earlier called the sphere of pure consciousness which, according to Husserl, is discovered in immanent perception after the transcendental *epoché* is carried out, and which we are to investigate by orientating ourselves toward the essence of pure subjective processes. Unfortunately, little is achieved by making this statement. In the *Crisis*, Husserl does not talk about 'immanent' perception, but he makes polemicel remarks against the distinction between 'external' and 'internal' perception, which he himself had made earlier. What is to be called in question, however, is not 'internal', but 'external' perception. The reason seems to be that what is given in external perception, i.e. the things of the life-world and perhaps the life-world itself, is according to Husserl's statement in the *Crisis*, something *subjective*.[2] After the *epoché* the life-world, according to Husserl, reveals itself as something 'subjective', no matter how 'objective' it appeared to us *before* the epoché. Of course the quotes around the word 'subjective' makes us uncertain how this word is to be understood, whether, and in what sense, the things of the world belong to the sphere of pure consciousness, and especially to the sphere of evidence which is indicated in the quoted sentence. For example, on p. 248 we read:

It is only through universal *epoché* that one sees, as a thematic field of its own, what pure egoical life actually is: as an intentional life, as being affected in this intentionality by intentional objects which are valid as appearing in this intentional life, as being directed toward them in many different ways, as dealing with them. Everything, every 'with which' of these dealings itself, belongs to pure immanence and must be descriptively graped in its purely subjective modes, in its implications together with all intentional implications contained therein.

Hence the intentional objects, being located in pure immanence, seem to belong to the sphere of pure consciousness, and are 'immanent'; they can be grasped in a specially orientated immanent perception. Hence it is not only what he earlier called 'adumbrations' (aspects), in whose experiencing things are given in perception, but also – though apparently only after the *epoché* – the things of the life-world which are now given immanently. Husserl seems to assert that what is often called 'external' perception, of which we (before the *epoché*?) falsely suppose that it immediately presents to us independently existing objects of the life-world, does not reach the 'in-itself' of these things and of the (life) world, but remains 'subjective'. On the basis of what is said in the *Crisis*, we must now also ask whether it remains relative? Does this 'relative' belong to the sphere of pure consciousness which, as was said in *Ideas*, Vol. I – is the sphere of absolute being and subsequently becomes immanently graspable and hence lies within the sphere of evidence? Let us set aside the problem how one and the same life-world, and the things in the life-world, can be given in 'external' (transcendent) and 'relative', as well as immanent, perception, and let us merely remember that Husserl in *Ideas I* emphasized that what remains after the *epoché* are not things but mere "thing-noemata"[3] as correlates of perception, indeed *thing-noemata* which are correlates of external sensory perception.

We know from the earlier writings of Husserl that these *noemata* are constituted as intentional unities of sense (*Sinnehinheiten*), which are constituted from a multiplicity of 'adumbrations'. If after carrying out the *epoché* we grasp them as fully constituted and still 'go back' (*Zurückgreifen*) to the various adumbrations (aspects) it is reasonable to ask whether this step does not already lead us, according to Husserl, 'behind' what is given immanently, and hence into some nonsense. But we take this step with Husserl himself, in order to understand how a manifold of different experienced adumbrations (visual perspectival views from various points of view and distances, tactile and auditory adumbrations, etc.) lead to *one* uniform thing-like (*gegenständlich*), 'noematic' sense. In doing so we inquire back to the associated acts of perception, which initially – especially before the *epoché* – belong to the so-called 'hidden' consciousness, to acts, that is, in which these various 'adumbrations' are experienced. On the basis of the experiencing of these adumbrations there arises a perceptual presentation (*Gegebensein*) of a determinate thing, or the constitution of a certain *thing-noemata* the manner of whose constitution corresponds to the manner in which these adumbrations are experienced and to their content. This going back and inquiring after the adumbrations and the acts in which they are experienced does not seem to be nonsensical at all (according to Husserl himself in *Ideas*, Vol. I), although it leads us 'behind' the sphere of the immanently and evidently given. But it is well known that this is not the end of the matter. Even the individual 'adumbrations' (aspects) are themselves certain 'completed' constituted structures which conceal a manifold of original 'unities of sense' which flows 'behind' them; they too can be drawn reflectively from their 'concealment' in order to experience and understand how it comes about that a unitary, and, as it were, stable adumbration (aspect) of a thing forms in the passive experiencing of a flowing, temporally stretched-out mass of 'unities of sense'. And here we can look back again to the associated subjective processes (to the simple 'having' of a sensation), in order to find out how the passing succeeding sensation-experiences of these flowing data organize themselves, as it were, into units and at the same time stabilize within the flow of growing time. But it is not the case that everything is simply sensed, all at once, finished at a moment, in a pure original impression. The narrow sphere of original impressions passes at once into retentive consciousness, but does not disappear abruptly as something no longer present, rather it remains as an echo and mixes with what is presently originally sensed, and modifies it more or less, depending on whether the retention is more or less lively and extensive, and depending on how active and intensive the original impressions are which presently occur and, finally depending on how their unfolding is impinged upon by protention. All this we discover in this way: in simply (unreflectively) having unities of sense which are the 'finished' constructional result of this entire complex and changeable process, we inquire back to the manifold of experience in which they are constructed, and in this way trace the entire process of constitution of the unities of sense. In this way we can learn to understand how stable and organized unities of sense can come about, and how their synthetic togetherness leads, and can lead to unified 'adumbrations' (aspects) of one and the same perceived thing. In this reflective procedure we step back, as it were, from the *thing-noema*, and also from the variously constituted adumbrations (although we do not lose sight of them) and concentrate upon the constitutional backgrounds of perception without committing any nonsense in asking back for, and inquiring into, these initially concealed matters of fact. Of course we go back from something given 'in evidence' to something which is initially concealed, and we grasp the dependency and the relativity of each of the higher constitutional levels upon their background, which is at first hidden, although the highest constituted objects are evident at first, and are undoubtedly given in the first reflection which is carried out after the *epoché*. This extremely complicated process, which can take place in very different ways of step by step constitution, as well as reflexive reduction to the lower and more concealed constitutional background and to the final time-constituting consciousness, was well-known to Husserl, and he did not desist from an investigation of this sphere. In his warning of the possible nonsense of 'inquiring back' he does not pay attention to the important difference in 'evidence' which depends upon the different constitutional levels, nor does he take into consideration that the transition from evidence on one constitutional level to the next lower, or more original, is not identical with the transition from evidence to non-evidence and to possible nonsense.

We have noted that the locution of inquiring 'behind the sphere of evidence' is metaphorical and hence can be understood in various ways. It is likely that this phrase was to express Husserl's concern that in the clarification of many facts within the 'sphere of evidence' we should not inquire back to something which is not 'transcendent' with respect to pure consciousness. But it is well-known that the expression 'transcendent' is also ambiguous and that the 'transcendent' can be sought in two different directions, as it were.[4] To begin with, 'transcendent' can mean 'not forming a real part or moment of pure consciousness, at the same time being a whole, which lies wholly outside of consciousness'. Husserl would conceivably answer to this that such a 'transcendent' does not, and cannot exist. I also do not want to assert that there really is such a transcendent, and I do not want to give arguments which would establish such an assertion. But this should not prohibit me from defining transcendence in this way, in order to inquire afterwards whether and in what cases something exists which is transcendent in this sense. However, I would agree with Husserl in rejecting 'transcendence' in the sense in which it signifies something which is absolutely unknowable (as for example Kant's thing-in-itself, in one of the possible interpretations of Kant's *Critique*). But if we hold to the sense of 'transcendence' given earlier then the transcendent is to be sought either 'behind' the meant intentional objects, or else 'behind' the act of consciousness or the ego of these acts. In my view, every intentional object is 'transcendent' in the indicated sense with respect to the conscious acts which determine (*entwerfen*) it. It is transcendent even vis-à-vis the manifolds of 'adumbrations' (aspects) in which it appears to the ego, so that it cannot be grasped in 'immanence' or immanent evidence. Nonetheless, its content and mode of being can be recognized with certainty, since it derives all its being and being-thus from the intentions of the acts which determine it.[5]

But if an object is to be 'transcendent' not only with respect to the conscious acts, but also with respect to the sense of the *object-noema*, then it must – if it is to exist at all – contain all its formal and material determinations *within itself* and must therefore have an amount of determinations *of its own* which effectively build it up, and are not merely assigned to it; all these determinations taken together are then the 'in-itself' of that object. On the basis of his analysis Husserl does not tell us in which sense he speaks of 'objectivity' or 'objective'. We must suppose that he would say of an object which is transcendent not only to the acts but also to the purely intentional objects, but which at the same time has an 'in-itself' in the just indicated sense, that it is 'objective' or belongs to the sphere of 'objectivity'. Husserl takes it to be a fundamental error of modern science and also of 'non-idealistic' philosophy to look for, or claim the existence of, such 'objectivities' in the world or 'behind' the things of the life-world which are given to the ego. What support does this important decision have? Husserl would perhaps answer in this way: if we are to assume an object which exists 'in itself', then only on the basis of a cognition which is accomplished in acts of consciousness. This means that for such an object we must uncover the correlate of 'subjectivity' in which its being and being-thus is demonstrated. But when this occurs, maintains Husserl, its being, as well as its sense of being derive from this subjectivity (the corresponding acts of consciousness) which are correlative to it. It then becomes a *cogitatum* and thus belongs to the sphere of immanence. That it is nothing but the intentional correlate of subjectivity shows itself after the execution of the universal *epoché* in the immanent evidence of transcendental analysis. The transcendental reduction shows it to be the object-pole of corresponding subjectivities of an ego-pole.

It is of course true that we can *know* the mode of being of such a transcendent object only from the mode of being which is assigned to, or bestowed upon it in a legitimately founded true cognition. But this does not mean that it can in this genuine sense *exist* only *if* we *assign* it this existence in this mode of being, that its existence *derives from* this assignment, that it *owes* its existence to it. But Husserl constantly expresses himself as if with all external, life-world objects the first were the basis of the second. But this holds only for purely intentional objects as they are created, for example, in poetry. But for this very reason they have no composition (*Bestand*) or formal and material determinations which are founded within *these objects themselves. In acts of consciousness they are freely and creatively composed by virtue of the intentions which are contained in these acts, but they are not made into objects which have being autonomously or in themselves, even if the mediation of language, for example, turns them into inter-subjectively accessible objects.*[6] *This freedom* of determining function is not present in legitimately executed acts with respect to the objects with autonomous being that are grasped in these acts. Also, it is not true that the act *of cognition can or may be freely creative in this sense*, as the poetic acts are. The entire effort of the cognizing subject is to form the sense-content of the acts of cognition in such a way that they *conform* to the mode of being and the determination of the object to be cognized, that they *obey* it, *subject themselves* to it. In this function of subjection, neither the act of cognition nor the cognition itself are free; they are not arbitrarily creative. Within the customary 'realistic' positivist epistemology (or psychology) this is taken to imply that the course and form of the process of cognition is causally determined by the independently existing cognized object. Husserl says justifiably that this is not permissible since it leads to a circle: an independent object (a physical thing or a real physical process, e.g. a series of electromagnetic waves) is presupposed so that existence can afterwards be conferred upon it upon the basis of an alleged causally determined process of cognition. Just this is the nonsense which Husserl feels whenever there is talk about objects that are 'in themselves'. However, one does not have to follow this road, and commit this nonsense, in order to uncover the conforming tendencies of the process of cognition, and to show that this process is not completely free and independent, but that its essence is determined at the very least by the fact that it grows from a *tendency to conform to something which it has not created itself*, which it can merely find, and that it is true to its mission only as long as it conforms to this noncreated something, and avoids everything that is merely *invented*. We can see in the manner which this process of cognition takes that it does not proceed freely, not arbitrarily, but that it is something where the ego is concerned, at every step, to subject itself to the noticed (*vorschweben*) object, to give it a hearing and to note only what is present in it, what it has learned from this object itself. Without having to take recourse to the real influences of the object in question, one can show in the course of cognition itself where and when cognition has subjected itself and must subject itself to the object, and where it has independently added something. One can find this out under the auspices of the transcendental *epoché* and does not commit any unallowable circle or nonsense, if one investigates the process of cognition from the outset under the aspect of its obedience and subjection to what the cognized object says. On the basis of the immanently graspable process of cognition, within the realm of 'pure subjectivity' – as Husserl says in the *Crisis* – we can distinguish between those manifolds of cognitional processes which subject themselves and conform to the objects to be cognized, and those other acts of consciousness which freely add something to the meant object, which assign and describe something to it although they have not given it a hearing and have concerned itself with it only very little, if at all. It is precisely the function of the so-called questions of legitimacy of reason which Husserl has treated in *Ideas*, Vol. I to carry out this distinction and especially to find out the principle according to which this distinction must be carried out in the concrete case. If one omits this and is concerned merely with the purely contingent factualness of the so-called processes of constitution, then one can indeed understand, in the individual case, how it happens that certain nexus of consciousness lead, for example, to the appearance of a material thing, or to the expression of the conduct of a person, but one can never know, under these circumstances, whether in these acts of consciousness one meets reality, an effectively existing object, or whether one misses it, or simply composes it poetically.

In this procedure we do not have to go beyond the sphere of immanence or evidence and do not have to naively presuppose anything, in order to assert that in cognition we do not have to do with mere phenomena but grasp an 'in itself'. Hence we do not have to take recourse, in the sense of naive empiricism, to causal connections between cognition and the *cogitatum*.

But, as I have pointed out, the inquiring back 'behind' the sphere of evidence can also be aimed in the direction of the ego itself. This is unjustified precisely if we presuppose, in the sense of the classical rationalists, that behind the ego, or, better, at its basis, there is a 'substance' and if we believe, moreover, that this transcendence must necessarily be presupposed, but that it is at the same time unknowable, that it has, for example, the form of an autonomously existing monad. One also commits nonsense if one asserts that it has the ability to carry out and achieve absolute knowledge of being, and that it can penetrate into the essence of any reality. Or, if one asserts that it could differentiate between genuine truth and falsity, and could decide, in each case, what has actually been achieved in a process of cognition. If we wanted to solve the questions for the legitimacy of reason, which constantly force themselves upon us in cognition, through such metaphysical hypostases, then Husserl would be correct in saying that we are inquiring back in a nonsensical way 'behind' the sphere of evidence. But this is not necessary in order to be certain of the objectivities and do not merely understand them. To avoid the nonsense it is also now necessary to make the unique, absolutely existing ego, which remains outside the reach of transcendental *epoché*, into a pure property-less ego, as Husserl initially attempted in *Ideas I*, or to weigh it down with the so-called 'habitualities', or finally, to degrade it soulless and mindless, to a pure function of the subjective being of intentional acts. In the pure essence of the transcendental ego as well as of pure consciousness we can see, whether it has independent being, or whether its being is dependent and requires of itself a mental subject, whose essence it predetermines.

Hence it is not only possible but quite necessary to ask the question of legitimacy respecting the objectivity or non-objectivity of the cognition or of knowledge in general. Actually, Husserl does it himself by putting this question with respect to the knowledge of the things in the life-world, of the life-world, and of the world in general; he not only puts this question but in a certain sense even *answers* it. For the thesis that it is a *mistake* to think that this knowledge is 'objective' i.e., that it reaches the being-in-itself of the world or of the things in the life-world, while it is *de facto*, as Husserl emphasizes several times, 'relative' and leads and can lead only to purely intentional objectivities, correlates of especially chosen acts of cognition, this thesis, which is stated several times in the *Crisis*, is itself nothing but an answer to the question of legitimacy. Incidently, this 'relativity' is discussed only in a few cases of so-called 'external' perception, which I have briefly discussed, but it is not demonstrated; hence, if this is all there was to be said about it, *nothing has been decided concerning the cognition of physical objects*. The *Crisis* contains the undoubtedly beautiful discussions concerning Galileo's rationalizing and idealizing of the life-world by means of the transition from the concrete data to the mathematical, and especially, geometrical limit-concepts, which can never be fully satisfied in concretely given things of the life-world. Still and all, mathematizing physical cognition as well as mathematical cognition itself, in spite of the attendent 'idealization' of nature, could be in itself 'objective' and could grasp at least mathematical being in itself. If we did not know Husserl's earlier discussions in *Ideas I* about the physical objectivities, and if we did not know that, according to Husserl, they are merely a higher level of 'objectivation', a higher level of constitution of special, unintuitive intentional objects, then we also would not know, that, according to Husserl, it is illegitimate to take mathematizing physical cognition as 'objective', and that it would be a grave error to take its correlates for 'true nature'.

No matter how this may turn out in the end, this entire consideration is not a renunciation in transcendental phenomenology of the consideration of questions for the legitimization of reason, it is not a renunciation of the attempt to 'secure objectivity', but it is a certain negative answer to the legitimacy questions for the truth value of cognition in 'external nature'. This negative answer is presupposed by Husserl when he declared the life-world and the physical world to be something 'subjective'.

The manner in which Husserl reaches this conclusion in the *Crisis* does not differ essentially from the way in which these problems were decided in *Ideas I* (even *before* the *epoché*). The only difference is that here everything is set forth in a much more stark manner, and without appeal to concrete analyses. We simply cannot doubt that Husserl in a life of hard work carried out a great many analyses of this sort. Part of these investigations have already been published after the *Crisis*. But so far it is not clear whether all these investigations have been carried to the point that a final result has been reached and the controversy concerning the capacity of world-cognition as well as the autonomous existence of the world has been decided, and the decision fully confirmed.[7] In the *Crisis* these numerous analytical investigations are, as it were, suppressed and all that is given is the main thesis which perhaps results from them. One gains the impression that in matters of 'idealism vs. realism' everything is decided from the outset, and even the either-or is not taken into consideration. One expects, in the unfolding of the methodological problems of transcendental phenomenology, that the decision in the idealism problem is not rendered simply *after* the execution of the transcendental *epoché*, but that it can be given only in the course of reductive considerations (*Reduktionsbetrachtung*), as the final result of difficult analytical investigations. When I wrote Husserl many years ago that the reason why I could not reconcile myself to his transcendental idealism probably lays in the fact that I did not know his many individual analyses on which he worked for many years, Husserl answered that it was in no way a matter of further analytical results but that above all one would have to experience the great transformation of the *epoché* or/and reduction. Thus it seems that the *epoché* is not the means whereby the problematic of the correlativity between world and world consciousness is uncovered, and which makes possible its investigation so that the results which are achieved in this way form the basis on which the decision can be made; rather it is a method which itself brings about the decision from the outset (namely that the life-world and perhaps the world in general is 'subjective') independently of how far the individual analysis of constitutional problems have been carried. But in this case one would have to beware of executing the transcendental *epoché*.

From this whole nexus of problems I should like to go on to consider a particular formulation which is repeated several times in the *Crisis*. As I have cited above, Husserl says that by executing the transcendental *epoché* the life-world (or the things of the life-world, or perhaps even the world in general) *changes into* a mere phenomenon, or that it *becomes* a phenomenon, that it *comes about*. This formulation, which is not merely an unguarded form of words has already left the decision of the whole question behind itself. Before the *epoché* neither the already given life-world, nor the things of the life-world with which we stand in various causal relations and which we recognize in psycho-physical processes, are given as mere 'phenomenon' but both are given as something that has a being which exists quite independently from pur cognitive experiences, and which continues to exist if we do not perceive it, or think of it. Put briefly, things are given to us as existing autonomously. We can change them or destroy them by bodily activities, but they prove recalcitrant when we want to attack them with our thoughts, perceptions, or feelings. Our consciousness proves powerless toward them. No matter what the experience, no matter what purely conscious changes we live through, they are in themselves what they are, and we cannot influence them in this manner.[8] Husserl will readily admit all this. But how can the execution of the *epoché*, which is, after all, a transformation that is altogether bound up with consciousness, have any real influence upon things? It does not matter at all, in this connection, whether we have consciously carried out that *epoché*, whose correct description is so difficult, or whether we have simply succumbed to an attack of psychasthenia. Things *cannot* thereby *change* or *become* phenomena. Only their mode of appearing changes through and through. For, in carrying out the *epoché we pass* from the real world or, if you will, the life-world into the *world-phenomenon*. Fundamentally, what took place was that *before* the *epoché* we grasped it, the world and the things, in *other* phenomena than the modes of appearance which we have *after* the execution of the *epoché*. Hence, carefully considered, matters are not as the sense of the just quoted locutions would require. Husserl was quite correct when he points out, in other passages, that *nothing* changes as a consequence of the *epoché*, and if something had changed, then the *epoché* would be nothing but a case of psychastenia, which it is *not*. For, before as well as after the *epoché*, the things of the life-world present themselves in the being-character of autonomy, as fully founded within their own self, and altogether independent of our experiences. *Therein* the epoché has changed *nothing*. The *epoché* has only this as a consequence that we leave, so to speak, simple reality and pass over to purely intentional objects which become *visible* in the *epoché*, in whose *content* the *phenomenon* is preserved as a *self-determinate, autonomous reality*; at the same time the intentional object becomes obvious. It comes into sight precisely because of the *epoché*. In the uncovering of this intentional object it becomes clear that the being-character of autonomy which appears in its content, is only a phenomenon and not true autonomy of being (*Seinsautonomie*) of the real object. Before the *epoché* the intentional object which belonged to the act of perception, was, as it were, 'transparent', one could look through it so that it was not itself visible. Because of its complete transparency we had, or at least thought that we had, converse directly with the real thing. The *epoché* has as its consequence that we have lost the immediate contact with reality and that we now have to do only with the intentional object and with a special content (*Gehalt*) with which it now seems to feign the real object which was earlier given directly. By investigating this content, we can find out what the real object in question was taken and believed to be before the *epoché*. There only remains the question whether it was taken to be such and so in a founded and legitimate manner, or whether it was merely an illusion which came about through a phenomenal intentional object with a certain content. When Husserl says that the tree in the garden can burn while, to use his phrase, the 'perceived tree as such' i.e. the purely intentional object in whose content the tree in the garden is given, cannot burn, he says nothing but that there is a heteronomy of essence between the two 'trees', which cannot be bridged, although the 'second' tree brings about the self-presentation of the 'first' tree *before* the *epoché*, and does not itself become apparent in its heterogeneity, while *after* the *epoché* it moves into the foreground and detaches us from the direct and immediate contact with the flammable tree. This heterogeneity between the real and the merely intentional object, (tree and tree-phenomenon) makes it impossible that because of the *epoché* the flammable tree in the garden *becomes* or *changes into* the merely intentional 'tree'. If Husserl seriously asserts this, it cannot mean anything but that he denies from the outset the heterogeneity between the two "trees" which he originally maintained himself, or he gives it up and holds from the outset that the real tree is a merely intentional object, or a noematic unity of sense (*Sinneinheit*) which has a sense of being (*Seinssinn*) in its content, which is forced upon the content of the intentional object by the corresponding act of consciousness, indeed by the act as it was carried out *before* the *epoché*. If this act were not from the beginning an act of sensory perception, but of poetic intention which composes intentionally a poetically phenomenal objectivity (*Gegenstandlichkeit*), then the object which occurs in its content (thing, man, etc.) would not show the sense of being (*Seinsinn*) of reality, but only a special modification of this sense of being (*Seinssinn*) as it is characteristic for objects depicted in literary works of art (hence of objects which are not real) and which deceptively appears in their content. But there is an important difference between the intentional object 'tree' which is merely depicted in poetry. It consists in this: as poets we are relatively free and independent of the surrounding a real world when we form these objects, and we do not have the tendency to conform to this surrounding reality and grasp it in *its true* sense. We merely want to form freely a new fictitious world which only occasionally depicts the real world and simulated it in esthetic garb, but does not have to do so. But poets are not knowers of the world and they are not philosophers who carry out the *epoché*; they are simply 'poets', who compose a poetic 'world' with linguistic means, in quasi-judgment. It is not our task as philosophers to preserve poetic freedom; on the contrary, we must give it up in order to discover how the content of our intentional perceptual experiences must be formed so that we are able to arrive at knowledge of reality, and can then, after the *epoché*, understand the reduction and function of cognition and investigate its capabilities.

Husserl has reduced all reality to purely intentional objects (*Gegenstandlichkeiten*) but he never inquired into the mode of being, and the peculiar form of, intentional objects; he consequently overlooked that they are essentially different in both these aspects from objects with autonomous being (particularly real objects). Thus he also did not see that purely intentional objects have a strange ambivalence of form; on the one hand, they have an intentionally formed content; as such they are supposed to be e.g., a tree, a man, etc. On the other hand, they have a structure which belongs to them *qua* intentional structures.[9]

All judgments which can be formed about purely intentional objects fall into two groups: those concerned with their content, and those concerned with their formation *qua* intentional objects, e.g. the judgment concerning their formal ambivalence. The content of purely intentional objects can be such that in it occurs the character of real autonomous being. But as purely intentional objects they have heteronomous being; hence all existential, formal, and material moments which occur in its content are merely meant, are merely assigned, but are not truly embodied (*verkörpert*), immanent, as it would be the case with an object of truly autonomous being, especially a real object (if indeed there is such a thing).

If Husserl had noted these formal peculiarities of purely intentional objects, he could have avoided certain contradictions, but he could also have opened up a different view of the problem in the real or life-world. The formal difference between the autonomous and purely intentional object would have made it impossible for Husserl to say that the objects of the life-world (or the real objects in general) *change* as a consequence of the *epoché* into 'object-poles' or *become* 'object-poles'. For precisely this is essentially impossible in both directions: the real things (things in the life-world, if indeed the life-world is the world in which we live and act) cannot become purely intentional objects, nor can the purely intentional objects become real things.

Another remark in conclusion. Husserl accuses modern science and certain schools of philosophy of succumbing to the fundamental fallacy of 'objectivism'. It is not easy to take a position vis-à-vis this thesis. In the *Crisis* Husserl did not carry out a further clarification of 'objectivity', or rather of its concept. But various locutions seem to indicate that objectivity is closely related to the "being in itself" of the particular object (*Gegenständlichkeit*). And in the main thesis, science is charged with committing the error of striving for and attaining objectivity in its knowledge, while it can never actually grasp the 'being in itself' e.g. of natural things. 'Objective' psychology, too, which is concerned with psychological occurrences of men living in the real world, is rejected as an aberration for just this reason. Instead, so-called 'pure psychology' is demanded and its problem situation, as well as the conditions for its conduct (among other things the *epoché*, which is indispensable for this psychology) are developed. It follows from this that it is considered identical with transcendental phenomenology. But what about this phenomenology itself? Is it also not supposed to seek 'objective' knowledge? Is it also incapable of advancing to the 'in itself' of pure consciousness, of absolute subjectivity? This seems to contradict Husserl's aims. After all, phenomenology is supposed to grasp the ultimate, the absolute of pure subjectivity in its own character of subjectivity. Hence it is to pursue 'objective' knowledge of 'subjectivity'. Thus, what is to be condemned is not 'objectivism' as such (in the sense here supposed), but only the objectivism of the positive natural sciences since, as Husserl asserts, they are not capable of reaching the being-in-itself of nature. If this is indeed true, then the special position of these sciences is responsible for the fact that they can be charged with 'objectivism', but the general tendency to gain 'objective' knowledge should not therefore be rejected. But is this certain? Does 'objectivism' merely contain the moment of being-in-itself, or is there a number of other moments as well, such as the special form of the 'objectivised' (*des objectivierten*), the not-being-correlated-with consciousness, and others, which are connected with it? And is there not a new subjectivity which is correlated with each self-comprehension (*Selbsterfassung*) of subjectivity, and does this not open up the possibility that this new 'subjectivity' cannot grasp consciousness in its in-itself, so that we can only grasp phenomena of consciousness and not consciousness itself? So that one would have to carry out a new *epoché* at every juncture, in order to discover and to realize that *every* tendency toward 'objective' grasps of subjectivity is only a delusion which must be renunciated?

There are passages in the *Crisis* which seem to indicate this. But then the dream of phenomenology as a rigorous science would have come to an end. On the basis of the *Crisis* these are only disturbing question marks which can neither be removed nor confirmed by reference to the assertions in the text.

Translated from the German text by Rolf George

NOTES

† We regret to announce the death of Roman Ingarden in June 1970.

1 After a philosophical evening with Husserl in October 1927, Oskar Becker said to Heidegger in my presence: "Actually, we are Dilthey-people". And Heidegger did not object.

2 Cf. p. 223; "of course it makes a difference in the manner of life-worldly experience, whether one experiences stones, rivers, or mountains, or whether one instead reflectively experiences one's experiencing of them, or other ego activity or some other activity... for psychology this may be an important difference and may lead to difficult problems, but does this change the fact that every thing about the lifeworld is obviously 'subjective'?"

3 It is true that in the *Crisis* Husserl does not speak about *noemata* and *noesis* but he also does not reject the dinstinction between 'thing' and 'thing-noema'; rather he

repeats his assertion from *Ideas*, in that the thing tree can burn down, but that a 'perceived tree' as such cannot burn down. In the *Crisis* Husserl does not use the terminology which was laid down in *Ideas I* in order to facilitate things for his readers, whom he obviously takes to be laymen; but this frequently obscures justifiable distinctions which had been carried out earlier. In my opinion, this has led to several misunderstandings, especially with those who have read *Crisis* without first having read *Ideas*, which they took to be 'dated'. Much damage was done by the fact that in France, for example, they first read the *Méditations Cartésiennes* and that Husserl became known from this vantage point. In addition, many have read Husserl only after Sartre's *L'Être et Le Néant* and have become acquainted with Husserl in an existentialistic light and have misinterpreted him.

4 Cf. my *Der Streit um die Existenz der Welt*, §48; 'Zusammenstellung verschiedener Transzendenzbegriffe', Vol. II/1, p. 224ff.

5 The contrast between 'transcendence' and 'immanence', which is here set forth, is of a formal or, if you will, structural nature. It has nothing to do with the epistemological contrast between evidence (especially apodictic or undubitable evidence) and non-evidence in whatever sense. These two contrasts overlap.

6 Of course I do not forget that the creative activity of the poet is also determined by various circumstances, but he is not influenced by autonomously existing objects which exactly corresponds to the intentional objects that he creates; he does not form the poetically and freely created objects under the pressure of such real objects. If they are so formed, then they are mere duplications or reproductions of reality and not a poetically formed new world.

7 In the Husserl Archive there is a sea of manuscripts which are related to problems of constitution; one should spend at least as much time in their study as it took Husserl to write them. But whatever one looks at, though it is often a very interesting contribution, will be a fragment which soon abruptly ends, and one cannot find the truly synthetic whole which leads to a final result after all the individual analysis. On the basis of Husserl's letters I expected the Archive to contain a new edition of the *Cartesian Meditations* which Husserl worked on for a protracted period around 1930. But I was shown a whole host of short manuscripts (in copied form) which contained several individual observations belonging to the theme of the *Meditations* but no large synthetic revision.

8 But already states of other persons seem to be a different matter; they do not appear to be quite as independent from our own experiences. This would have to be investigated further.

9 It was no accident that I chose the product of literary art as the topic of the book which I published in 1931. I had supposed that literary productions and the objects which are depicted in them are purely intentional structures (*Gebilde*) and that they differ in their mode of being as well as in their form from the mode of being and form of real objects in such a way that the latter must not be reduced to the former. My inquiries have confirmed this supposition, and this was the first step in the dispute with transcendental idealism. In that book I made for the first time a distinction between the content and the structure of purely intentional objects (l.c. §20). But the formal difference between (purely intentional) objects of autonomous and of heteronomous being was first conclusively demonstrated with the *Streit um die Existenz der Welt*.

DALLAS LASKEY

INGARDEN'S CRITICISM OF HUSSERL

It is a matter of history that few of Husserl's former students from the Göttingen and Freiburg era were willing to follow him into the perspective of transcendental phenomenology.[1] To be sure, as Husserl remarked in a letter to Roman Ingarden,[2] a whole generation had passed by and former students had moved on to their own special areas of interest; now they were unable or unwilling to comprehend the new development. While Ingarden had moved to a formulation of his own in *Das Literarische Kunstwerk*,[3] he was nevertheless considered by Husserl as one of the latter's most faithful and devoted students.[4] Perhaps because of this respect and trust, Ingarden was asked to comment on Husserl's *Cartesian Meditations* and he obliged with critical remarks on the first four meditations.[5] Husserl was so impressed by these that he demanded that they be included along with the text in the German edition. The richness and scope of Ingarden's remarks have not been fully appreciated by the English-speaking world, no doubt in large part to the fact that they were not translated. I propose to comment on some of these in so far as they illuminate the nature of the difference between the two phenomenologies. At the same time they will bring into sharper focus the development of what has been called 'The Second Phenomenology'.

It is not my purpose here to present all of Ingarden's criticisms, or to sketch however briefly the main features of his phenomenological perspective. Nor is it my intention to delineate in any way the basic characteristics of Husserl's transcendental phenomenology. On the contrary, I wish to deal with just those criticisms which not only indicate basic difficulties in Husserl's formulation, but which stress the reasons for Ingarden's particular turn to the ontological perspective. This was *not* an arbitrary gesture on Ingarden's part but based on reasonable grounds. It is precisely because of the difficulties within Husserl's transcendental phenomenological perspective that 'the way out' is via the ontological or metaphysical perspective. To show this I will deal with the following criticisms: (1) the attack on the foundations (or the 'Problem of the Beginning' as Ingarden

Tymieniecka (ed.), Analecta Husserliana, Vol. II, 48–54.

called it), (2) the rejection of the transcendental turn, and (3) the shortcomings of Husserl's doctrine of constitution. Each one will be dealt with briefly in turn and will be followed by a brief summary and conclusion.

I. THE PROBLEM OF THE BEGINNING

In the *Cartesian Meditations* Husserl attempted to establish the foundations of knowledge in an absolute and final sense, free from all prejudice and unwarranted assumptions. Such a concern for absolute foundations characterized Husserl's writings throughout his life [6] and stemmed in part from a desire to overcome the strife of systems. The latter developed because their fundamental terms were not perfectly or completely anchored in experience, and as a result had the status of assumptions or posits; this could only lead to skepticism and relativism. Husserl considered it necessary for civilization and the future of mankind that a truly universal philosophy be developed from absolute foundations.[7]

This ideal for philosophy was introduced in the First Meditation as a 'provisional hypothesis'. Such a guiding ideal could not be established immediately but could be determined in the future:

> Thus, by immersing ourselves meditatively in the general intentions of scientific endeavor, we discover fundamental parts of the final idea, genuine science, which, though vague at first, governs that striving.[8]

It was not established in any single experience but in a series of such.

Ingarden is quite right in insisting that the idea and value of the absolute foundation is not established within the scope of the *Cartesian Meditations*; nor is there any apodictic evidence available. Such a conception could only be achieved through a transcendental-phenomenological reduction and this is not possible at this point. Thus we have what Ingarden calls 'the problem of the beginning'.[9]

A second and more serious criticism follows from the first. It is now claimed that the phenomenological descriptions of the scientific intendings will not lead to anything like the establishment of the ideal of scientific knowledge. How could such an idea be justified in a series of single concrete experiences?[10] Because Husserl was unable to complete the process within the scope of the *Cartesian Meditations*, Ingarden doubted the validity of the project.

Suppose now that it is admitted that Husserl was not able to establish his ideal. What follows from this? Would it necessarily follow that his particular phenomenology was doomed? A more charitable interpretation might hold that the ideal, introduced by Husserl, was a guiding thread pointing to an infinite task. M. Farber, for example, has argued that the ideal of a presuppositionless philosophy does not refer to every assumption in every sense of the term, and therefore has affirmed that 'Husserl's ideal of philosophy was held to be realizable in an infinite historical process'.[11] On the other hand, Ingarden chose the more drastic alternative of reinterpreting the goal and task of phenomenology. Its task is not the solution of all the great issues of philosophy but to aid in the correct formulation of such issues. This he could do by seeking apodicitity not in transcendental subjectivity but in the noematic structures of our lived experience. By criticizing the ideal of transcendental phenomenology, Ingarden sets the stage for his own more modest and more realistically geared phenomenological perspective.

II. THE REJECTION OF THE TRANSCENDENTAL TURN

Husserl's introduction of the phenomenological *epoché* revealed the *ego cogito* as transcendental subjectivity and represented a direction that Ingarden was unwilling to take. Through this radical epoché Husserl discovered the vast realm of transcendental subjectivity which was the ground of all judgment and meaning. The *ego cogito* now became the acceptance basis of all objectivities whatsoever.

> Anything belonging to the world, any spatiotemporal being, exists for me – that is to say, is accepted by me – in that I experience it, perceive it, remember it, think of it somehow, judge about it, value it, desire it, or the like... The world is for me absolutely nothing else but the world existing for and accepted by me in such a conscious *cogito*. It gets its whole sense, universal and specific, and its acceptance as existing, exclusively from such *cogitationes*.[12]

Husserl then concludes that the being of the pure ego and his *cogitationes* is antecedent to the natural being of the world. Natural being is secondary and continually presupposes the realm of transcendental being.[13]

Ingarden charges that Husserl's claim goes beyond the evidence which can be brought forward for it, and to make matters worse, involves a hidden metaphysical assertion. The conclusion "the existence of the world presupposes the existence of the pure ego" requires the assumption of an *identity* between 'the natural existence of the world' and the 'existence of the world comprehended by me'. But how can this identity be established at this level of Husserl's exposition? Ingarden charges that Husserl has in fact made a decision involving a metaphysical assertion, for Husserl has literally stepped outside the realm of transcendental subjectivity in asserting such an identity or equivalence. But this metaphysical assertion is not made manifest and if challenged, it is difficult to find evidence for it. One could infer then, that it was a characteristically idealist assumption which may have been the consequence of the way Husserl interpreted intentionality – or possibly the result of the way he interpreted the Cartesian ego. In any case, Ingarden was critical of the direction taken by Husserl and was loath to follow.

III. THE SHORTCOMINGS OF HUSSERL'S DOCTRINE OF CONSTITUTION

The remaining criticisms of Ingarden that are considered here all stem from some difficulty arising from Husserl's concept of constitution and provide further justification for Ingarden's own perspective. The concept of constitution runs into trouble on a number of counts, but perhaps most outstanding is its failure to cope with the distinction between transcendent and the transcendental. Ingarden first discloses the difficulties and then proposes a method for their resolution.

At the close of the First Meditation Husserl stated that the concept of the transcendental and its correlate (the concept of the transcendent) must be derived phenomenologically.[14] Transcendence is part of the intrinsic sense of anything worldly, notwithstanding the fact that the sense of anything worldly is gained from the grounding acts of our experience.[15] Thus the very meaning and significance of the world comes from my experiencing. In this perspective then, the Ego who bears within himself the world and who is necessarily presupposed by it (as the correlate) is properly called transcendental in the phenomenological sense.

This account is not quite as innocent as it seems. Ingarden asks whether, if transcendence belongs to the intrinsic sense of all worldly things, it would follow necessarily that everything which is not distinguished by this transcendence is therefore not worldly?[16] There seems to be no clear evidence for this, and once again Husserl is seen involved in another metaphysical assertion. The root of the difficulty lies in the distinction between what is constituted and the constituting activity; this has not been examined carefully enough. For does it follow that the mere fact of being constituted is the condition for being called worldly? Are there no other elements involved?

The solution for the difficulty comes in the form of a radical proposal to adopt a *new principle*:[17] – out of the *contents* of lived experiences and only out of such *contents* can knowledge be achieved. Ingarden is proposing a new form of constitution in which the contents of our experience play a significant role. Constitution, therefore, belongs not on the side of the active consciousness but on the side of the object which is constituted. This has the particular virtue of supplying the kind of apodictic evidence which was lacking with Husserl's constitution. It has also the advantage of accounting for the transcendental in terms other than a constituting consciousness and would thus avoid the kind of metaphysical assumption covertly made by Husserl on this matter.

If Ingarden's revision of the constitution principle is not adopted, further difficulties may be faced in the constitution of 'abiding properties' of the *Ego*. Let us now turn to this question. In the *Meditations* Husserl claims that the Ego is not an empty pole of identity, but, on the contrary, according to a law of 'transcendental generation' it acquires abiding properties. By its own active generating, the Ego constitutes himself as the identical substrate of *Ego*-properties... as a fixed and abiding personal *Ego*.[18]

The question now may be raised whether the 'abiding properties' are themselves mere intentional correlates of the corresponding systems of unified lived-experiences or is their existence absolute? Such a question is considered seriously by Ingarden on both epistemological and metaphysical grounds. Are they constituted exclusively from the noetic experiences or is there a material element or component in each experience which is a factor in such constitution? Ingarden claims that Husserl is not clear on whether there is or is not a content that plays a role in the constitution of such abiding properties. Again Ingarden is calling for a material element or component of content which is a factor in constitution.

Another variation on the same theme is found in questioning the identity of the I of the natural attitude and the I of the transcendental realm. How can there be one and the same constituting pure I and constituted real I, when the properties of such are mutually excluded by one another and could not exist in the same unity? This is indeed the heart of the problem of transcendental phenomenology and one for which Husserl had no clear answer.

Ingarden advanced another interesting argument in which he claims that Husserl was guilty of a *petitio principii* which invalidates his whole endeavor in the *Cartesian Meditations*. Husserl has to assume the specific nature of the transcendental *ego* in order to carry out the phenomenological reductions leading to apodictic knowledge. However, it is only through such phenomenological reductions that the transcendental ego can be revealed in the first place. Thus the whole endeavor is trapped in a vicious circle from which it is difficult to break out.

The argument used here is merely a form of a more general difficulty alleged to lie in every theory of knowledge. The heart of the difficulty lies in the distinction between the activity of knowing and that which is known. As long as the distinction is not overcome, one is always hung up on either an infinite regress or a petitio (assumption of truth or validity). The way out of this dilemma was revealed in Ingarden's study of external perception, wherein various grades or levels or organization are discovered. In the course of his exposition it became clear that there was another element of constitution which could be isolated and intuited apodictically.

IV. CONCLUSION

Ingarden's criticism revealed certain difficulties in Husserl's phenomenology which stemmed from his transcendental reduction and his concept of constitution. In most of the cases it was noted that Husserl was guilty of an implicit metaphysical step which could not be substantiated within the context of his method. Rather than attempt to patch up the doctrine within the perspective delineated by Husserl, Ingarden chose to move in another direction and to a phenomenology in an ontological perspective. His own studies of perception had brought him to the awareness that in order to account for constitution one had to know precisely *what a definite constituted X is supposed to be*[19] as a starting point for further considerations. This led to a study of that *X*'s form, its material constitution, and its mode of being. Only then could one see how a particular *X* was constituted from all possible multiplicities of consciousness. "The ontological investigation of the object must precede the constitutive phenomenological perspective".[20] This is not to say, however, that the ontological is the final consideration. But his position is radically different from that of Husserl who tried to derive ontology and metaphysics from the perspective of constitutive phenomenology.

NOTES

[1] Herbert Spiegelberg, *The Phenomenological Movement*, Vol. 1, Martinus Nijhoff, The Hague, 1960, p. 170.
[2] Roman Ingarden, *Edmund Husserl. Briefe an Roman Ingarden.* Martinus Nijhoff, Den Haag, 1968, p. 63. Hereafter referred to as *Briefe*.
[3] Roman Ingarden, *Das literarische Kunstwerk*, Halle 1931.
[4] *Briefe*, p. 30 *et passim.*
[5] Edmund Husserl, *Cartesianische Meditationen und Pariser Vorträge.* 2. Auflage Martinus Nijhoff, The Hague, 1963. Hereafter referred to as *Bemerkungen.*
[6] *Briefe*, p. 40.
[7] *Briefe*, p. 42.
[8] Edmund Husserl, *Cartesian Meditations* (translated by Dorion Cairns), Martinus Nijhoff, The Hague, 1910, pp. 12–13.
[9] Ingarden, *Bemerkungen*, p. 203.
[10] Ingarden, *Bemerkungen*, p. 206.
[11] Marvin Farber, *The Aims of Phenomenology. The Motives, Methods, and Import of Husserl's Thought*, Harper Torchbooks, New York, 1966, p. 42.
[12] Husserl, *Cartesian Meditations*, p. 21.
[13] *Ibid.*
[14] Husserl, *Cartesian Meditat ions*, p. 26.
[15] *Ibid.*
[16] Ingarden, *Bemerkungen*, p. 211.
[17] Ingarden, *Bemerkungen*, p. 211.
[18] Husserl, *Cartesian Meditations*, p. 66.
[19] *Briefe*, note 50, p. 166.
[20] *Ibid.*, note 50, p. 166.

FRED KERSTEN

ON UNDERSTANDING IDEA AND ESSENCE IN HUSSERL AND INGARDEN

The following discussion seeks to determine in broad outline the basic structure and significance of the analysis of Idea and Essence in the work of E. Husserl and R. Ingarden.

The on-going course of daily living gears us into and confronts us with a world of changing things, affairs and events under perchance similar but varying conditions, from varying and variable perspectives. But the presented world and things, affairs and events in it are not, as such, chaos: in the midst or at the margin of that course of living certain 'non-changes' with definite-indefinite limits are also co-presented. To be sure, the changes, variations, alterations, the factual contingencies may be expressed by the laws of functional regularity, by belief in the uniformity of nature or in the permanent possibility of 'sensations' and the like. Yet those laws and beliefs are external to the matters of fact experienced and thought about; they do not express the 'essence' or 'inner What' of the things, affairs and events encountered. A given thing or set of things change; *this* or *those* things change; *a* thing or *some* things change. Accordingly there is a limit to change, to Becoming. There is indeed no more 'pure' becoming than there is 'pure' contingent factuality. The correlative of contingent change is, then, necessity. Individual matters of fact are not meant merely in their individuality, but also *as* individuals, as instances of specific and generic kinds.

In Husserl's words: "If we say that every factuality can, 'according to the essence peculiar to it,' be otherwise, we have already said, then, *that it belongs to the sense of that contingency precisely to have an essence and hence an Eidos which can be purely apprehended, falling under truths of essence at different levels of universality.*"[1] Of course, it is not the prerogative of the philosopher to see this with original insight: "The truth is," Husserl says, "that everyone sees 'ideas,' 'essences,' and sees them, so to speak, continuously; they operate with them in their thinking and they also make judgments about them. It is only that, from their theoretical 'standpoint,' people interpret them away."[2] The chief philosophical

Tymieniecka (ed.), Analecta Husserliana, Vol. II, 55–63.

problem at hand is to analyze, clarify and explain what in truth everyone sees in confronting a matter-of-fact world.

A concrete example, taken from Husserl, will help us explore this problem – namely, the logical presentation of 'All *A*'. For that presentation to occur originaliter it is not sufficient to survey in fact all *A*, even were it the case that all the particulars in question were finite in number. Rather a further and founded act of consciousness at the end of the survey must be made explicit along with "All *A*".[3] In other words, only with the original presentation of the kind as well as the individual *A*'s can one judge that this individual is an instance of 'All *A*,' and then the fact itself that this is an *A*.

As the example suggests, the individuals presented in the survey do not appear simply as mere, bare and actual data. Instead they are also presented and may be explicitly grasped as eventual examples, possible varieties, so to speak, of something permanent and invariant, namely 'All *A*'. To say that, of course, is not to eliminate or divest them of their character of real things given in actual experience. But if what has been said is the case, it follows that the presented actual *A*'s of the survey need not have any special primary in the presentation of 'All *A*'.

We can clarify this further with a different example – the Idea 'material thing at large'. This is exemplified equally by golden mountains and winged horses, by trees and stones. Seen as to content, the Idea is exemplified by products of the imagination *no less than* by the facts of actual perceptual and apperceptual experience. Clearly to apprehend the Idea as exemplified by imaginary or actual cases in no way eliminates the status of those cases *as* imaginary or actual.

Taking both examples together, we may also observe that in the process of regarding individuals as possible examples of Ideas certain features and structures are themselves presented unaltered throughout, but exhibited in, that process. All members of the series of *A*'s surveyed, actual and imaginary, are 'congruent' in a certain respect, exhibit an invariant form and content regardless of all other differences. Moreover the manifold of possible examples need not be exhaustively surveyed; though the process may be broken off after a given number of steps, it can of course be carried out *ad libitum*. Indeed, precisely consciousness of the *ad libitum* continuation makes possible the reference of the Idea not only to all those examples that have been or might have been or will be considered, but also to more examples *ad libitum*. In short, the Idea is grasped and objectivated as an identity in contradistinction but nonetheless with reference to possible exemplars (actual and imaginary) within a given domain of greater or lesser universality. In view of the reference to more examples *ad libitum*, Idea is not exhausted in its extension by imaginary and actual varieties. And just as individuals as possible instances of Idea may be judged about in original ways, so too Idea *qua* Idea may be grasped, objectivated and judged about equally originally regardless of its being non-individual and of whether or not it has possible exemplifications.

Finally, so far as 'specification' of Idea is concerned it may be said that if the contents of and relations among Ideas hold *a priori* for matters of fact, this is because actual things and events must be considered purely as possible cases which happen to be 'realized'. To regard the actually real as purely possible is, of course, not the same as regarding it as imaginary. Rather the actual is one possibility among others. Accordingly the complexity of 'specification' may be formulated as the *a priori* limits of the essentially possible (which may or may not be actualized or 'realized'), the essentially impossible (which cannot be actualized or 'realized'), and the essentially necessary (which must be actualized, 'realized') in any possible realm of being whatever.

In the foregoing we have, largely implicitly, considered Idea and Essence as objective correlates of specific syntheses in consciousness. And although the temporal syntheses and objective correlates are mutually irreducible, nonetheless they do not exclude each other.[4] Correlatively, then, what has been said so far allows of distinguishing between three different evident intendings:

(1) The meaning and believing in an individual as of a co-intended kind.[5]

(2) The meaning and believing in the kind itself as exemplified in and by co-intended imaginary or perhaps individual instances.[6]

(3) The meaning and believing in the kind itself for its own sake, i.e., regardless of whether or not it is exemplified in or by anything at all, unrestricted to any given or likely co-intended instances.

Leaving aside any consideration of the relations among these intendings, and using Husserl's terms in a loose sense open to revision, we may call the intendings in question 'sensuous intuiting,' 'categorial intuiting' and 'ideational intuiting' respectively. For our word 'kind' Husserl has sometimes used the highly ambiguous words 'Eidos,' 'Wesen,' and 'Idee.' In the passage cited earlier from the *Ideas*, Husserl's word for 'essence' not only signifies 'pure essence' as the Whatness of a real individual in its own being peculiar to it, but it also signifies 'essence' understood as the highest category as well as its particularizations down to full concretion. This maze of terminological and to some extent analytical confusion was challenged by Jean Hering[7] who distinguished between 'essence' (*Wesen*), 'essentiality' (*Wesenheit*) and 'idea' (*Idee*). These distinctions serve as the point of departure for Roman Ingarden's clarification of Idea and Essence begun in his essay 'Essential Questions'.[8]

In the light of the *noetic*-noematic distinctions now made, and hopefully without doing too much violence to the thought of Husserl, Ingarden and Hering, I would like to submit the following as a schema for clarification and analysis of Idea and Essence: (1) In the narrower sense of the term, 'essence' signifies what is co-intended to in and through intuiting a real or ideal individual as exemplifying and, perchance, actualizing, What and How it is. (2) 'Idea' signifies what is presented originaliter in contradistinction but with reference to the real or ideal individual as to What and How it is. (3) The essence of something is exemplified *in* that something, and hence is inseparable from it. (4) The essentiality of something is, in contrast, independent and non-individual, though exemplified (perhaps better: embodied) *by* something (for instance, the 'species' Red in contrast to the 'red' of the red thing). And (5), in each case, Idea, Essence and Essentiality, the *a priori* complexity of 'specification' must be analyzed.[9]

We cannot develop the schema further here. Instead, something else must be made explicit. In each case of intuiting distinguished there is a 'positional' moment; not only are things meant and intended to but, in corresponding ways, they are also believed in. Stated universally, objects of all kinds are presented in and through systematically concatenated mental processes, and through those processes objects not only disclose their properties, qualities, relations, but also themselves as existent in specific modes (really or ideally) – though existence is to be understood as a 'thetic' or 'belief character' rather than as a property of the objects. As a consequence, we distinguish between what is meant and intended to on the one hand, and, on the other hand, that object as it is intended to. For example, I am presently seeing the desk on which I write, and while seeing it I may pay attention to certain features of the desk – its size, shape, color and so forth. But all the while seeing it I am (implicitly) believing in it as existing. By objectivating the noematic sense of the seeing of the desk, its rectangularity, color, shape are made thematic. But by objectivating the character of believing, the mode of existence is made thematic. The seen and meant desk is believed in as existing, as inserted into a spatiotemporal context – the object as it is intended to. This distinction in objectivating points to one in different dimensions of consciousness, such that if one dimension varies, the other varies – if the desk is presented as absent, then it may be believed in dubitatively, and so forth.

This distinction between presentationality and positionality allows for the development of an ontology in which the Idea, Essentiality and Essence and their relations can be analyzed both as regards their respective presentational and positional modes. Shortening and oversimplifying his account of the situation, we can refer here to four basic ontological concepts mentioned by Ingarden as expressing the fundamental modes of positionality: (1) absolute supra-temporal existence, (2) extra-temporal, ideal existence, (3) temporal existence, and (4) purely intentive existence.[10] Instead of elaborating these concepts further, we can use them as a framework within which we may discuss Ingarden and Husserl together. (Limitation of space prohibits direct consideration of the various concepts of Essence and Transcendence beyond what has been said.)

The fundamental phenomenological situation to be clarified and explained is the correlation of contingent change and necessity. Reformulated this means that on the basis of grasping and objectivating the 'kind' as well as the individual, the factuality or contingency itself is grasped and objectivated.[11] (For example, on the basis of judging that this an instance of color, one proceeds to judge that this is a color.) When regarded as clarification of this basic situation, the four concepts of modes of existence are immediately seen not only to include (with radical revision) the Husserlian explanation given, but also to surpass it.[12] To evaluate this confrontation between Husserl and Ingarden in the most radical way, we may focus our attention on the fourth mode of existence: purely intentive existence. Within the context of *Das literarische Kunstwerk* (1930), this would seem to hold only for the intentionality of a work of art which is partly at the mercy of certain significational acts of consciousness. The implication is that what has been said about the constituting of Idea and Essence would not hold for works of art,[13] and that Husserl's account of the nature of intentionality would have to be revised. However, in *Der Streit um die Existenz der Welt* (1964f.) Ingarden seems to universalize this special case and hold that what is true of the intentionality of works of art is true of all intentionality.[14] Idea and all cases of ideal existence, then, would be radically transcendent to consciousness in a manner distinct from purely intentive objects. But can this mean anything more than that Ideas and cases of ideal existence, as noematic correlates, are irreducible to the modes of consciousness of them?[15] If it means anything more than this the danger arises of (implicitly) holding that, for instance, the perceptual noema 'tree' is numerically distinct from the 'real tree.'[16] In general, can one conclude from the fact that Idea stands out as self-identical in contradistinction to actual and imaginary *ad libitum* exemplifications that, as regards its 'belief character,' it is somehow autonomous and non- or 'meta-'intentive? To say, for instance, that the perceptual noema, 'desk,' and the Idea, 'material thing at large' are not the same both as regards presentationality and positionality no more asserts that Idea is not a noematic object than does denying the existence of the desk in any way assert that the desk is rather an object of the imagination.[17] In this connection, it also may be asked whether or not the distinction between presentationality and positionality cuts across all four modes of existence mentioned; or whether or not the fourth mode of existence is not really a mode of existence at all but rather a (limiting?) case of *mere* (pure?) presentationality. This raises the whole problem of 'passive Doxa' which Husserl mentions in his last writings.

Finally, it must be noted that according to Ingarden Ideas, like real individual objects, lie outside our sphere of influence.[18] But to assert this may very well be to maintain that an imputed real relation of consciousness, as a process in the world, to its objects does not hold in some cases – perchance only in the case of intentive objects are objects within our sphere of influence. But to consider the matter in these terms is to assert implicitly that there is an objective analogue between intending to something and that something purely as intended to, *and* that real relation, for instance: consciousness and that whereof it is consciousness as real things in time. Ingarden seems to say that if one denies the latter relation in a given case (namely in the case of the first three modes of existence), then one *ipso facto* denies the other relation.[19] As one Husserlian phenomenologist, Dorion Cairns, has put the matter, *in the world* consciousness and its objects are not only related *as*, for instance, perceiving and perceived purely as intended to, but *also* as having objective relationships as objects in time. The two relationships must not be regarded as analogous, and since consciousness only discovers itself in and of the world so far as it is consciousness of the world, the latter is not only fundamental but it is also *sui generis*.[20] (This, of course, in no way implies that consciousness is concretely apart from the world.) Only by referring Idea and Essence back to the constitutings of them can one avoid even implicitly asserting that analogue. As a consequence, to clarify and explain what in truth everyone sees and operates with a constitutional inquiry must always be fundamental to and precede an ontological one.[21] To the extent that the controversy over realism and idealism assumes the truth of that analogue, the controversy is vitiated by a constitutional inquiry into the correlation of contingent change and necessity. But we cannot close without observing that to vitiate the controversy is not equivalent to resolving the philosophical problem of the existence of a real world. Rather that problem only then emerges as a problem in the first place. And even if it is granted that the "transcendentally 'reduced' standpoint" is the correct one, it still is legitimate and not inconsistent to ask, as Ingarden has shown,[22] whether or not that 'standpoint' provides the basis for answering all questions about the existence of the real world, especially ontological and metaphysical ones; ultimately whether or not that 'standpoint' includes other ones as well. Only in this light can one avoid accepting as philosophical and genuine what is not derived from 'evidence,' only in this way can one examine the whole range of 'evidence': where this is still lacking one's judgments cannot claim any final validity must, at best, be regarded as possible intermediate states on the way to final validity.[23]

NOTES

[1] Edmund Husserl, *Ideen zu einer reinen Phänomenologie und phänomenologischen Philosophie*, Vol. I, §9, p. 9.

[2] *Ibid.*, §22, p. 41. (Translation by Dorion Cairns.)

[3] Edmund Husserl, *Logische Untersuchungen*, 4 Vol. II, 1, p. 173. See Dorion Cairns, 'An Approach to Phenomenology', *Essays in Memory of Edmund Husserl* (ed. by M. Farber), Harvard University Press, 1940, p. 12. See also Aron Gurwitsch, *Studies in Phenomenology and Psychology*, Northwestern University Press, 1966, pp. 246ff.

[4] Aron Gurwitsch, *op. cit.*, pp. 137f.

[5] Cf. Roman Ingarden, *Der Streit um die Existenz der Welt*, Vol II, 1, Niemeyer, 1965, pp. 241ff.

[6] Cf. *Ibid.*, pp. 246ff.
[7] Jean Hering, 'Bemerkungen über das Wesen, die Wesenheit und die Idee', *Jahrbuch für Philosophie und phänomenologische Forschung*, Vol. IV, Niemeyer, 1921.
[8] Roman Ingarden, 'Essentiale Fragen. Ein Beitrag zu dem Wesensproblem', *Jahrbuch für Philosophie und phänomenologische Forschung*, Vol. VII, Niemeyer, 1925.
[9] See Hering, *op. cit.*, pp. 498f., 530; Ingarden, 'Essentiale Fragen', pp. 168, 171, 175ff., 225ff., and *Der Streit um die Existenz der Welt*, Vol II, 1, Niemeyer, 1965, pp. 232, 241ff, 246ff.
[10] Roman Ingarden, 'Les Modes d'existence et le problème 'idéalisme-réalisme', *Proceedings of the 10th International Congress of Philosophy*, Amsterdam, 1949, pp. 349f. Cf. *Der Streit um die Existenz der Welt*, Vol. I, Niemeyer, 1964, §§14f.
[11] See above, p. 2.
[12] Cf. Ingarden, *Der Streit um die Existenz der Welt*, Vol. II, 2, Niemeyer, 1965, pp. 394ff.; also cf. A.-T. Tymieniecka, 'Le dessin de la philosophie de Roman Ingarden', *Revue de Métaphysique et de Morale* (1955), 42, 48, 49.
[13] Cf. A.-T. Tymieniecka, *Phenomenology and Science in Contemporary European Thought*, Noonday, 1962, pp. 26f. But see also for possible solutions to the problem from a purely Husserlian standpoint, Dorion Cairns, 'The Ideality of Verbal Expressions', *Philosophy and Phenomenological Research*, Vol. I, pp. 455ff., and Fritz Kaufmann, 'The Phenomenological Approach to History', *Philosophy and Phenomenological Research*, Vol. II, pp. 163f.
[14] Cf. Ingarden, *Der Streit um die Existenz der Welt*, Vol. II, 1, Niemeyer, 1965, §§46ff., especially pp. 216ff.
[15] Cf. *ibid.*, Vol. II, 2, pp. 375ff.; and Roman Ingarden, *Das literarische Kunstwerk*, Niemeyer, 1960, p. 121.
[16] Cf. Mikel Dufrenne, *Phénoménologie de l'expérience esthétique* Vol I, PUF, 1953, pp. 269f.
[17] Cf. Husserl's distinction between 'existential predications' and 'predications of reality' in *Erfahrung und Urteil*, Claassen, 1954, §74.
[18] See Ingarden, *Der Streit um die Existenz der Welt*, Vol. II, 1, Niemeyer, 1965, p. 253; *Das literarische Kunstwerk*, Niemeyer, 1960, pp. 121, 362; A.-T. Tymieniecka, 'Le dessin de la philosophie de Roman Ingarden', *Revue de Métaphysique et de Morale* (1955), 42.
[19] Cf. Ingarden, *Der Streit um die Existenz der Welt*, Vol. II, 1, Niemeyer, 1965, pp. 253, 256f.
[20] Cairns, 'An Approach to phenomenology', pp. 17f. Contrast with Ingarden, *Der Streit um die Existenz der Welt*, Vol II, 2, Niemeyer, pp. 323ff., especially p. 370.
[21] Ingarden has suggested that the order of procedure is to first become aware of what is being investigated as to form, material composition and mode of being. The resulting ontology, however, is not the last word; it is to be followed by a metaphysical and epistemological inquiry which provides an ultimate control of the whole analysis, wherein Husserl's constitutional analysis is retained. See Edmund Husserl, *Briefe an Roman Ingarden* (ed. by Roman Ingarden Martinus Nijhoff, 1968) p. 166; also *Das literarische Kunstwerk*, Niemeyer, 1960, pp. x, xii. In shorthand terms, an eidetic ontological analysis is then fundamental to and precedes a constitutional analysis. It seems to me that this would be the least radical evaluation of Ingarden's philosophy as regards Husserl's.
[22] See Roman Ingarden, 'Bemerkungen zum Problem "Idealismus-Realismus", *Jahrbuch für Philosophie und phänomenologische Forschung*, Festschrift, Niemeyer, 1929,

pp. 168f., 175f., 190. Cf. also Alfred Schutz, 'The Problem of Transcendental Intersubjectivity in Husserl', *Collected Papers*, Vol. III, Martinus Nijhoff, 1966, pp. 83f., 90.

[23] See Edmund Husserl, *Cartesian Meditations* (translated by Dorion Cairns), Martinus Nijhoff, 1960, §5, p. 13.

DISCUSSION

Klibansky: Whatever their differences are, both Husserl and Ingarden hold that philosophy is a rigorous discipline; and both have in their writings attempted to conform to this standard. To be true to their spirit, our discussion should not be lacking in rigour. Hence two points should be clearly distinguished in our approach: The first concerns the proper understanding of Husserl's and Ingarden's thought; the second concerns the validity of their positions. May I recommend that these points be kept well apart?

With regard to the first point – the right interpretation –, Mr. Laskey drew attention to three fundamental questions which gave rise to Ingarden's criticism of Husserl. I propose that each of these questions be taken up in turn and that, subsequently, we address ourselves to the second speaker's presentation of the part which the basic concepts of idea, essence and existence play in Ingarden's writings.

Once these questions of interpretation are clarified, we shall be better equipped to enter into the more interesting and more important discussion of Ingarden's attempt tp shift the emphasis of the phenomenological approach to the side of ontology.

We should thus first examine whether Ingarden's criticism regarding the problem of the beginning is justified. We should then consider whether there are indeed hidden metaphysical assumptions in Husserl's writings, underlying above all his transcendental reduction. Finally, we should decide whether Ingarden's strictures on Husserl's concept of constitution are valid.

I propose that we discuss these questions in this order and then give Mr. Laskey the opportunity of replying.

Questioner: I have a question about the problem of the beginning. In the text it reads: "In order to have an insight of the possibility and necessity of the phenomenological transcendental reduction one has first to have already practiced it." (Ingarden's letter to Husserl) My problem is that if you can reach the formulation of the concept of the transcendental

Tymieniecka (ed.), Analecta Husserliana, Vol. II, 64–77. *All Rights Reserved.*

phenomenological reduction you have only shown its possibility. You cannot say you have shown its necessity, but you have certainly shown its possibility, as a possible experience. But you don't need to have experienced the transcendental reduction in order to show that the transcendental reduction is possible as such.

Van Breda: In a discussion not long ago with Foucault, the proponent of the French structuralism, it appeared that it is accepted now that the human being, possessing a ready made language, receives all that he knows about the world he is living in. He receives all that in the structure of his mostly, only not so well-known, mother tongue; in particular, in the structures of the linguistic way of expressing scientific truth. Let's take that for granted, although not for one moment do I believe it completely. Then we will understand that we have to stress the mediation of language in all that we say about our knowledge. Concerning what Husserl means by reduction, I recall discussions with the great master, Ingarden, along the line of the dichotomy between objectivity and subjectivity. We receive all that we see from the world in which we live. Maybe not immediately; also to a great extent in the framework of our language, but before all, in the framework of our consciousness. That is, one cannot, of course, make an abstraction from the human *ego*, or the intersubjectivity of the human *egos*. But it is necessary, first of all, to reduce the complete knowledge we have about our world – daily life, *life-world*, (scientific life, cultural life) to the framework of the consciousness. And the problem is to reduce to that framework, just as the linguistic philosophy does now. However, when Husserl claims the return to 'subjectivity,' that does not mean to reduce all experience to a subjective world in the classic sense, but to the subjective framework, where all is caught in the mediation of the consciousness. The letters from Husserl to Ingarden are always saying in effect, "But my friend why don't you see that I am not against the sciences, not against the real world."

Küng: There seems to be a tendency to scale down Husserl's idealism by saying that he doesn't really deny the existence of the world, that Husserl stresses the importance, for instance, of the body in our experience. I think, one should make a distinction between the order of knowing and the order of being. Certainly in the order of knowing the first thing we acknowledge is that we find ourselves in the world – the *noema* of the world and the *noema* of our bodily experience are given and so world and body are primary in the order of knowledge. Husserl, however, tries to find through reduction what is primary in the order of being. In the order of being, I think, the *noema* is for Husserl a construct, a product of the *noesis*, and therefore the 'sense data', which are the material part of the *noesis*, are in the order of being primary to the *noema* while the *noema* is produced on the basis of this material part of the *noesis* through the *noetic* act. I think it is only Ingarden's insistence which stresses that in the order of being there is a real world existing from the start as an autonomous referent and as a referent which guides our processes of constitution, to which these processes have to adapt. The constitutive processes according to Ingarden are not like the imaginations of a poet, which are completely free. When we constitute the natural world then we have to follow, or to adapt to, something else which is primarily really existing. For Husserl, by contrast, I think the whole noematic world is only a product of subjectivity. Of course it belongs to the subjectivity to have the inclination of producing a world and to have the inclination of living as an embodied consciousness in a world. But it seems, according to Husserl, that it might have been that there hadn't been any world, in the sense that this inclination hadn't been able to be fulfilled, couldn't have been fulfilled, and yet consciousness could have been there. Therefore, this inclination of projecting a world and considering one's self as an embodied consciousness in the world is only an inclination.

A similar thing holds with respect to language. Father Van Breda mentioned that one could also say that the phenomenological reduction is a reduction to the framework of language. Probably in an analogical way one would say that it is a reduction to the basic framework of our embodied experience in the world. However, neither for Husserl nor for Ingarden is language primary in the order of being. It is, of course, primary in the order of knowledge, in the sense that we have grown into a language which we find from the start. But again in the order of being the meaning intentions embodied in a language are secondary; the primary thing is what may or may not fulfill these meaning intentions. So again, the inclination of producing a language may be primarily there, but the language as an actual structure of embodied meaning intentions is something secondary and also something which can vary, which is not unique. There are different languages. Ingarden himself has written explicitly on the role of language in phenomenological analysis, very early in fact, before the end of the first World War, when he was explaining phenomenology to the analytic philosophers of Poland. There he gives a very detailed account of his procedures in making phenomenological analyses, and he stresses that as a matter of fact we start with the meaning intentions which are given with our mother tongue. Then we have to become critical and it may be that these meaning intentions of our mother tongue are not the best possible ones, that they do not exactly fit reality, and we may modify our language and so go from our ordinary language to a technical language which might prove to be more accurate than the ordinary language. (I have elaborated on this in a paper for the meetings of the American Philosophical Association in St. Louis, which will be published in the *American Philosophical Quarterly*.) Then a final remark concerning the question of idealism-realism: I think it is extremely important to make a rigorous distinction between 'sense' and 'denotation', or 'sense' and 'referent' to use Frege's terms. My own paper will be on this subject.

Questioner: My observations are inspired by the observations just made by Dr. Küng. I wonder why we shouldn't look at this problem formulated in a more fundamental way and consider the possibility of solving it not just by phenomenology itself but by bringing up what I consider a more basic question, namely, that of the relation between phenomenology and dialectics. As Father Van Breda has explained, whether one takes the subjective on the one side or the objective on the other side the contrary relation between the two should be viewed from a dialectical point of view.

Gadamer (Professor Gadamer's contributions to the debate were, with his permission rewritten and edited by the editor of *Analecta.*): I am very pleased with the contributions to this point since they give us the occasion to discuss the fundamental points of Ingarden's criticism of Husserl's thought. Hearing Ingarden's paper, I cannot accept any of the critical points he is raising. Since several of the people present are convinced about the correctness of Ingarden's views, we have an opportunity to enter into a debate. Of course, we all have the greatest respect for Roman Ingarden and regret that he cannot be here, for it would have been easier for me to carry on a controversy with him in person than just with his ideas.

The first question which arises is: What does Ingarden mean by the distinction he proposes between intentional object and 'purely intentional' (heteronomous) object? From the transcendental point of view I cannot see the point of such a distinction. Upon what principle could such a distinction be drawn? Although I am the last member of the Marburger School of Neo-Kantianism, which has influenced the transcendental self-interpretation of Husserl, I am aware of the limits of a transcendental point of view. Nevertheless, I cannot find in Husserl's text any of the contradictions and difficulties pointed out by Ingarden.

Secondly, the difficulty raised by Ingarden concerning the pregivenness of the life-world, we must take into consideration the role of the world-horizon for our experience of the world. If from the point of view of the transcendental constitution we accept the validity of the world-horizon as a point of reference with respect to which all possible forms of constitution emerge and can be ultimately accounted for, then the pre-givenness of the world-horizon offers the final explanation of our world-experience. As we know, the notion of the world-horizon was discovered by Husserl in his later period. However, we have to keep in mind that the world-horizon of the transcendental constitution varies with different cultures, epochs, etc.

Thirdly, one never understands a thinker unless one can appreciate his sound points. Polemics dealing merely with contradictions are always quixotic. In fact, I believe that what leads Ingarden to his criticism is his own interest in the work of art and in the origin of this specific type of entity. Concerning the production of the work of art Husserl has merely differentiated between intentions constitutive of real objects, which would be existence positing acts, and intentions constituting works of art, which would not posit their real existence but would contain 'neutralisation modifications' of the existence positing acts. In other words, the work of art, in this very vague way Husserl treats it in passing, appears as a neutral mode of givenness, a neutral representation of the real thing. Ingarden, I believe, was interested in the ontological creativity of the human being and did not realize that Husserl's quest was oriented quite differently. In fact, what Husserl sought was not whether and how any mode of the ontological status would be created, but how the constitutive intentionality explains all the modes of givenness. Nevertheless, there is no contradiction between Ingarden's and Husserl's different approaches; in fact, what Ingarden describes as a transcendent and autonomous ontological status of real things of the external world, is the same as what Husserl describes as the meaning of a transcendent thing by means of transcendental reduction.

However, for Husserl to accept Ingarden's statement that there are exclusively intentional objects on the one hand, and independent objects on the other hand, would mean to abandon the methodological coherence of the transcendental analysis. Now I would like you, my colleagues, to teach me how one can defend Ingarden's position from the viewpoint of his own approach. Nobody can, I think, defend this attack on Husserl.

Laskey: I think on this question I wouldn't mind arguing on the other side. It would be more comfortable. But I will try to be consistent here. It seems to me that Husserl laid down very severe, very rigid, conditions which he was unable to fulfill. I am taking this to be Ingarden's interpretation. And in failing to fulfill these he commits himself to what Ingarden calls "metaphysical assertions" which can in no way be defended within the context of constitutive phenomenology. On the grounds of this criticism Ingarden makes a decision for his own way of approaching philosophical problems. Using the language here that Professor Küng mentioned before – the order of knowing and the order of being – Ingarden is claiming that the initial clarification in the order of being is prior to the epistemological investigation. It seems to me that this is just exactly the opposite of what Husserl says. Husserl says that you have got to do your ontological work and that ontology and metaphysics legitimately come out of transcendental subjectivity. But what I have tried to do in my paper is to show the route by means of which Ingarden comes to his own different position.

Tymieniecka: I do not believe that Ingarden needs anyone to rescue him. With a life work of 15 volumes he is secure for posterity. However, I cannot fail to take up the challenge of Professor Gadamer, having by no means an intention to teach him anything. I could learn a lot from him myself. But I confess that I could not understand Prof. Gadamer's criticisms. In view of Ingarden's work, the criticisms do not hold. To answer them we have to go back to the fundamental issues of phenomenology and to Ingarden's basic divergence from Husserl.

Let us recall that the starting and the crucial point in the whole phenomenological enterprise is the claim which Husserl made concerning the self-justification of cognition. This means that we should not start to establish any philosophical explanation of cognition of man and the universe unless we have an apodictically certain cognition; 'apodictically certain' means a cognition, which in addition to the evidence that it possesses itself, has still to be justified by other means. That is the thesis of the *Ideas* where Husserl took up the idea of the eidetic insight. According to these initial and basic views of Husserl, we do possess a certain realm of cognition which spreads over the whole universe of human discourse, of all possible cognition within which we can cognize objects immediately, with an absolute evidence as they present themselves, as they are given; but in addition this type of cognition would guarantee the justification of itself. Here lies a considerable problem. Husserl has given as a justification for the eidetic insight the nature of the *eidos*, the essence which was supposed to be initiated and self-justifying as such but the nature of which he never further elaborated. Consequently, the apodicticity of the eidetic insight was missing. The evidence of cognition was there but the justification of it which he claimed was not provided by Husserl himself. At this point, as I have shown it, in particular in my book *Essence et existence*, Ingarden, following the impulse from Jean Hering (who made the fundamental distinction in the realm of what we call essence, *eidos*, and formal structure in general), took up the issue and elaborated a complete theory of the nature of the *eidos*. Ingarden's structural analysis of the *eidos* reveals the validity of the eidetic insight. The idea is supposed, in the first place, to exhibit a content, and the *eidos* which Husserl claims that we immediately view, is this content. But to present this content the idea has to possess a structure *qua* idea as such. There is a question of its mode of existence and consequently of the possibility of the relation of the singular cases to the universal idea, such that this content which the idea exhibits is present both in the idea as an abstract thing and in the singular cases fromwhich we abstracted it. This question I have tried to elucidate myself beyond Ingarden. By means of these elaborated distinctions, Ingarden provided for the apodictic certainty of the eidetic insight. Then we come to a second point which Professor Gadamer did not mention, but which would have clarified a great deal.

Husserl himself proposed an ontological inquiry restricted strictly and uniquely to the eidetic insight of the whole universe of discourse and still completely, let us say, foreign to the epistemological explanation in terms of consciousness. Since the universe of discourse is divided into various regions, according to the types of objects that are exhibited there, we may divide such a fundamental study into various 'regional ontologies' (e.g. regional ontology of objects of nature; regional ontology of created objects, like works of art; regional ontology of mathematical sciences; and so on). The investigation of regional ontologies has been undertaken by various of Husserl's students from the Göttingen School – I mention only Hedwig Conrad Martius, who undertook it in the realm of nature, within the domain of biological studies; and, of course, Max Scheller, who undertook to in the realm of ethics, Pfänder in logic, and Reinach in law. The project of regional ontologies was the first idea of phenomenology as the 'universal science' I was speaking of this morning. The universal science as it was first proposed was to be grounded precisely in this absolute apodictically certain cognition. However, it was in this form abandoned later by Husserl himself. As Professor Laskey mentioned, each of the other disciples of Husserl has developed his own trend of thought. Ingarden, however, took up again the whole project of the fundamental ontology. How did Ingarden take it up? How did he come to separate himself from Husserl?

Along with the regional ontologies which Husserl proposed as dealing with the concrete sections of knowledge he mentioned an ontological investigation of the *object as such* and furthermore of various *types of objects*. The expression 'object as such' means the object of our cognition. Does this mean that ontological study in the analysis of the types of objects as such should in the Kantian tradition be based upon an epistemological inquiry? Ingarden, on the contrary, reproaches Husserl for having turned with the transcendental constitution towards an epistemological justification of the eidetic realm. Having adopted Husserl's claim of making a radical start in philosophy, Ingarden made this start on the apodictic foundations which he gave to the eidetic insight, at the level of ontology. The superiority of the ontological point of beginning over Husserl's constitutive genesis lies, according to Ingarden, in its ultimate radicalism. That is, if we assume the eidetic insight together with the justification of its possibility in ideas as a type of object as the basis of an inquiry, there is no need of any further justification in order to unfold the inquiry into the structures of being. Thus we may fulfill Husserl's quest for presuppositionless philosophy. We see then how Ingarden, having taken up both of Husserl's basic postulates of a radical beginning for a presuppositionless philosophy and its apodictic foundation, sought it in the ultimate ontological status of ideas. Husserl, by contrast, never disclaimed the validity of the eidetic insight – all his work in the transcendental constitution is carried on by means of the eidetic insight acknowledged by him as absolutely necessary, not only for inquiring into the states of consciousness, but first of all, as he explains in the second *Cartesian Meditation*, for uncovering the field of consciousness. But missing the possibility of the radical beginning in the *eidos* he made the transcendental turn as a further probing, further searching justification of epistemological presuppositions of the eidetic structures in which object of cognition appear.

Thus the major point of divergence between the two may be summarized as follows: Husser llaid the foundation for phenomenology as an ontological foundation, but having failed to give its final justification in ontology itself sought this justification in the nature of consciousness, namely, in the way in which these structures are given to us. Ingarden went further into the analysis of the eidetic insight and gave a final ontological foundation as a radical beginning to proceed towards an ontological edifice for phenomenology. What did Ingarden accomplish in this ontological inquiry? What use could we have for the analysis of formal structures of types of beings? Here, coming to Professor Gadamer's last point of confrontation of both thinkers, I would interpret Ingarden's endeavour as basically discovering the structures, and principles of the fundamental *rationality* of the human mind rather than as structures of the constitution which, of course, follows them. Indeed basic rationality, established in the fundamental types of structures does not contradict at all Husserl's constitutive analysis. To the contrary, insofar as they are performed correctly, since the constitutive analysis is as I have tried to establish in my own work guided by the *eidos*, the phenomenological ontology of Ingarden presents all the a priori possible types of beings that could be constituted. The question of the pregivenness of the world, the second point of Professor Gadamer's criticism, is then for Ingarden first to be elucidated at the level of what is ontologically a priori possible. Here we come to Professor Gadamer's first and crucial objection. Differentiation of various types of beings in the structural analysis within the content of ideas, reveals radical differences among them. In particular, there are two which are structurally radically opposed: the type of the real being, – that is the 'contingent' type of objects – and the ideal being, e.g. ideas as such, mathematical objects. Both are however autonomous in their nature with respect to the constitutive consciousness. But in between them lies the *heteronomous* type of being: the 'intentional' object, and the *purely intentional* object. The intentional object shows two variations: (1) intentional object of cognition, through which we cognize the real object of the world, or mathematical objects, and then (2) *purely intentional* objects such as we *create* in works of art.

The radical difference between a 'real' being ('contingent' being) and 'purely intentional' being – for instance a work of art – the difference between this table and Shakespeare's *Hamlet*, consists in one absolutely fundamental feature: transcendence. There is a certain parallel between Husserl's constitutive account of the transcendence of the real being and Ingarden's. Yet ontological analysis brings to light a new radical point for which constitutive genesis does not seem to account: existential autonomy of the real being. Real beings show a very special self-sufficiency, possess within their own structure all the elements necessary for them to unfold and carry on their action and passion, with their regulations, and no cognitive transformations of any sort can change or influence their course of nature in any way (e.g., this table possesses its four legs within its eminent structure and the four legs are certainly not dependent upon my cognition of them, all the features of the table belong eminently to the table no matter in what conscious acts we cognize them).

By contrast, the analysis of Hamlet shows a strikingly different situation. To limit ourselves just to one essential feature: In the text of the drama there is only a certain limited number of features of Hamlet or Ophelia or of the situation given. Other features are open to be completed by our imagination at will (e.g., we don't know what color hair Hamlet had, how tall he is etc.). If the hero of the drama had been a *real* Danish prince, he would have had a very special constitution which would be eminent to himself, about which we could change nothing. Even if he would dye his hair, his real hair would be blonde or black and no imagination could change it. The same may be said concerning his inner being, his disposition, character, and so forth. But the Hamlet which we have in the work of art is totally different. Shakespeare meant to indicate only certain of his features, whereas all the others we invent to complement those which have been indicated. On the contrary, concerning the structure of the real living prince, all is present, there is nothing to be complemented. However, an even more striking distinction between the structure of the real being and that of a work of art is that whereas all the features of the real contingent being are contained eminently within the compass of its structure, that is not the case even with those features of Hamlet which Shakespeare meant to put into the text. Shakespeare could have made his Hamlet melancholy, foolish, or joyful, could have made Ophelia a prosaic type of girl, or an exalted, poetic one, a foolish or very womanly one, etc. He had the freedom to choose all the characteristics of their physical appearance or phychic dispositions because he, as the author, was establishing the existence and the being of Hamlet in *purely intentional acts of consciousness*. In opposition to what we had in the case of the real contingent being, there was nothing which he had to take account of as 'being there'. He had, to the contrary, the freedom to establish all the features of the protagonist in the play in the intentional acts drawing basically upon his imagination and depending ultimately upon his will. In other words, while the real contingent object or being as it appears in the structural analysis has this autonomous type of structure and of existence (that is, we know him as possessing all his features within himself), the work of art owes its features to the intentional acts of the author, reposes entirely in its content upon these acts, thus is heteronomous with respect to the intentional consciousness or 'purely intentional'. If in Husserl such distinction as Professor Gadamer insisted upon is not possible, it is because Husserl reduced all the objects of cognition, or of the constitution, to intentional objects; the real object is represented by Husserl in constitution as transcendent, yet this transcendence is supposed to be an existential mode constituted in a special set-up of conscious acts. Thus this table with all its transcendent features is existentially reduced by Husserl to the status of Hamlet, that is, of a purely intentional object. If this were true, that the real object could be completely assimilated to the intentional object, constitution of reality would be like the creation of a work of art. It is precisely Ingarden's objection to Husserl that this is not true. The structure of the real object shows that it cannot be reduced to the status of the purely intentional object of constitution and that it demands to be taken in its own right.

Klibansky: If Ingarden were here, I am sure he would be pleased by the way in which his differences with Husserl have been brought out and justified. The challenge can be summed up in the following four major points: First, Husserl claims apodicticity. This claim according to Ingarden is not fulfilled. Secondly, Husserl claims to be presuppositionless.

However, he introduces hidden metaphysical assumptions. Thirdly, this was not raised so much by Miss Tymieniecka but in Mr. Laskey's presentation, there is a failure to distinguish in the analysis of constitution between what is constituted and the constituting activity. And, fourthly, the point brought out by Miss Tymieniecka at the end, of Husserl's deficiencies, or alleged deficiencies, in the analysis of the structure of real contingent objects. Now these are four very important points. I am sure that this conference would have achieved something if these four points were clarified.

I will repeat them: First of all the claim of apodicticity which, according to Ingarden, has not been fulfilled by Husserl. Secondly, the claim to be presuppositionless (*voraussetzungs-los*) while hidden metaphysical assumptions are introduced. Thirdly, insufficient analysis of the notion of constitution and failure to distinguish between what is constituted and the constitutive act, itself. And, fourthly, the deficiencies, or alleged deficiencies, of the analysis of the structure of real contingent objects.

Strauss: The idea of the *epoché* as it is expressed by Husserl gives me frequently the impression that our belief, as he says, in a real world, the faces of the real world, is a kind of pollution. If we only could avoid that pollution we would restore a condition like that before the fall. I am afraid this is a completely mistaken attitude. First of all the *epoché* did not transform our consciousness into a pure consciousness. If there were a kind of 'pure consciousness', the 'pure consciousness' could not use our language. After all, everything which we know from Husserl was expressed in human language. In the process of reduction, one does not reduce one's living being but remains, in what one sees, not a consciousness which has intentional data, but the seeing, living creature. In the *Ideas* I, Husserl says something like, "If the whole world would disappear and everything with it, *my* pure consciousness would still remain." With all great respect for Husserl, I think this is indeed a mistake which needs correction. It finds a kind of spontaneous correction if we speak not about *mine* but about *our* body or about *our* consciousness. If any combination of pronouns and nouns is wrong it's that of 'my' consciousness. The consciousness is personal but in *my* consciousness I experience other creatures with whom I share a common world. The development in modern days – I'm thinking of Merleau-Ponty – which emphasizes that, after all, our experience is that of a living creature is perhaps the real correction of Husserl's description.

Van Breda: Husserl first of all says that consciousness is unthinkable without the world. The philosophy of the human body of Merleau-Ponty stems in great part from the texts of Husserl on the consciousness of human beings. Husserl describes also how we are conscious of our own body. He continually takes care to avoid a possible idealistic interpretation.

Tymieniecka: It is indeed quite an important point that Father Van Breda has raised and it would be quite precious if Mr. Claesges were to answer it, since the constitution of the body is a very special difficulty in Husserl's transcendentalism. It is absolutely true that he insists on the existence, transcendental existence, of the body with which we live. What matters is how he conceives of it.

Questioner: Reading Husserl's texts we may have the impression that *what is fundamental* to the constitution is something like the 'sense data'. But the fact is that Husserl emphasizes also that we not only see something but we are conscious of seeing completely... of what we see, the concept we know, something about our own body, our eyes, the way we are moving, and so on. So we are conscious of what he calls 'kinesthetic intentions' – we are conscious of our own body and that alone makes possible the views about, or the possibility to be conscious of, sense data. It is a dialectic process.

Sokolowski: I think that kinesthesia are a small segment of sensory experience. In physiology and when Husserl speaks about *kinesthesis* himself, I don't think it's a very happy term – it means the sensory experience of the motions of lips and of joints, and so forth. But I would say that all experience of these motions of joints is preceded by the activity of motion. And first of all of being awake. There is no basic consciousness without being awake and being awake is one of the basic data. We have to thank our bodies for it. When we are awake we are in the possibility of establishing a relationship with the world and opposing the world. In getting up, it's the antigravity which we can perform which sets up opposition to objects. We also then may experience in other complex ways the motivity of our bodies. I would even say that motion and the antigravity are fundamental for all sensory experience. That is a neglected area. Locomotion is only secondary to the antimotion against gravity, which is permanent.

Questioner: I am prompted here by Dr. Strauss' remarks just to say a few words on what I will later present to the Conference, and that is this: For me, Husserl is illuminating because of the concept of transcendental *ego*. I think evidence shows that the infant is awake, arrives at awakeness before being aware of body, before being aware of self as a corporeal being. Husserl's term, the 'transcendental ego', is a term that applies precisely to this phenomenon of awakeness prior to the awareness of self as a bodily being. The whole human being is not transcendental ego. But this is an aspect of human being that I think is reflected by Husserl certainly far more clearly and concisely than by any other modern philosopher.

Klibansky: Would Professor Sokolowski care to comment on and clarify the question of the body?

Sokolowski: I will treat that problem indirectly by way of the problem of reduction and transcendental consciousness in general. The problem of reduction is an attempt to raise a question about everything. In fact, in that sense it's really the philosophical enterprise *par excellence* that has been done by every philosopher. It is not that Husserl wants to transform himself into some type of being which would be different from a normal human being, but that he can raise questions even about the mobility of a person. The fact that he is able to raise such questions means that there is a dimension of his own being which is, to use a metaphor, behind his own being. Even the bodiliness of human consciousness can be questioned. This is precisely what the reduction does. Of course, he does not say that one does not have a body anymore or that one is somehow restored to a state one had before one was embodied, like the neo-Platonic approach. Perhaps I would agree with Father Van Breda that the problem of reduction is also largely the problem of language and not only of consciousness, because you are speaking within it. It generates its own type of verifiability and evidence, because of the type of language it uses.

JOSEPH J. KOCKELMANS

PHENOMENOLOGICO-PSYCHOLOGICAL AND TRANSCENDENTAL REDUCTIONS IN HUSSERL'S 'CRISIS'

The problem on which I wish to focus attention in this paper can be briefly formulated as follows: Is it true that towards the end of his life Husserl abolished the radical distinction between phenomenological psychology and transcendental philosophy and, therefore, the distinction between the phenomenologico-psychological and the transcendental reductions, also? This question must be asked because Husserl himself seems to suggest such a view in the concluding sections of *Crisis* and in other manuscripts of the same period, and in addition this notion is further corroborated by Fink in his outline for the continuation of *Crisis* as well as by Drüe in his book on Husserl's conception of psychology.[1]

It is a well-known fact that the distinction between the two reductions, as well as the distinction between phenomenological psychology and transcendental phenomenology, gradually developed in Husserl's thought as a further specification of the two different types of phenomenology which he developed successively between 1901 and 1907, that is to say in the second volume of *Logical Investigations* and in *The Idea of Phenomenology*[2]. It is also well-known that an *explicit* distinction between the phenomenologico-psychological and transcendental reductions appears rather late in Husserl's work. To the best of my knowledge this distinction is explicitly found for the first time in the *Encyclopaedia Britannica* article, the *Amsterdamer Vorträge*[3], *Cartesian Meditations*, and in various manuscripts written between 1925 and 1931.

Since time does not permit me to enter into historico-critical investigations and to critically substantiate the two statements just formulated, I suggest that we first turn our attention to two vital questions: (1) In the years between 1925 and 1931 how did Husserl specifically conceive of the relationship between phenomenological psychology and transcendental phenomenology on the one hand and the two corresponding reductions on the other? and, (2) What does Husserl precisely claim in this regard in the final sections of *Crisis*? Once we have established Husserl's view in

Tymieniecka (ed.), Analecta Husserliana, Vol. II, 78–89.

these two different phases of his development we must then turn to the conceptions of Fink and Drüe in an attempt to reach our final solution.

The question concerning the precise relationship between phenomenological psychology and transcendental phenomenology constitutes one of the central themes occupying Husserl's reflections between 1925 and 1931. This is evidenced not only by the lecture series *Phenomenological Psychology* (1925) and the investigations which were meant to serve as a preparation for the *Encyclopaedia Britannica* article, but also by certain lectures, as for example the *Amsterdamer Vorträge*, passages in *Cartesian Meditations* and many manuscripts produced during the same period. In order to briefly characterize Husserl's view during this particular period of his life I should like to take my point of departure in the *Amsterdamer Vorträge* (1928).[4]

Referring to phenomenological psychology Husserl says here that the first problem is to clearly delineate the subject matter of such a pure psychology and to explain how one can gain access to this subject matter. As Husserl sees it, the subject matter of phenomenological psychology is the pure psychical which is taken in this science as completely isolated from and independent of the physical with which it is nonetheless essentially interwoven. One gains access to this domain not by a process of negative abstraction, but by a special intuition. And such a special intuition can occur only in the realm of reflection. What is called now 'phenomenological experience', Husserl says, is a very peculiar type of reflection the nature of which is to be determined more accurately.[5]

In order to materialize a pure, phenomenological experience, Husserl maintains, and to be able then to make consciousness in its pure and unique Being the universal theme of my research, I must put between brackets as being not psychical all that is covered by the general thesis of our natural attitude, according to which the real world about me is at all times known as a fact-world that has its Being out there. In so doing the thesis of the natural attitude is still experienced as lived, but I make no use of it any more in order exclusively to focus my attention on the psychical as such.[6] In performing this reduction I obviously cannot deny that consciousness is and remains intentional, consciousness of something, but as long as I decide to remain within the realm opened up by the reduction I take that which I experience not as things in a real world, but exclusively as that which is intended by consciousness, thus as unities of meaning, as phenomena.[7]

Phenomenological psychology is thus impossible without a reduction. At first sight this conception seems to mean that pure, that is phenomenological, psychology in the final analysis will be able to unravel all transcendental problems and thus will become identical with transcendental phenomenology. We must remember here, however, Husserl says, that phenomenological psychology is only and exclusively interested in the psychical as such, that is to say in 'souls' and 'communities of souls' which as such are encountered in the natural world which psychology therefore must still somehow presuppose. In other words as a method of phenomenological psychology the reduction serves *exclusively* to isolate and determine the purely psychical. Although it is reduced to phenomena in the phenomenological sense of the term, the psychical itself which constitutes the immediate subject matter of psychology is and remains the psychic life of concrete human beings within the real, spatiotemporal world, or inasfar as this psychology is eidetic, within a *possibly* real world. It is only and exclusively in this sense that in phenomenological psychology 'the' world is always pre-supposed as a constant world at hand and that this psychology, too, is still a 'positing' science which therefore basically *remains within the realm of the natural attitude.*[8]

Comparing phenomenological psychology with empirical psychology on the one hand and transcendental phenomenology on the other, Husserl says that empirical psychology as a positive science studies the psychical as belonging to the always pre-given, spatio-temporal real world which for this psychology remains the only valid one. As an eidetic science phenomenological psychology seeks the *logos* of the psychical. Its thematic basis is not the real world as such, but merely a possibly real world, a world which is thought to be pre-given as real. The phenemenologico-psychological reduction forms a necessary method by means of which it becomes possible to delimit and determine the psychical (especially in its character of being intentional) by means of an inactivation of all the transcendental positings which are found in this intentional life. Eidetic psychology, however, remains a *positive* science in that it retains the validity of the natural apperception of the world, at least inasfar as the psychical itself is concerned. As soon as this latter aspect is inactivated we enter the realm of transcendental phenomenology.[9]

In *Cartesian Meditations* Husserl presents a similar view. In the fourth meditation, in speaking about the transcendental *ego* from an eidetic point of view, he says that what has been said about the transcendental *ego* can to a great extent be maintained in phenomenological psychology. "To the transcendental *ego* there corresponds then the human *ego*, concretely as the *psyche* taken purely in itself and as it is for itself, with is psychic polarization: I as the pole of my habitualities, the properties comprised in my character. Instead of eidetic transcendental phenomenology we then have an eidetic pure psychology, relating to the *eidos* 'psyche', whose eidetic horizon, however, remains unexamined."[10]

In *Crisis* Husserl's explanation of the difference and the relationship between *empirical* and phenomenological psychology does not contain essentially new insights, but his attempt to combine his conception of the life-world with his view on the meaning and function of the phenomenologico-psychological reduction seems to entail that this reduction becomes identical to the transcendental reduction. If this is indeed true, then there can no longer be an essential difference between phenomenological psychology and transcendental phenomenology. Therefore let us consider this idea more carefully.

Psychology, Husserl is saying here,[11] deals with the psychical, and phenomenological psychology is concerned with the *eidos* of the psychical. Since all psychic phenomena are intentional phenomena the specific theme of eidetic psychology concerns the essence of the human soul seen exclusively as the subject of intentional life. Eidetic psychology studies the individual soul from the perspective of its intentional life. Now every soul is intentionally connected with other souls in a community of souls; that is why phenomenological psychology in the final analysis is concerned with a community of pure subjects. In order to be able to bring to light the *eidos* of the pure ego and the community of the pure subjects the psychologist has to perform the phenomenologico-psychological reduction. This reduction places between brackets all the psychologist's existential convictions and judgments concerning the subject matter of his study. He is required to assume the position of that of a disinterested observer and witness to the intentional life of the pure subjects. This reduction includes all souls, all subjects, and the research psychologist is here no exception. Thus he must become a disinterested observer of himself as well as of other subjects. Unless he performs this reduction consistently the phenomenological psychologist will never reach the essentially uniform and in itself absolutely closed 'innerworld' of the pure egos and the universal unity of intentional life which precisely is the subject matter of his science.[12]

Husserl states again that the universal reduction is merely the necessary and sufficient condition which must be fulfilled if one is to achieve access to the essentially intentional, psychic entities. For only by means of this reduction does it become possible to see the genuine meaning of the life of the pure egos as an *independent* theme of investigation. Only within the sphere of the reduction can this life be seen as intentional life in which these subjects in various forms and modalities are directed toward and concerned with purely intentional objects. The task of psychology consists precisely in descriptively and adequately uncovering these manifold intentional relationships in all their immediate and mediate implications, inasfar at least as all of this belongs essentially to the immanence of the pure egos. Husserl finally specifies again that all persons, including the observing phenomenologist, are in this way transformed into phenomena; this transformation reveals these egos as pure subjects of pure, intentional, conscious life.[13]

After all of this, which is still in complete harmony with what is found in the earlier publications mentioned, Husserl quite unexpectedly continues now saying that it should be clear from what we already know on the subject that a full treatment of these ideas necessarily leads phenomenological psychology toward transcendental phenomenology, because the phenomenologico-psychological reduction necessarily turns over into the transcendental reduction.[14] The reason for this is that the descriptive psychology of intentionality has to deal with the problem of the life-world as the intersubjective 'product' of a quasi infinite community of pure subjects whose lives are mutually interconnected, and that a logical development of this problem requires that we take the last step and perform the transcendental reduction. For we want to solve the problem arising from those acts and systems of acts by means of which the absolutely isolated and pure ego is able to experience other egos and to enter into contact with them so that in their intersubjective cooperation an intersubjective world can emerge, but this can obviously be done only after the phenomenologist has made it possible for himself to experience himself as *absolute ego*.

In all sciences, psychology included, we search for objective knowledge. That is why in dealing with human subjects scientifically, we try to achieve objectivity by excluding what is merely subjective. But this means that in order to achieve objective knowledge we must eliminate ourselves, even as that functioning subjectivity in whose functioning the ontological meaning 'world' precisely originates. As scientist who searches for intersubjectively valid truth I thus have to admit that (even in my investigations which aim at the intentional constitution of the world) there are other subjects with me whose being real as subjects I must admit on the basis of empathy and with whom I know myself to be communalized. But when I perform the reduction and apply it to myself and to the world then, at the same time, this reduction affects all other men just as it affects the world in general, so that these men and the world in general become mere intentional phenomena for me. A radical and consistent reduction thus necessarily leads to an absolute and unique *ego*, the *ego* of the absolutely growing-lonely psychologist who now no longer is a real human being in the world, but a pure subject, the pure subject of his universal and pure intentionality. But this *ego* is an apodictic, a transcendental ego.[15]

In Husserl's view the surprising result of these reflections can be summarized briefly as follows. A pure psychology within the realm of the natural attitude seems to be impossible; there is only a *transcendental* psychology or phenomenology. Husserl adds, however, that upon closer consideration one will see that this conclusion, although indeed containing a good deal of truth, on the other hand cannot be entirely correct.[16]

Further explanation of this remark does not help Husserl to transcend his ambiguous conclusion. For on the one hand he defends the thesis that phenomenological psychology and transcendental phenomenology cannot be completely identical,[17] whereas on the other he argues that the difference between the reductions cannot be maintained.[18]

Max Drüe has collected a whole series of texts which seem to aggravate the difficulties.[19] I shall limit myself here to a brief indication of his most important considerations in this regard.

First of all, Drüe says, after performing the phenomenologico-psychological reduction the psychologist becomes gradually aware of the fact that his empirical ego is merely a mundane fixation of the transcendental ego which in the realm of the transcendental precedes him and, furthermore, that each transcendental ego is an individualized materialization of the transcendental subjectivity. But this latter insight leads the psychological reduction over into the transcendental reduction and, thus, phenomenological psychology changes into transcendental phenomenology.[20]

In a similar sense Husserl writes that the separation between the pure, phenomenologico-psychological and the transcendental attitudes which is necessary at first, is later transcended and overcome by the phenomenologist who tries to come to a higher level of self-understanding. Understood in the proper manner phenomenologico-psychological reduction is nothing but the transcendental reduction which has not yet arrived at a complete self-understanding.[21]

Another argument can be developed along the following lines. From 1925 on Husserl always maintained that we must make a clear distinction between the empirical-psychological *ego* and the pure *ego* of phenomenology.[22] We must furthermore make a clear distinction between the pure *ego* of phenomenological psychology and the transcendental *ego* dealt with in transcendental phenomenology.[23] The empirical-psychological *ego* is described by Husserl as "I, this man in the world".[24] The phenomenologico-psychological ego is determined as "the human *ego*, concretely as the *psyche* taken purely in itself and as it is for itself, with the psychic polarization: I as pole of my habitualities, that is the properties comprised in my character",[25] and elsewhere as "pure psychic subject with a pure psychic life".[26] In comparison with the transcendental *ego* this phenomenologico-psychological *ego* is still a mundane *ego*.[27] Husserl goes on to explain this as follows. Phenomenological psychology studies the fundamental types of psychic phenomena. It deals only with psychic entities and events as 'unities of meaning' as they appear to me in my intentional acts. Psychology is interested only in the essences of the psychic entities and events, and describes them in intentional analyses. The necessary presuppositions of such a study are first the eidetic reduction from facts to essences and, then, a reduction from 'objective and mandane things' to 'unities of meaning' which are given to me (who is and remains here a real psychological entity in the world); this last reduction is what is called the phenomenologico-psychological reduction.[28] If one reflects upon this more carefully it soon becomes quite clear that there is an inconsistency in the attitude of the psychologist. On the one hand he performs the phenomenological reduction, and on the other he refuses to perform the reduction consistently so as to include his own ego. It is this inconsistency which leads him to transcendental phenomenology by reducing his mundane *ego* to the transcendental *ego* and then, in the final analysis, this transcendental *ego* in turn into the transcendental subjectivity, which (as Husserl explicitly says) can no longer be understood as 'personal'.[29] In other words, a clear distinction between phenomenological psychology and transcendental phenomenology cannot be consistently maintained.

In another section of his book[30] Drüe turns to certain reflections made by Fink in a first draft which he prepared for Husserl in 1936, suggesting some ideas for the remainder of *Crisis* which at that time was still incomplete.[31] In this outline Fink maintains that phenomenological psychology necessarily debouches into transcendental phenomenology. Nonetheless, Fink claims, there is and remains a difference between the two even where the psychologist indeed passes completely through the road which leads from psychology to transcendental phenomenology. In other words phenomenological psychology is not only different from transcendental phenomenology inasfar as and as long as it is conceived of as a 'preparatory phase' of transcendental phenomenology; but even where psychology has passed completely through the way and already has debouched into transcendental phenomenology, even then there is still a difference between the two. Fink conceives of this difference in terms of the extension of interest spheres: phenomenological psychology is interested only in a limited realm of problems; in other words within the domain of transcendental phenomenology, phenomenological psychology has a rather limited task and scope. Drüe agrees with these considerations and conceives of them as the final words one can say on the issue. Personally I feel that these reflections do not give an adequate idea of Husserl's final view on the topic. I am convinced that there is a way of bringing together all the relevant texts (Husserl's manuscripts as well as Fink's project for the continuation of *Crisis*) into a harmonious synthesis, but this requires that we interpret all of these statements in a different way. In order to make my position as clear as possible I should like first to make some introductory remarks and then to formulate my view in a few concise statements.

As I see it, Husserl never defended the view that phenomenological psychology and transcendental phenomenology are to be *identified*. First of all, we have a very impressive list of texts, written between 1913 and 1936, in which he explicitly claims that both disciplines are *essentially* different in that psychology is a mundane science to be developed in the *natural* attitude, whereas transcendental phenomenology moves in the *philosophical* attitude. Furthermore, the meaning and scope of the second part of *Crisis* taken as a whole is certainly in conflict with the thesis which completely identifies psychology and transcendental phenomenology since psychology (that is empirical as well as phenomenological psychology) is described there as a road leading toward transcendental phenomenology. Thirdly, if Husserl had been of the opinion in 1936 that the views he was developing at that time were in conflict with the ideas he had developed earlier in *Phenomenological Psychology* and *Cartesian Meditations*, he most certainly would have indicated this somehow. And finally Husserl always maintained that regional ontologies are not to be identified with transcendental phenomenology, whereas phenomenological psychology is consistently referred to as the regional ontology of the psychical.[32]

On the other hand I do subscribe to the view that a consistent development of phenomenological psychology will ultimately and of necessity lead to transcendental phenomenology. Phenomenological psychology is introduced by Husserl as the eidetic science which must lay the foundations of empirical psychology. But obviously phenomenological psychology is not the ultimate founding science; it itself is to be founded ultimately in transcendental phenomenology just as every other regional ontology. The relationship between this particular regional ontology and transcendental phenomenology, however, is of a very peculiar type, not only because of the close relationship between the subject matters of phenomenological psychology and transcendental phenomenology, but also and even particularly because of the fact that in phenomenological psychology a phenomenological reduction is necessary in addition to the eidetic reduction. This must be specified in greater detail.

Phenomenological psychology is a regional ontology. As such it remains in the natural attitude and thus outside the realm of the transcendental reduction. It is interested in bringing to light the *eidos* of the psychical, and in so doing it uses the eidetic reduction in order to make the transition from fact to *eidos* possible. Further it employs the phenomenologico-psychological reduction in order to separate the psychical from the physical with which it is essentially connected. However once this subject matter has been clearly demarcated, in phenomenological psychology as such there is no longer any need to maintain this reduction. Then the psychologist may return to the natural attitude and attempt to describe the intentional activities of the lives of the pure egos. Husserl follows this procedure exactly in *Ideas* where after introducing the phenomenological reduction and explaining its goal and scope he continues: "We start now with a series of reflections within which we are not troubled with any phenomenological *epoché*. Directed to the 'outer world' and without forsaking the natural standpoint we reflect psychologically on our ego and its lived experiences... *on the essential nature of the consciousness of something...*"[33] In other words, phenomenological psychology as a science of the natural attitude remains oriented toward the real world, and as eidetic toward a possibly real world; and its goal is to clarify the eidetic structures of real and possibly real egos and their conscious or psychic lives. Looking upon phenomenological psychology from this point of view there are no serious problems.

However if we turn to the question of how this particular science is precisely related to transcendental phenomenology, then it becomes evident that the position of the phenomenological psychologist is ambiguous. This ambiguity can be approached from different perspectives. One could say for instance that inasfar as the psychologist himself is a human soul, a human subject, he belongs in principle to the subject matter of phenomenological psychology, and as such he is obviously objectified. As investigator, however, he is not an objectified ego but precisely the objectifying subjectivity. By returning now to the natural attitude the psychologist is not able to justify his objectifying function. Such a function of the ego necessarily presupposes the transcendental reduction. One could say also that the moment the psychologist takes into consideration that a subject is not merely in his own world, but also in an intersubjective life-world and tries systematically to account for the intersubjective validity of the world for everyone, he must realize that he can do so only if he occupies a privileged position in regard to all other egos as well as in regard to the world. But such a privileged position can be obtained only by means of the transcendental reduction.

The question which remains is whether these and similar considerations necessarily entail that phenomenological psychology debouches into transcendental phenomenology in such a way that phenomenological psychology cannot be maintained as an independent discipline essentially distinguished from transcendental phenomenology. The answer in my view is obviously in the negative. Such a consequence follows only if one as able to show that phenomenological psychology's task consists precisely in giving the *ultimate* explanation of the 'meaning and Being' of being.[34] There is no doubt that in Husserl's view the goal of phenomenological psychology consists in laying bare the *eidos* of the psychical; the ultimate explanation of the 'meaning and Being' of being remains the task of transcendental phenomenology. When the phenomenological psychologist first performs the phenomenological reduction and then returns to the natural attitude, he is completely justified in doing so in view of the goal he wishes to achieve. But from the viewpoint of someone who desires to give an *ultimate* explanation, justification, and foundation the psychologist's behavior is inconsistent. However, even though the transcendental phenomenologist sees this inconsistency, he cannot blame the psychologist for one important reason, namely that the psychologist never pretended that his particular methodical procedure was ever meant to solve the problems of transcendental phenomenology. His sole legitimate claim is that his procedure enables him to attain the specific goal he originally set for himself.

NOTES

[1] Edmund Husserl, *Die Krisis der europäischen Wissenschaften und die transzendentale Phänomenologie. Eine Einleitung in die phänomenologische Philosophie*, Martinus Nijhoff, The Hague, 1962, pp. 238–69; pp. 514–16. Max Drüe, *Edmund Husserls System der phänomenologischen Psychologie* De Gruyter, Berlin, 1963, pp. 227–45.
[2] Edmund Husserl, *Phänomenologische Psychologie*, Martinus Nijhoff, The Hague, 1962, pp. 20–46. Joseph J. Kockelmans, *Edmund Husserl's Phenomenological Psychology*, Duquesne University Press, Pittsburgh, 1967, pp. 87–137.
[3] Cf. Edmund Husserl, *Phänomenologische Psychologie*, pp. 237–301, 302–49.
[4] Cf. for what follows: *Ibid.*, pp. 303–8, 312–5, 321–4, 331–47.
[5] *Ibid.*, pp. 303–8.
[6] *Ibid.*, p. 312.
[7] *Ibid.*, pp. 312–3; *Krisis*, p. 247, p. 257.
[8] *Ibid.*, pp. 335–6.
[9] *Ibid.*, pp. 338–41; 343–6.
[10] Edmund Husserl, *Cartesianische Meditationen und Pariser Vorträge*, Martinus Nijhoff, The Hague, 1963, p. 107.
[11] Cf. for what follows: *Krisis*, pp. 238–69.
[12] *Ibid.*, p. 243.

[13] *Ibid.*, pp. 254–7.
[14] *Ibid.*, p. 259.
[15] *Ibid.*, pp. 259–60.
[16] *Ibid.*, p. 261.
[17] *Ibid.*, p. 261.
[18] *Ibid.*, pp. 261–3.
[19] Max Drüe, *op. cit.*, pp. 240–2.
[20] *Ibid.*, p. 241.
[21] Ms. K III 1, p. 134; quoted by Drüe, *Ibid.*, p. 241.
[22] Edmund Husserl, *Cart. Med.*, pp. 64–5.
[23] *Ibid.*, p. 107.
[24] *Ibid.*, p. 64.
[25] *Ibid.*, p. 107.
[26] Edmund Husserl, *Phän. Psych.*, p. 342.
[27] *Ibid.*, pp. 343–4.
[28] *Ibid.*, pp. 335–4 (passim).
[29] Edmund Husserl, *Krisis*, pp. 259–60; Ms. B I 13, IV, 26, quoted by Drüe, *op. cit.*, p. 242.
[30] Max Drüe, *op. cit.*, pp. 243–5.
[31] Edmund Husserl, *Krisis*, pp. 514–6; 557.
[32] Cf. Edmund Husserl, *Die Idee der Phänomenologie*, Martinus Nijhoff, The Hague, 1950, pp. 14–5, 79; *Ideen zu einer reinen Phänomenologie*, 3 vols. Martinus Nijhoff, The Hague, 1950–1952, Vol. I, p. 39, pp. 140–4; *Phän. Psych.*, pp. 46–51.
[33] Edmund Husserl, *Ideen*, Vol. I, p. 74.
[34] Edmund Husserl, *Formale und transzendentale Logik. Versuch einer Kritik der logischen Vernunft*, Niemeyer, Halle a.d.S., 1929, p. 13, p. 148.

ANDRZEJ PÓŁTAWSKI

CONSTITUTIVE PHENOMENOLOGY AND INTENTIONAL OBJECTS

There is, according to Husserl, a radical difference – "die kardinalste, die es überhaupt gibt" (*Ideas*, Vol. I, p. 77) – between the real world and pure consciousness. But a peculiar shift in the way in which this difference was described occurred as early as in the *Ideas*, Vol. I. In his letter to Husserl from 1918[1] Roman Ingarden writes:

> Niemand hat vielleicht die Wesensheterogenität (zwischen Sein und Bewusstsein) so stark betont (wie Sie in den *Ideen*). Indessen... ändert sich im Laufe der Untersuchung diese ursprüngliche Stellung, so dass im Grunde die Wesensunterschiedheit eigentlich geleugnet wird, oder in die Scheidung zwischen Noesis und Noema umgedeutet. Dies geschieht einerseits dadurch, dass die Realität... *nur* als Vermeintes "und darüber hinaus ein Nichts" gefasst wird, anderseits aber das Wesen des Bewusstseins auf das Noema ausgedehnt wird. Denn im Grunde ist – nach Ihrer Auffassung – z.B. das Ding als Korrelat einer unendlichen Wahrnehmungsmannigfältigkeit nichts anderes als ein noematischer Sinn... Wir kommen am Ende zu der Gleichung: Ding = ein eigentümlich gebautes Noema-Bewusstsein.

Thus, an inner contradiction in Husserl's position seems to be present: either reality is given by means of perceptual adumbrations (*Abschattungen, Ansichten*) and is essentially different from pure consciousness in its structure and mode of existence (cf. *Ideas*, Vol. I, pp. 76ff.) (and to be that it must have, in itself, a genuine essence of its own), or it is nothing but a noematic sense-content existing in pure consciousness (cf. *Ideas*, Vol. I, p. 93). According to the very sense of experience, says Ingarden in the letter quoted, what is given is real only when it is not my imagination; while 'imagination' means here more then 'being presumed (vermeint) without sufficient motivation', and 'reality' more than being presumed with sufficient motivation in previous experience: it means that in every experience involved, there appears something that exists 'in itself'.

The question is, in the first place, that of the sense of 'transcendent'. In the same letter, Ingarden distinguishes two involved meanings of the word: (1) being essentially (in itself) different from pure consciousness, (2) surpassing what is actually given (a) in a single perception, or (b) in any finite series of perceptual syntheses. Transcendence in the sense (2), if

Tymieniecka (ed.), Analecta Husserliana, Vol. II, 90–95.

taken as a noematic feature only, does not, according to him, affect the point at issue; but if by 'the sense of a real object' the sense incorporated in the object is meant, the object, in order to exist and to transcend the data, must be 'in itself'.

These were ontological questions. The parallel epistemological problem is that of substantiating a 'position' (*Setzung*) of an incorporated sense of reality or of reality itself. Here, a new concept of transcendence emerges; but, as Ingarden stresses, a mode of existence should not be confused with the validity of a 'position'.

These early considerations have led Ingarden to further investigations and to the development of his later views as stated in his mature works. But we may already see from the above that the points at issue are the crucial concepts and axioms of philosophy, the very sense of the phenomenological attitude, and the goals we set to ourselves when we adopt it and those which, on the other hand, phenomenology is fit to realize. The root of the controversy is a peculiar duality in Husserl's own position. On the one hand, he stated the "principle of all principles", i.e. that "jede originär gebende Anschauung eine Rechtsquelle der Erkenntnis sei"; on the other hand, he adopted the rule of transcendental constitution which, more and more radicalized, led him to treat every object, even an immanent one (in the sense of *Ideen* I), as transcendent in the widest sense of the term in relation to its consciousness, and 'constituted' in it; this constitution is understood as some sort of creation (the terms used: 'erzeugen', 'schaffen', 'Produkt', 'Gebilde', 'Stiftung'), and pure inner consciousness is said to be the source (*Ursprung*) of all transcendent being. The difference of attitude, underlying this duality is, by and large, that between the so-called München-Göttingen School of 'ontological' phenomenology and transcendental phenomenology. But Ingarden's approach to the problems involved is original and based on such a broadly laid foundation of investigations that it seems worth while to sketch the main lines of his way towards a solution; it goes without saying that such a sketch may only be regarded as an invitation to a more detailed study of Ingarden's own works.

In his contribution to the Krefeld Colloquium Ingarden mentions some important points where Husserl's views do not seem quite clear to him. Some of them are: first, what is precisely meant by Husserl by the 'relativity' of the real world in relation to pure consciousness? Secondly, Husserl, although he has laid foundations of formal ontology, did not analyze the form of an individual object and did not ask whether an object being thus 'relative' would not have to possess a specific form differentiating it from an 'absolutely' existing one; thirdly, Husserl treated our cognition of particular real objects in the same way as that of the real world at large, and he did not, at least in his works published during his lifetime investigate whether the world as a whole region of being can have a form quite similar to that of a single thing. Neither did he discuss the problem whether pure consciousness is really a whole region of being or merely a single individual. When, as Husserl made it clear later, the world is said to be constituted by the acts of many transcendental *egos* (cf. *the V Cartes. Meditation*), the situation becomes radically different from that in which only one stream of consciousness is involved. But Husserl never succeeded in showing in a satisfactory way what makes it possible for him to regard other streams, transcendent in relation to his own, as not constituted by it.

To make up for these deficiencies Ingarden undertook, in the first place, his investigations concerning the ontological status of literary works. They seemed to be the best examples of purely intentional objects, objects which are typical products of a free activity of human consciousness. At the same time he tried to analyse different possible modes of existence, forms of different kinds of objects and of the whole regions of objects. These analyses have led him to his existential and formal ontology as it is presented in the *Controversy about the Existence of the World.*

In this paper, I shall only try to discuss some implications of Ingarden's analyses of intentional objects and of his views on the proper place of their investigation, as these implications regard the problem of transcendental constitution and the closely related idealism-realism issue.

In his existential ontology, Ingarden distinguishes different 'moments of being' which are predicated about objects when these are said to possess this or that mode of existence.

According to Ingarden, the 'relativity' of real being to pure consciousness, in spite of the already mentioned vagueness of the concept, means for Husserl in the first place that real things are *determined* by the conscious acts in which they are apprehended; i.e., that these determinations are only, so to speak, ascribed to them from the outside. This is the moment of existential *heteronomy*. But its peculiarity is that a heteronomous object, a purely intentional one, does not, for plain common sense, exist at all. Thus e.g. a hero of a novel, although he is described as real, is but a product of the author's fancy. He is determined by the printed text of the novel, but only by it; in himself, he is a pure nothing. The positive feature of heteronomy is just this circumstance of the properties in question being ascribed to the object.

As Ingarden shows, this existential moment is connected with a specific form, characterized by a peculiar two-sidedness of structure. Such an object is endowed with two separate series of properties having two separate subjects, respectively. There is, first, its 'content' – all the ascribed characteristics, together with the form and mode of existence which e.g. Hamlet, as Shakespeare's character, is described as possessing. On the other hand, there is its determination as something created by such and such acts, or corresponding to such and such a text: its 'intentional structure'. The subject of the latter properties is the proper intentional object, that of the former ones is only a central point, a pole of the content. But both are heteronomous, there is nothing autonomous in a purely intentional object.

Can reality be regarded as 'purely intentional'? And, if not, how are Husserl's ideas of transcendental constitution and of pure consciousness to be understood in order to be acceptable? It does not seem that these questions have been answered by Husserl himself or by other transcendental phenomenologists quite satisfactorily.

As may be seen from the letter quoted above, Ingarden does not think it possible to give up the concept of reality as something existing 'in itself'. His reasons are: first, this sense of reality has its source in our experience, reality is *given* as existing just in this way. Were it but an illusion, we would not be entitled to assert that the real world exists 'for something else only', but merely that it does not, as reality, exist at all.

The second reason is that the proper sense of the transcendental investigation is determined by the assumption of this sense of 'reality'. The goal of this investigation is, in the first place, the attainment of an absolute, indubitable knowledge. If every act of consciousness were to create its object, the very idea of such a knowledge would be denied; because, by cognition we understand nothing else than seeking and apprehending the object as it is in itself. Thus, the difference between a noematic sense and its object cannot be discarded. And, correlatively, we must distinguish among our acts those which are creative and those which, although in a way they may be quite active, aim only at the discovery of what is actually there.

Surely, for Husserl, a realisation of such a true knowledge is possible in immanent perception. And, in this case, it is clear that where we are confronted with intentional creation, and where this creation is disclosed in transcendental investigation, what is thus revealed is not a cognition at all; and, therefore, there is no need for any changes in our idea of knowledge.

If we ask about the validity of our knowledge of essences, the situation becomes even more difficult. In his discussion of *Formale und transzendentale Logik* Ingarden speaks about a peculiar conflict between two discordant assertions of Husserl's: on the one hand, the whole phenomenology should be, and is, 'eidetic', it tries to discover essences. And this is, indeed, the starting point and a presupposition of the whole transcendental inquiry; constitutive investigation is said to be a 'Rechtsbetrachtung', an examination of the validity of experience (*Ideas*, Vol. I, part IV); on the other hand, we are told that a real object has no genuine essence at all, that it has no determinations in itself, that its being is 'relative' to the conscious acts. But, asks Ingarden, which set of these contradictory 'eidetic' assertions is true – that of the ontological, or of the transcendental ones? And, above all, can the results of *a priori* cognition be false?

It would, of course, be possible to deem all the results of ontological investigation as the *content* of purely intentional objects, corresponding to creative acts of consciousness (in the sense explained above). But we are, in fact, sure that there is an essential difference between, say, poetic creation and genuine experience; we do not regard poetic imagination as able to create authentic reality. On the contrary, we think it to be necessarily dependent on a long process of genuine experience.

As pointed out by Ingarden, the equation of real objects to noematic senses is not accepted by Husserl as a valid metaphysical conclusion from previous constitutive investigations, but occurs at the beginning of the inquiry. Were it a temporary demand only, and were we allowed, after investigating the problems of cognition, to come back to the objects, the idealistic solution would not necessarily follow. But Husserl does not allow this. And so we are left with noematic senses alone.

This point is closely connected with the problem of the sense and range of the phenomenological reduction and with that of the proper method of epistemology. The expressly stated function of the reduction is, in *Ideas*, Vol. I, the disclosure of a special region of individual being; a region which can be known in an 'absolute' way. Thus, a peculiar duality of this function arises. On the one hand, the reduction seems to be a purely methodological device, which, indeed, allows us to avoid a *petitio principii* in epistemology. On the other hand, it plays the role, just mentioned, of revealing to us a being of a special sort. But does it not, as applied by Husserl in the latter capacity, lead automatically to an idealistic conclusion? Such, indeed, seems to be the case. For, as Ingarden remarks, even in the discovery of pure consciousness any assertions concerning, e.g., the difference of the mode of existence between consciousness and reality should not be assumed. To be methodologically correct, this discovery cannot, by itself, imply anything about reality. In epistemology, Ingarden concedes, the phenomenological reduction is necessary. But not in ontology or in metaphysics. Here, on the contrary, the inquiry must be free of it. Otherwise, a properly substantiated solution, both a realistic and an idealistic one, becomes impossible.

As may be seen from this argument, the problem of heterogeneity and of mutual relation of reality and pure consciousness has not been solved by Husserl quite satisfactorily; and his ideas about reality, intentionality, pure consciousness and transcendental constitution are in need of further elucidation and inquiry. This inquiry has been, indeed, undertaken by Ingarden in a manner at the same time extensive and systematic. Thus, a broadly laid confrontation of his position with the transcendental one seems both promising and interesting.

NOTE

[1] Published in Polish in Ingarden's *Studies on Contemporary Philosophy*, Warsaw, 1963. From among the seven studies dealing with Husserl included in the collection, only the contributions to the Krefeld (1956) and Royaumont (1957) colloquims, a small part of a discussion of Formale und transzendentale Logik (*Kantstudien*, 1933) and a radio speech have been hitherto available in languages other than Polish. The quotation is from the German original.

ALPHONSO LINGIS

HYLETIC DATA

The concept of sensation arose out of the empiricist enterprise taking the psychic as a sector of Nature, intelligibly integrated into the totality of Nature by laws, and understood by the exposition of the relationships between its simplest component elements. Husserl surely did not believe that consciousness could be treated as a sector of Nature among others, and rejected not only the analysis of the psychic into the atomic sensations of classical empiricism, but also of the analysis of the psychic into the molecular or molar sensations of Gestaltism.[1]

And yet Husserl himself resurrects the concept of sensation under a new name, *hyletic data*. All the ancient ambiguity of the term is maintained: under hyletic data he classes together indiscriminately the exteroceptive sensations corresponding to the various sense organs, the interoceptive kinesthetic sensations, the affective sensations of pleasure, pain, tickling, and other sensile feelings, and the sensile phase of tension and release, impulses, drives, volitions, and inclinations. Like the empiricist sensations, they are passivity and matter (*hyle*), and are of the nature of 'impressions.' They are a discrete multiplicity.[2] When they vary, the appearance of the intended transcendent object varies.[3] And they are said to *resemble* those very characteristics of the transcendent objects which they mirror within consciousness – the sensation for white (*Empfindungsfarbe*) differs from the sensation for black as white differs from black on the things.[4] Husserl goes so far as to affirm that the visual sensations have extensity (*Ausbreitung*) which depicts (*darstellt*) the physical extension (*Ausdehnung*) of physical objects.[5] The consciousness then must be a sort of place that can contain these somehow extended sensations.

But consciousness is not consciousness *of* these sense data. Unlike the mental representations of empiricism, the intentional objects of consciousness are not formed out of this matter. Thus we do not reach this matter by abstracting from the form of objects of consciousness, but rather by a kind of reflection back from its objects to the consciousness itself.[6] This reflection discovers the hyletic data as *Abschattungen*, shadows of the

Tymieniecka (ed.), Analecta Husserliana, Vol. II, 96–101.

things cast back upon the sphere of consciousness itself. Intentionality passes through them without coagulating them into mental tableaux, nor synthesizing them even ideally.

Thus Husserl deliberately neutralizes the naturalist psychology, and more profoundly the representational epistemology and the matter-form metaphysical analysis of things, from which the notion of sensation comes. What role will it then play in his phenomenology?

The idea of sensation would seem to be the simple contradictory of the idea of intentionality. The idea of sensation belongs to mental activity conceived as immanence, whereas the idea of intentionality is mental activity conceived as transitive, as transcendence. Intentionality means that the consciousness is directly open upon the things themselves, and Husserl affirmed that intentionality is not a property or trait of a substance called mind or consciousness, but that intentionality is the very essence of consciousness. Whence Sartre understood the Husserlian concept of intentionality to be the radical liberation of consciousness from its mental contents – "It was all we needed to put an end to the soft philosophy of immanence, where everything takes place by compromise, protoplasmic exchanges, by a tepid cellular chemistry".[7]

And yet the genuine Husserlian concept of intentionality requires and presupposes sense data. The concept of intentionality is the difficult conjuncture of, on the one hand, the notion of mind as a being open to what is other then itself, essentially an intuitive immediacy, and, on the other hand, the notion of mental activity as synthesis, as com-prehension. It is simultaneously immediately transitive orientation, ec-stasy, and com-prehending, synthetizing constitution. It is simultaneously intuition and *Sinngebung*.

Intentionality is constitution, *Sinngebung*, because it is the recognizing of a presumptive, an ideal unity across a flux of data. It constitutes objects not by subsuming a sensible diversity under a form or a category, but rather by ascribing a diversity given across time to ideal identity. Thus the hyletic material in the mind itself does not coagulate into a tableau obstructing the ray of intentionality; rather intentionality moves unto the ideal identities which are its objects by *taking* these shadows *as* adumbrations of something intelligible, something meant, the *Sinn*. Intentionality posits, presumes, or proclaims (*meinen*)[8] identity. Intentionality then is not a relating of one real being to another (for example, the subject with the object); it is a relating of givenness to identity, of flux to ideality. The 'movement' of intentionality is a movement from the sensible to the ideal; it is this 'movement' or 'tending toward' that synthesizes. Thus the ray of intentionality, passing through the hyletic thickness of the psychic unto the transcendent objects meant, is called by Husserl an *Interpretieren*, a *Deuten*. As in Aristotle, the mental act is the hermeneutic act.

But not all hyletic data function as analogues, *Darstellungen*, for aspects of objects. Kinesthetic sensations are hyletic data that do not depict profiles of transcendent objects. But Husserl does not treat them as representations or motor images of corporeal movements recorded or reflected in the immobility of subjectivity. Kinesthetic sensations function not as representations, but as motivations. They make felt, within the sphere of consciousness, the relativity of each perceptual profile to a certain orientation. They reveal that the intentionality that means or intends an object across hyletic data is also a movement of orientation with regard to that object. Even if the object is a centaur I but imagine, kinesthetic sensations reveal that centaur in a certain orientation with regard to the zero-point with which the ego coincides.[9]

Husserl says that there is a 'freedom' in kinesthetic sensations: "The course of kinesthetic sensations are free processes, and this freedom in the consciousness of the course of the sensations is an essential part of the constitution of space."[10] This means that the kinesthetic sensations do not reveal the object as simply oriented with regard to a factual zero-point, but rather a freely displaceable zero-point, a freely mobile subject. Conditionality, the 'if...then...' mode, stamps the orientation given across kinesthetic sensations.[11] Thus the experience of the object as an ideal unity of a succession of perspective *Abschattungen* is also the experience of the object as oriented with regard to a mobile zero-point. The I think is an I can. The ideal unity of the object implicates the free mobility of the subject.

It is then by animating the kinesthetic sensations that intentionality is revealed to be itself already turning in the very space it will constitute by constituting objects. In moving from the sensible to the ideal it moves in space.

The kinesthetic sensations reveal the consciousness to be somewhere. Its hyletic substance is passivity and matter. Hyletic data are neither actively produced, nor passively received by consciousness; it is rather encumbered with them: it lives them. But they are the very passivity of the consciousness with regard to itself, they are the consciousness laden, affected with itself. To live oneself is not to objectify oneself; it is to be encumbered with oneself, to be affected with oneself. The Husserlian consciousness is not just representation, or an epistemological function: it is life, *Erlebnis*, experience lived through.

The carnal presence (*leibhaft*) of the things themselves answers to the sensuality of the intentionality of consciousness. It is across this sensuality of consciousness, and not in its *Sinngebung*, that consciousness *approaches* the things, that it vists them, and comes into contact with them. The meaning-constitutive intentionality of consciousness implies distance; it is going beyond the given unto the ideal totality, it is objectification: it posits an ob-ject before itself, across a distance. Distance, interval, is continually both opened and traversed by the very enactment of intentionality. But then proximity, the contact with the thing itself, is not accomplished by the idealizing intentionality that objectifies it and thus posits it at a distance; it rather consists in the materiality of the sensations, which are lived through and not themselves envisaged from a point of orientation, and which, across intentionality, are found to be the adumbrations of the things within the substance of the consciousness itself. Because consciousness not only objectifies the things but *approaches* them there is such a thing as sensuous cognition, a cognition for which the objects are not just intersections of relations, but sensible proximity, dense and carnal presence, *Leibhaftigkeit*.[12]

On the most radical stage of the retrogression of phenomenological analysis, the hyletic materiality opens upon the constitution of time and the *Urkonstitution* of the *ego*. Husserl conceives consciousness to upsurge uninterruptedly in the living now as a sensorial impression. An impression that endures. And that passes. Its passing is a modification, a mutation of the primal impression. But already the impression that emerges anew and primal in the living now vacated by the passing impression feels, within itself, the sensation now passing. This inner adhesion to the passing impression is the very tension of the present, the primal tension of intention. This way for the sensing of sensation to retain the impression that passes is the very production of temporalization. Time and intention are born in this dephasing.[13]

But the primal phase of the living now itself is conceived by Husserl as an impression produced by spontaneous generation. For the primal impression can be conceived neither as produced nor as received by a preexisting proto-consciousness; it "is the absolute beginning... the primal source, that from which all others are continuously generated. In itself, however, it is not generated; it does not come into existence as that which is generated but through spontaneous generation. It does not grow up (it has no seed): it is primal creation."[14] To arise, by spontaneous generation, and to endure by affecting oneself with the coalescence of mutations one undergoes in passing, is to exist as impression. The materiality of what is hyletically given is now discovered to consist not in formlessness (or in pre-formed determinability), but rather in its impressional nature.

Thus life arises in the autogeneration of impressions, and life becomes consciousness and intentionality across the modifications of impressions, and their retentional and protentional tensions. And consciousness forms one flux by virtue of the coalescing of hyletic mutations. And the flux of hyletic data gives duration, endurance, to the intention aiming at the objects, which thus does not move with the instantaneousness of ideal operations.

Thus the Husserlian doctrine of hyletic data is not a simple regression to empiricist naturalism. For Husserl conscious life (*Erlebnis*) is intentionality and sensuality, ideal movement of objectification of things, and movement of approach to the things in their carnal presence.

NOTES

[1] Cf. for example *Formale und Transzendentale Logik*, Halle, 1929, p. 252; *Ideas*, Author's Preface to the English Edition, New York, 1931, pp. 16–17.

[2] *Ideen I*, Halle, 1928, p. 203. Aron Gurwitsch understands the distinction between *hyle* and *morphe* classically as a distinction between determinability and determination. Thus for him "The hyletic itself is 'formless stuff,' merely present and nothing more," and he consequently argues that "Data devoid of all articulation, hyletic data in the strict sense, do not exist at all.", *Studies in Phenomenology and Psychology*, Evanston, 1966, pp. 254 and 256. The present paper aims to show that for Husserl the "materiality" of hyle is not the classical "determinability", and that a phenomenon is not a compound of hyle and morphe in the sense that for Aristotle an existent is formed matter.

[3] *Ideen*, Vol. I, p. 203.

[4] *Logische Untersuchungen*, Halle, 1922, Vol. III, p. 101.

[5] *Ideen*, Vol. I, p. 162.

[6] *Ideen*, Vol. I, pp. 203, 65.

[7] *Situations*, Vol. I, Paris, 1947, p. 33.

[8] Emmanuel Lévinas, *En découvrant l'existence avec Husserl et Heidegger*, Paris, 1967, pp. 218–22.

[9] *Ideen*, Vol. II, The Hague, 1952, pp. 56–7.

[10] *Ideen*, Vol. II, p. 58.
[11] *Ideen*, Vol. II, p. 57.
[12] *Ideen*, Vol. II, p. 153.
[13] E. Lévinas, *En découvrant l'existence avec Husserl et Heidegger*, pp. 151–6.
[14] *Zur Phänomenologie des Inneren Zeitbewusstseins*, The Hague, 1966, p. 100. Engl. transl. by James S. Churchill, *The Phenomenology of Internal Time Consciousness*, Bloomington, 1964, p. 131.

DISCUSSION

Klibansky: I thank Mr. Kockelmans for his stimulating paper. I would ask if there is anyone who would defend Eugen Fink with regard to the interpretation of the relation between phenomenological, psychological, and transcendental phenomenology?

Kockelmans: If you read in the *Phänomenologische Psychologie* – in the tests which are added to the main text – what Fink suggests is that there is no essential difference between phenomenological, psychological, and transcendental phenomenology. And Fink formulates his view in this way: The phenomenological psychology and transcendental phenomenology differ in their extension of interest in these spheres. It is indeed against this which I argue. I feel that this is too limited a distinction between the phenomenological psychology and transcendental phenomenology. But we have to realize Fink only suggested this to Husserl as a possibility for further research. I doubt very much whether Fink, for instance, today still sticks to his idea. I do not say he doesn't, but I could easily imagine that over the years he has come to the conclusion that probably another approach to this problem would be more appropriate.

Van Breda: You quoted, and that is absolutely exact, the text of Fink; but there is also a new edition of another paper prepared by Fink for the continuation of *Crisis* and that publication is in the *Edmund Husserls Briefe on Roman Ingarden*. Ingarden makes a number of explanations and historical notes. That means also that Ingarden was always trying to find out the definite systematic work that Husserl was speaking of between 1926 and 1936. We don't have that systematic work. The reason is that Husserl didn't write it. It is interesting from the historical point of view to read also the answers of Husserl. We don't have the letters from Ingarden to Husserl. We have only some summaries Ingarden possesses and these we plan to publish.*

Husserl was working, after *Ideas*, Vol. II, on several regional phenomenologies. And we will have one of them in the publication of Claesges here present: 'Space and Thing'.

Tymieniecka (ed.), Analecta Husserliana, Vol. II, 102–127.

Gadamer: First of all I must acknowledge that the main line of your paper is in complete agreement with my own view. Indeed, when we examine Husserl's later thinking it is obvious that all his crucial concerns converge in the intention of finding ever new ways towards the transcendental reduction. One among them, particularly meaningul if we seek to avoid possible mistakes, is to take the start from the transcendental psychology.

I have, however, a minor question. It appears that we have in psychology a striking paradox. You said that the psychologist is in the situation of an observer or of an 'objectifying' investigator, and that he should change this position into that of a transcendental *ego*. But, and I think it is also Husserl's view, the observer and the observed person stay in interaction, hence the well-known difficulty arises concerning introspection, namely, that by observing himself one is changing himself. In order to avoid this paradox, one has to realize that what is involved in this pursuit is not the empirical *ego* but the functioning of the transcendental *ego*. Therefore the objectifying attitude might be not only enough for the psychologist but also legitimate.

Kockelmans: I completely agree with you. But you know when I brought up the question of objectification I brought it up in connection with the main theme of *Crisis*, which is an investigation into the life-world. If the psychologist has to account for the life-world, not only his own but the life-world for everyone, or at least a possible life-world for everyone, then he has to have a privileged position in regard to the other *egos*. That is the same thing that Husserl developed in the *Fifth Cartesian Meditation*, where he explicitly says, "If I am going to account not only for the world which is mine but for our world, then I have to account for the constitution of the other ego before I am able to account for the constitution of the world for everyone." That is only possible on the basis of a privileged position you adopt by putting yourself in the transcendental realm, in the absolute realm, and that is the meaning of the reduction. But as long as the life-world question was not brought into the picture, Husserl never mentions this question of objectification. If I could have read my paper completely you would have seen that there is an allusion to this problem. But since it was not of importance, seeing the lack of time we had, I just didn't read it.

Erling Eng: I would like to amplify an agreement I have with Professor

Kockelmans, but also add something that may change the complexion of the discussion. I think it is significant that he added that when one enters the *life-world* then the situation somewhat changes. The important thing is that when Husserl talks about psychology it is largely a psychology of perception. And as long as one stays out of the sphere of action and pursues the phenomenological reduction in the sphere of perception there are not too many problems with phenomenological psychology.

It is when one enters the sphere of the drives (*Triebe*), or *pulsions*, that one encounters difficulty. Significantly enough, Husserl, in one of the supplements to *Ideas II*, discusses the sensibility (*Sinnlichkeit*) and specifically makes a very striking observation concerning the drives. He says the drives are *eidetic*. This means that even in my experiencing of my drives, or perhaps we could use the French word that has sometimes been used for Triebe, these *Pulsions*, I occupy a privileged position as pure *ego*. It is significant that Husserl in the later period of his development devoted many pages to passive genesis and passive constitution. Driesch makes use of much of this material, some of which has not even been published in the volume on the passive synthesis. And this is extremely important because as a psychologist encounters the evidence for passive genesis in the lives of individuals, also in himself (the concrete example being the way in which Freud, if I may mention that name here, encountered relics of his own earlier experience in the course of working with patients), the question is: "What is his position with regard to these?" In my paper at the end of this Conference I will go into the element of *pathos* in the Aristotelian sense associated with a phenomenological reduction, and I submit that it is by moving through that phase of *pathos* that one recovers a transcendental position. In order to talk of passive genesis or of passive intentionality – a very important feature of Husserl's later thought – one must have accomplished a transcendental reduction in some form. In my field of psychology often this is accomplished in a mythological fashion. The pure *ego* takes on a particular mythological form. But there is a transcendental reduction present and I submit that the psychologist is again and again put in this position we are in, sheer self-defence, because of his involvement in what he is dealing with. In the course of being a phenomenological psychologist, in order to save himself, the psychologist must accomplish that transcendental reduction and in his work from then on what he does implicitly is somehow to communicate by his own stance the possibility of this transcendental reduction to the person or persons with whom he is in contact.

Strauss: I don't feel in a position to continue the text of the exam, not sufficiently informed about that. But I remember that Brentano, Husserl's teacher, in his *Psychology from an Empirical Standpoint*, has one chapter on love and hate and I would say in this phenomenological account of the human being there is no possibility of love and hate. Pure *egos* never love and pure *egos* never hate. Pure *egos* don't kiss, pure *egos* don't even die. They don't eat, they don't move. This is the kind of psychology which is far remote from, let me say, even the experience of practice of everyday life which really is profound after all and which we all exercise, including the animals. I think we have some kind of understanding not with the pure *ego* of our dog but with the dog in his attitude as an experiencing being. And I think that is the whole approach, if I understand it quite right. If I read about a pure *ego*, I have grave doubts at several points whether this is really a possible scientific basis, a philosophical basis, for psychology. Let me mention one more thing. On the basis of empathy, I must admit, there are subjects with whom I know myself to be communicating. Now I think empathy would mean that if we are in the world we really meet things which are not inanimate. But for some strange reason, we endow them by this process of empathy with psyche and so on. The historical fact is that the discovery of dead things is laid in history. I have the feeling that the terms 'the pure *ego*' and 'the *ego*' and 'our *ego*' are unfortunate. That is not the basis, cannot be the real basis, for scientific psychology. I think we have to start from our experience of living together with all other living creatures and try to understand them and ourselves in a kind of communication, in a world which we share together. There is no word about motion and so on in this whole thing. I think that Husserl talks about *Wachleben*, but there is no *Wachleben*, there are only creatures which are awake or asleep. But *Wachleben* in general doesn't exist. I mentioned already this morning the *kinesthesia* which I think is also a rather poor attempt to reach from Husserl's rather abstract position to the concreteness of life and past life.

Kockelmans: I am not going to argue with that, which probably just concerns the language to be used. But I feel when we talk about Husserl and Husserl's phenomenological psychology we have first of all to realize that indeed he was not a professional psychologist. Secondly, he never claimed to be. Thirdly, when he speaks about phenomenological psychology he is not functioning as a psychologist but as a theorist who is talking about how it eventually could be or should be developed. Obviously you can say, "He doesn't talk about emotions, he doesn't talk about passions, he doesn't talk about guilt, he doesn't talk about being good or bad." Husserl would agree that a good psychology should talk about all these things but would point out that it is not his profession to do so. As far as pure *ego* is concerned, Husserl would probably have said that certainly *egos* as pure do not love, do not kiss, do not die, do not do any of these normal things. But you have to make a very careful distinction between the psychologist who is doing the investigation, who is trying to account for the epistemological position he adopts, and, on the other hand, the human subjects about whom he is talking: I grant you wholeheartedly, that is the reason why I put it all between quotation marks. When Husserl talks about the subject matter of psychology as something that is pure psychic then indeed he makes a mistake. But I believe in 1913 one was still allowed to make that mistake. Today it would be a very grave mistake to say so, but I could quote you quite a number of authors writing at almost the same time who still were talking about 'soul' as a normal concept. That was allowed in psychology. The psychic, the pure psychic even, are terms that have been used. Concerning the main thing: We can point to a great number of difficulties which we meet reading Husserl's lectures on phenomenological psychology. We also can point to obvious shortcomings. We can point to clear mistakes he made, most of which were historical mistakes. But nonetheless Husserl saw – in my view – one very important thing. If we start talking about man in psychology, we have a choice between telling stories or being scientific. And if one is scientific one has to use empirical methods. That leaves the whole realm of the meaning of man's behavior out of consideration. The question is whether or not we are going to attribute this realm of meaning to philosophy and create the disastrous consequence that psychology is not going to become a science. From such a decision it follows that psychology will remain divided in a multiplicity of schools which maintain all kinds of different things on the basis of so-called, most of the time badly understood, 'philosophical presuppositions.' Or are we going to aim at psychology as a purely descriptive science in the sense of Brentano, with the result that we do not as philosophers perform empirical investigation, but make *eidetic* statements about human behavior in Husserl's sense, statements true for all human beings in similar conditions. That doesn't mean that in a particular situation there will be no variables in man's behavior, nevertheless we can say something descriptive about man in general. When one starts developing such a descriptive psychology one can make mistakes. But then it is up to an empirical psychologist to point out what is wrong. If it is convincing one shall change one's attitude immediately. This is the important thing: On the one hand, in such a dialogue in which we describe the meaning of man's behavior in such a way that we can stay out of the multiplicity of *egos* we do not perform a reduction, what Husserl called a transcendental reduction. That means we do not get involved in philosophic problems. On the other hand, we do not claim that there cannot be any truth other than established by means of empirical procedures. Should we allow for truth, Husserl points to an enormously important thing. He didn't develop it but it could be and should be developed.

Klibansky: The challenge can be summed up in the following four major points: First, Husserl claims apodicticity. This claim according to Ingarden is not fulfilled. Secondly, Husserl demands the absence of presuppositions, but in fact he introduces hidden metaphysical assumptions Third, Husserl fails to distinguish clearly between the constituting activity and what is constituted. Fourth, Husserl fails to give an adequate analysis of the structure of real contingent objects. These were the four main initial thoughts. If I may be allowed, I would introduce a fifth point which seems to me the main point of Ingarden, and to which yesterday it seems the discussion did not do justice; namely, the charge made by Ingarden that if we take *Crisis* seriously and interpret it in a way we think Husserl interpreted it then the dream of phenomenology as the rigorous science would have come to an end. It seems to me that this Conference should not be allowed to pass without having answered this question: Is phenomenology still a rigorous science? Does it aim at being one? And if so then Ingarden's challenge must be successfully met.

De Boer: I will try to say something about the paper by Poltawski. I think that in this paper some artificial oppositions in the thinking of Husserl are made. For instance, on page 2, you read: "On the one hand, he stated the 'principle of all principles', i.e. that '*jede originär gebende Anschauung eine Rechtsquelle der Erkenntnis sei*'; on the other hand, he adopted the rule of transcendental constitution which, more and more radicalized, led him to treat every object, even an immanent one (in the sense of *Ideen*, Vol. I), as transcendent in the widest sense of the term in relation to its consciousness, and 'constituted' in it." I think there is no opposition – real opposition – in the thought of Husserl between 'principle of all principles' and idea of constitution, because Husserl himself emphasizes every time that the great difference between his idealism and the classic German idealism is that the German idealism was mythological and operated with constructed concepts. In comparison with this classic idealism the phenomenological idealism is not mythological but scientific because it is descriptive and bases its propositions on evidence and intuition, on givenness (*Gegebenheit*). In my opinion, there is no opposition between the principle of intuition, on the one hand, and, on the other hand, the theory of constitution. I think then that it is not correct when, on page 6, we read that Husserl made no distinction between creative acts and acts in which we are active, and intuitive expressions, perceptive acts. In addition to this, on page 7, that in the theory of constitution, Husserl lacks a theory of justification (*Rechtsbetrachtung*). It is a typical characteristic of Husserl's idealism that it is *Rechtsbetrachtung*. All these sentences and propositions about the constitution are based on intuitive experience. I cannot see that there is an opposition between the fourth part of *Ideas*, Vol. I and the theory of constitution.

Gadamer: We may now discuss the problems raised by Ingarden in their own right, rather than from the point of view of the justification of his criticism of Husserl. Firstly, I agree completely with the remarks of Mr. De Boer. He is right in defending the coherence of the two points as he describes them. However, let me now take up again the four questions formulated by Professor Klibansky (I keep the fifth one for my own conference).

Concerning the first point, that is the question of apodicticity, I believe that Ingarden has the same concern as Husserl, neither of the two is satisfied with what Husserl calls later 'the record of evidence'. Both have tried, in different ways, to find a more adequate foundation for apodicticity. The theory of Ingarden develops the Husserlian theory of fulfillment of intentionality of the *Logical Investigations*. The elaboration of the relationship between idea and *eidos* seems to be a new form by which Ingarden hopes to differentiate an empty, vague supposition from a real, authentic and apodictic statement. Husserl's own elaboration of the transcendental thinking is the consequence of his discontent with his earlier version of the *a priori* lacking the basic guarantee of apodicticity. The transcendental *ego* is clearly meant as the ground for a hierarchy of evidences and as a guarantee for the apodictic evidence. In this context, I do not believe that we may speak about metaphysical presuppositions. What we may ask, however, is whether the transcendental *ego* and its constitutive function are a sufficient foundation for any form of philosophy.

As far as the second question, about the presuppositions, is concerned, it certainly poses the most serious challenge to the Husserlian thinking and it would be foolish to deny that there are none. I will mention only three, emphasizing that I have discovered them through the critique of *Hyle*. The first of the presuppositions is, I believe, that of an absolute givenness in presence. Husserl makes, in fact, a continuous effort to account for the apodictic self-givenness in the temporal flux of consciousness, making a presupposition, that the absolute self-givenness must be in the form of presence. His description of sensory perception follows from this presupposition. In point of fact, when in his later thought he defends his theory of foundation against the critique of Heidegger, he formulates the problem of foundation by identifying it with the first and original form of self-givenness. The first primordial givenness is seen by Husserl in the sensory perception; all modes of cognition that would follow – variation, critical evaluation, abstraction – can be legitimate only with reference to this primordial givenness of a *hyletic datum*. Hence in the evolution of knowledge this first basis of our world experience has to be protected from distortions.

Regarding the third question, namely what the constituting act and the constituting activity are, there seems to be a clear equivocation in Ingarden's thought between constitution and a forming activity. Ingarden assumes the work of art as merely a fiction because intentional acts constituting it would be producing a form that has no claim upon the transcendent reality of the world, being merely a human creation. But Husserl does not conceive of the meaning of the constitutive intentional act in this sense of production. There, I think, lies the basic difference between Husserl and Ingarden.

However, among many other problems, there still remains the foundation of logic. I think I am right in saying that Ingarden defended the identity of mathematical objects against the theory of Husserl. His main argument, I believe, was that mathematical objects cannot be conceived as the result of human activity, that enumerating is not the origin of the number. It seems impossible to admit that all logical and mathematical entities would depend upon the interrrelations among meanings. I think, then, that instead of a critique that would be the soundest point of Ingarden's interpretation of Husserl. Indeed, Husserl could not and did not mean that logical entities depend upon the mental activity constructing them, especially since Husserl pointed out that infinite numbers cannot be enumerated and following this paradigm Husserl unfolds the various problems of self-givenness. If one asks how the self-givenness of the infinite number could then be stated, the Husserlian answer is that it is constructible but it does not depend in its nature upon the actual construction.

Coming to the fourth question, that of the real objects, I cannot but repeat, that Husserl's theory of transcendence is sound. In the case of transcendent things there must be a difference between intentional objects and real objects. It is the intentionality of grasping, of perceiving the transcendent thing that poses the transcendence. What distinguishes then Ingarden from Husserl is the consequence of the equivocation of intentionality, of the intentional act and object in Ingarden's thought.

I will still make a general remark rejoining what seems to me the most important question of the presuppositions of Husserl. There are many doubts with which the Husserlian thought is afflicted concerning, for instance, the constitution of the human, living body, perception of the living, human body, etc. Husserl has strange ideas about what he believes is given first of all, namely the extended entity. Although I believe that the conception of transcendence in Husserl is quite sound, yet the question arises: Why and in virtue of what can we say that 'we are perceiving a thing'? Professor Strauss alluded yesterday to the fact that a living being is prior to the dead object. In this respect Husserl fails to account for this distinction between living, acting, perceiving real being and the real object as simply a transcendent, extendent body. In the *Crisis*, Husserl describes the form of idealisation involved in the mathematized natural science of Galileo; and yet I believe that that would be the only reason to conceive of the body as first of all an extended corporeal being. Of course that would be the consequence of accepting mathematical idealisation as the first and basic ontological approach. But Husserl rejects it and so it is still difficult to justify his basic statement about the body.

Ingarden is not the only one to criticize the consequences of Husserl's thinking. In the first place, it was Scheler who criticized Husserl, and in a certain sense they agree with Ingarden on the point that the Husserlian approach cannot grasp anything but *eidoses* and essences. Husserl's suspending of all positing consciousness and thematizing of essences alone precludes the possibility of describing real things. Therefore Scheler points out that we should inquire into the origins, the genetic prehistory of the very possibility of the suspension and of the eidetic insight. It recalls the old struggle between Fichte and Schelling. Scheler desires to ground the possibility of the eidetic insight in the structure of lived experience, in the physiological and biological foundation of the human mind. That is the great question of anthropology, to show how the human being can detach himself from his congenital engagement in life, integrating him in the world. However, I do not think that it is possible to follow Scheler's proposal perfectly, although he is right that transcendental reduction should find a counterpart in the known transcendent points of view. We can describe, in fact, the extent to which the transcendental reflection depends upon a cosmological, realistic basis. Scheler's demand, nevertheless, is not unjustified as long as all forms of the intellectual detachment from reality are not sufficiently explained by the deduction of reality itself from the transcendental reflection. In the Husserlian thinking, the transcendental *ego* has no ontological but only a *rational* claim.

Van Breda: First of all I should like to express my agreement with Mr. Gadamer. But now I will say something to the disciples of Ingarden. I won't give direct answers to the four questions. I feel a little unhappy because of the problem proposed by Ingarden. We are missing, I fear, the essential points that Husserl could contribute to the Americal way of life and to the Canadian way of life. There is a fundamental problem Ingarden saw. Just as for the Aristotelian tradition, also for Ingarden, and not only for Ingarden, for Husserl and his disciples in Göttingen, the central problem was reality. For most of the neo-Platonists the problem of reality is the problem of the reality of ideas. Ingarden, asking the question, "What is reality?", preceded it by the question, "What is the *ontological* status first of ideas and the *ontological* status of reality?" But if you are sitting down on that problem you are missing another point. Ingarden's approach to this question is by the route of analyzing all types of beings. I don't say Husserl refused that route, but Husserl was not interested in that kind of route. He was constantly asking himself: "What is the genesis for me of the real significant world we are living in?" This genesis must be self-explanatory. The way to follow up the constitution is intentional analysis. What Husserl really brought to contemporary thinking is a method. The problem we are discussing here is whether transcendental truth has something to do with the transcendental entities. But if you follow that you will miss the most important points in Husserl. To repeat, the real problem for Husserl is, "What is the genesis of the significant world for me now?" I am living in space and time. I have a daily space, I have a scientific space. I am living in time. I can think of this time as Einstein did. I can also think of it simply as the one who is taking care of the kitchen does. What is the contingent history of that? The present moment we are now living in? I don't say Husserl will give the solution, but at least he gives the first ascent toward a solution. That is what I want to say before all. This approach is the only possibility of getting something real from Husserl. The problem of idealism–realism does not concern Husserl ultimately. Now to pose another problem: The significance of transcendence. How does the object come to my consciousness? If you will add, "It is in my consciousness", then it must be also something autonomous. I think that things are autonomous notwithstanding that I try to change them in many cases. I take them to be autonomous because I take them with the significance of transcendence. There Husserl goes much farther than Ingarden. Husserl never will concede that the world of Hamlet is a world not coming from the autonomous external world. Consider the world of a car. Cars do not grow on trees. And we make cars just as we make pieces of art; I think there is a value in what Ingarden says. But I think Husserl will not concede that it is something autonomous – that freedom you were describing yesterday.

I think personally, also, that the freedom of Othello, to quote another example, and it is quite a famous one, is extremely relative. There are a lot of things stemming from the transcendent world in Hamlet. You have only to see the difference between Shakespeare and Joyce. Notwithstanding the fact that the world of Joyce is something almost pathological, and let's say a case for a psychologist, there is a lot of transcendency in it and I don't think it is independent. The question is the extent of the freedom of the one who is creating poetical works, the one who is making a statue the one who is making Flemsh paintings. In the end you depend, for instance, on the tools you use. The question is then about the meaning

Husserl gave to the problem of transcendence. Is he speaking about the transcendent world in the sense of Ingarden? But if you stay with that problem you might overlook the real contribution Husserl made with his analysis of language.

Tymieniecka: Gentlemen, if I am to pick up again the glove thrown by Professor Gadamer, it is certainly not to represent Professor Ingarden, because he does not need to be represented, but to continue the philosophical debate with Professor Gadamer, and go deeper into the problems themselves. Before that I would like to discuss the points with which Professor Van Breda has reproached me. We have in the first place, I believe, to make a sharp distinction which in scholarly work is often blurred; whether we want to consider Husserl as textual scholar and exclusively as textual scholar, namely, in his very terms and in his very own perspectives as has been done in the last 15 years, or even say in the last 40 years, or whether we can, as reflective and independently thinking philosophers, reach towards the philosophical significance of what he says. I would not deny the great richness and value of Husserl's constitutive analysis, as well as that of Merleau-Ponty, who contributed to philosophy enormously. As I mentioned in my lecture, they have analysed human consciousness in its cognitive-constitutive function to an extent that has never been even anticipated in history. However, a philosopher, who after all is searching for the truth and pursues it by all available means, cannot be satisfied with the mere descriptive analytic approach. What really interests a philosopher is the import of all this analysis for the progress of philosophical reflection. From such a point of view we cannot simply remain within the universe of the thinker, attempting to unfold these descriptive ideas further, but we must discover what their final significance for the philosophical quest is. Within the ultimate philosophical perspective the problem of idealism–realism is unavoidable. On this point I am in radical disagreement with Father Van Breda and I would claim that the whole history of Western Philosophy backs my statement, that the crucial philosophical question is *whether the world exists; whether we exist; in which way the world exists;* and in which way *we exist in relation to the world* It is impossible for a philosopher who really wants to get to the bottom of things, as Husserl himself says in one of his letters, to live in a meaningful way without satisfying himself regarding the crucial questions concerning, "What can we know and how can we know it." The final significance of these extensive analyses of phenomenology is what does it really contribute to the solution of this fundamental problem? We are forgetting also in our discussion what Ingarden, with some reason, pointed out in his lecture, that Husserl cannot be interpreted uniquely on the basis of his *latest* writings. He has been unfolding further his thought from the *Philosophy of Arithmetic* to the end of his life. In *Ideas*, Vol. I, Husserl explicitly makes the points which Półtawski mentioned in his paper, namely, that transcendental consciousness, at which we arrive by reduction, is the only absolutely existing substance, that *nulla res indiget ad existendum*. Whereas the world (and Husserl says it in many ways) is to some degree, of course, necessary for consciousness (if the world did not exist at all, then probably certain constitutive functions of consciousness would have to be modified), nevertheless consciousness could exist with these little modifications, as it were, without the existence of the world as such. If such a crucial state ment is made in a central work, has never been disclaimed no matter how extensive developments and nuances have been introduced in the later stages until the very last one concerning e.g. the constitution of the body, in his latest manuscripts, can we overlook its importance? There remains also the question: If in the constitutive process we arrive at an exfoliation of the human universe and of the human body as *relative* to the constitutive processes, we must ask in which way it is relative. To what degree is it relative? Another point of Professor Gadamer's discussion which is very important, concerns the transcendental consciousness itself. If we encompass constitution in its complete unfolding as it is, we will not fail to come to the conclusion that actually the constitutive processes are all meant as *universal types* of consciousness as it is constitutive of the human being. The question is: "Why should such a consciousness be given to us and not another type of consciousness?" There we come again to the question of the relation of consciousness to this world which it constitutes. Is it pure accident that there is such a type of consciousness, with such a world that it itself projects and no other? Or maybe there should be as Leibniz would say "sufficient reason" for such or other consciousness in the world. These are questions which I think are totally unavoidable if we really think philosophically through what Husserl *says* without being totally absorbed by what he *does*. Furthermore, concerning the question of apodicticity, I will refer again to Ingarden, with whom however on this point I am not in total agreement. It is not a question of whether we are to agree with Husserl or with Ingarden, but rather we are challenged to probe a little deeper into the problems themselves. Concerning apodicticity, it seems to me that from yesterday's debate till today we met halfway with Professor Gadamer. I have a difficulty, however, with the claim of Husserl which you are raising, namely, that once we have started to follow the flow of the constitutive genesis with the attitude of reduction leaving out all the contingent features, and have reached the field of the pure consciousness – with all data given with absolute evidence – then this field of pure consciousness as such becomes the justifying agency towards whatever we discover within the field. However, Professor Gadamer, there are two points here which seem to me very difficult indeed. One of them is pointed out in Ingarden's paper; the other would be mine. First, if we want to reach the apodictic evidence as justification of our cognition, Husserl himself says that this justification should be kept at every step. At every step of all procedures. The question is of priority, since what is at stake for all of us here is how we should start philosophy at all? Where should we start? If we demand that what we want to find out has to be certain, the matter becomes difficult; we have to find out where one can obtain certainty; surely not from all sources of cognition. The question is, what is the *right absolute beginning*: The cognitive processes or the structures of objects already given? Secondly, coming back to the fourth point specified by Professor Klibansky, Professor Gadamer has mentioned something important. Should there be found important in the specific structure of the real object as given – not as constituted, that's irrelevant – certain specific features that oppose it to an intentional object, as Ingarden tries to establish, then, of course, the question is how can we establish this difference? As I have already pointed out, Ingarden does it on the basis of structural difference. Professor Gadamer dismisses the distinction by saying that Ingarden is making within intentionality an equivocation by giving to intentionality the measure of "production". Thus, the constitutive intentional object, les us say the constitutive intentional object of the external world, would be confused in Ingarden's analysis with the created work of art.

Switching the issue to this line of consideration I will mention first that Ingarden has not treated the creative process at all. Then I answer Professor Gadamer's objection no more by Ingarden's thought but from my own research expounded earlier, which deals with the question of the constitutive *a priori*. My thesis is that if we contrast the process of constitution of an external, real object with that of a work of art, the respective processes themselves exhibit radical differences. They follow different guidelines in constructing the *noema* of their respective intentional objects. If we consider, first, the classic description of external perception we proceed by a series of perceptual glimpses performing each time a noetic act in which a fragment of the *noema* is present while other aspects are only anticipated. It has been acknowledged in Husserlian thinking that the perceptual process follows a continuous line in which the intrinsic sequence of anticipations leads towards the completion of a *noema* and thus towards the intentional object. But I am coming here to my point: how can we account for this continuity of anticipations? How can we account for the fact that we actually perform a series of individual instances of perception, which line up into a continuous series? The fact that it occurs because we walk around the table, look underneath, knock at it, feel its surface, etc., points out that there is an *a priori* guideline that we follow, which is not inherent in the process itself: the *eidos* which is transcendent to the process in a quite specific sense and in respect to which the perceptual glimpses perform the coalescing structuration of the noema of the intentional object. Ingarden asked after such a guide-line but did not find it.

In my research into the creative process in art, while contrasting the constitutive perception with what I have called 'creative perception' we find that the constitutive process in which the intentional object of a poem, a drama, a painting is formed, is of quite a different nature. Instead of the *automatized*, spontaneous course the unfolding of the sensory perceptual series takes, in the creative perception this course has to be established upon *deliberation*, *discrimination* and *choice*. Consequently, it is no more guided in structurizing the noema of its intentional object by the established *a priori* of the *eidos*, but according to a different principle. It is the spontaneity and automatisation with the pre-established *a prioric* guideline of the constitution of the real, transcendent things, which differentiates its intentional object from the one of the 'freely' and deliberately accomplished creative perception, which appears then really at the will of the performer.

However, this is just an opening of a vast array of problems concerning the creative function of consciousness which, as I have shown in my book *Eros et Logos*, throws a new light upon many issues basic for phenomenology.

Küng: I have several remarks to make to what Professor Gadamer and Father Van Breda said. First let me also say something about the paper of Professor Lingis. I would like to keep the distinction straight between the *noesis*, the *noema*, and the *referent*. In the paper on the *hyletic data*, the distinction, of course, between the *noesis* and the *noema* is very important. It has to be stressed that the intentions are a part of the *noesis*, a matter of the *noesis* and not the *noema*, and as such the speaker tried to stress that somehow they are extended. But I think for Husserl clearly the *noesis* is not extended at all. But, of course, the spatial world is constituted on the basis of the *hyletic* data in the *noesis* and so it seems to be necessary to say they have something analogous to spatiality in them. I would like to remark that an analytic philosopher, namely, Wilfred Sellars, is thinking in very much the same line. He is a materialist, and so ultimately there will be brain processes; nevertheless, he would say: "If we have a sensation of something red, of a *noema* which includes the property red, then there must be an analogous property in this mental act." To the spatiality of the *noematic* world there must be some analogical space in the *noesis*; Sellars would also say that to a sensation of something red there must be property analogous to red in the mental act – or to the brain process. Concerning what has been said by Father Van Breda and Professor Gadamer, first I must apologize to Father Van Breda that I still insist on the problem of *realism–idealism*. Of course, if according to his taste it is boring I am sorry to take up his time, but he said he would be glad if the abscess would be emptied and so some relief may come from it even to him. I think that the problem of realism and idealism is not an unimportant problem. Especially, again, in the contemporary philosophy, particularly in philosophy of science, this is a very live problem. Not in the sense that many philosophers of science would be idealists, but it is very much debated today what kind of Kantianism one should adopt. If phenomenology wants to be relevant not only to phenomenologists but to philosophers of science in general it must be clear on this issue; it could certainly contribute very much to the discussion of it in the philosophy of science. Secondly, I think the problem was also important for Husserl himself. He wouldn't have thought it was only something less interesting. It seems that he had a very emotional commitment to his idealism. Only he found it important to stress that the real world after all is only a noematic world, and that if it is a bad world it's not so serious because the spirit is what counts and the spirit is that which is absolutely real. In this sense, I think for Husserl himself the question of realism–idealism was extremely important. Consequently, it is clear that one cannot leave out the question of realism and idealism while discussing the genetic development of knowledge. I think the two things are not separate pieces without any connection. But if our own answer to the realism–idealism problem is already given, that vitiates the genetic account and I think in Husserl this shows up in all the difficulties which he has in trying to explain how in constitutive genesis we come to know other people. Indeed, if you don't accept the real world and with this a real body then it becomes very mysterious how we do relate to other people. Thus in this existential framework, the question of realism–idealism plays a great role. Toward the clarification of the issue of realism and idealism, I think it is absolutely crucial to add a further distinction between the *noema* and the *referent*. There are four things at least to be distinguished. First, the *noema*, when we refer to a real object. Secondly, the *noema*, for instance when we refer to a poetical figure like Hamlet. Thirdly, the real object *itself*, it there is one. Namely, the table itself. And fourthly, the poetical object *itself*. Thus the *noema* and the *referent* are not identical. Sometimes there is only a *noema* and there is no *referent*. If you think about a centaur going around in this room, then you may have an intentional object but there is no *referent* to which your intention refers. There is a *noema* but no *referent*. Now with respect to poetical objects, the distinction between the *noema* and the poetical object leads to sophistications which are not of great value in themselves. But it's a different matter with the distinction between the *noema* of the real world and the question of whether there is a *referent* beyond the *noema* of the real world. And there, I think, is the difference between Husserl and Ingarden. Both would accept that there is a *noema* of the real world but Husserl seems to deny that there is a *referent* which corresponds to this *noema*. He seems to hold that there is only the *noema*. In a way you can put the poetical object and the table on the same level. You can say both are *referents*; you can talk about the real table or you can talk about Hamlet. But although you can say whatever you like about the table, only *some* things are true. So also you can say what you want about Hamlet, and if you make Hamlet the *referent* of your discussion then there is, of course, a limitation of your freedom, namely, something is true about it and other things are false. Freedom is only in the creation of Hamlet. Shakespeare was free to create his personality in any way he wanted. But, if it is already created by Shakespeare and written in the text and then you want to get to know it, you have to adapt to what this poetical figure in the text is. It has some objectivity as such. Even the creation of Hamlet by Shakespeare has some objective limitations. If he wants to create a fiction of a person then he cannot without contradiction say that at the same time the person is an inhuman animal. So there are limitations on the freedom of poetical creation, which has been, in part, brought out before.

Strauss: The first paper heard this morning encourages me to appear here as a *persona dramatis*. I imagine I attended this meeting by accident. I came in from the life-world as a nobody, and was surprised by many things I was listening to. I had the impression in my naiveté that I saw running through his first paper a white thread which was never mentioned, and this was time. Let me take up again the problem of Hamlet. Hamlet may be performed by a group day after day – let me say for one week at least. When I see Hamlet I realize, although I see it now with *noema* and *noesis* just as I see the stage, that what is going on there on the stage does not belong to the actual present. I see that present but it does not belong there. It happened in no time, or it happened somewhere in the past. But suddenly we see place, characteristics, the structure of the play at this very moment with our own eyes just as we see everything else. Nevertheless, we realize it does not belong in that *frame* of present. So the question then would be: "How do we experience that present which makes it possible to have this kind of negation in the artistic presentation?" This perhaps could be connected with a reference to another topic mentioned here – a novel may tell the story of someone, perhaps not very differently from a newspaper giving a report about something which happened several generations ago or which happened yesterday. And let's suppose that what we read in the newspaper is correct. For those are the facts. In any case, we are confronted with a dual aspect of time. What I experience as past I do experience in the present. When I say I did, I say it now, but I refer to something which is no longer present. Our experience is limited to this moment and to this ever-changing moment. The Hamlet can be repeated. That's one temporal structure. Events in reality cannot be repeated. And I am not saying that history doesn't repeat itself. But suppose that I was fascinated by Hamlet and went to see it a second time. This second time is something in my history and this repetition which I performed for myself is different from the things on the stage. It seems to me then that one of the basic problems of our experience is actuality. This actuality extends to past and future. There is no *retention* in all these things. We have an order of time which is the continuity of reality. And beyond that we have an order of things conceived, temporal structures, which are not really rooted in that primary reality. I think this has some implication for this whole problem of the absolutely given and of the authentic intuition. If there is such a thing, is it not limited to a moment? If Husserl or anyone, after figuring out the noematic structure, describes it is the given still present? Or did a change already occur within our flow of experience, which is something that we cannot avoid but which of course, makes experience different from, let me say, knowing the Pythagorean theorem? This knowing is really timeless and there I grasp a structure which is not repeated, and independent from any temporal change. These logical structures, the mathematical structures, have a different *atemporal* structure. So I would think that even if there is this kind of intuition that Husserl insists upon it cannot be reported. To quote the final line from Faust: "Alles Vergängliche ist nur ein Gleichnis". We translate it: "Everything that perishes is but a simile". I may perhaps use the opportunity to say something more about and against the *kinesthetic sensation*. It was said that in Husserl's opinion sensory experience also implies some motion. But all motion, if intended, is to project 'somewhere' where I am not yet. It is an intention between my 'here' and places 'over there' which are nevertheless accessible to me. The eyes are seeing from 'here over there'. I don't see my 'here', I see with my eyes 'over there'.

Questioner: I see "here there"?

Strauss: No, you don't see 'here', you see 'there'. Everything which is seen is 'over there'. It is 'where I am not eyt'. This thing 'here' is 'where I am not yet'. I can reach it. It's 'in reach' but it is a character of the visible object, a temporal structure of the visible object for me that I am not yet 'over there' but I can reach it. I think that the *kinesthetic sensations* are only in the effectuation of those intentions which, of course, are intentions of my own live presence.

Claesges: Husserl has distinguished between two ways to phenomenology: The Cartesian and the ontological way. The Cartesian way would be, as is shown in the *Crisis*, through the transcendental phenomenology of the life-world, to which belongs also the way through phenomenological psychology, which is also a regional ontology. I believe that a 'mundane' ontology, that is 'naive ontology' insofar as it would not take into consideration the transcendental standpoint, would get confused in contradictions which go together with the relation of the universal and the particular, whereas it is precisely the aim of Husserl's transcendental phenomenology to avoid these contradictions, these *apories*, which are ancient. Husserl himself believes that the problem of reality can be raised and should be raised. He refers to it once, I believe in the *Crisis*, as the problem of metaphysics. Husserl had a strangely narrowed conception of metaphysics and the task of this metaphysics is to solve the problem of reality or, as he calls it, of 'facticity'. However, he is of the opinion, and I believe that this is essential and decisive to his thought, that this problem may be approached only when, on the way towards ontology *through the transcendental phenomenology*, these *apories*, into which a naive ontology would *eo ipso* fall, are avoided.

Pietersma: Just a few remarks in reaction to what Mr. Küng was discussing. We are talking constantly about the thing and the *noema*; we say sometimes 'there is' just a *noema*, sometimes 'there is' a *noema* and also a thing. We seem to be talking from the same standpoint – we just put two 'things' together: thing and *noema*. That seems to be confusing because what we end up with if we are not very careful is what is often called 'the double-object theory of knowledge'. It seems in every case we have two objects; in every case we can say 'there is' a *noema*, and thank goodness, 'there is' also a thing. What do we mean when we say 'noema'? In British and American philosophy, we had a long period in which we talked about 'sense data'. It seems to me that this discussion could be relevant to our present discussion. After Moore and Russell introduced the term 'sense data', they made the assumption that *there are* sense data. And before long they started to put sense data together as things are put together. What was forgotten is that 'sense datum' is in a sense the mode of appearing of an object; when we talk about sense data we don't talk about the object, we talk about its modes of appearing. I am just wondering whether when we talk about *noemata*, we are not talking similarly about modes of appearing, or modes of givenness, if you like. That is perhaps closer to the sensory terminology. And when we talk about modes of givenness and say *there are* modes of givenness, we talk specifically from, let's say, the transcendental point of view. If we keep that in mind, that we are talking a particular kind of language, that this existential operator is used by us from a specific point of view, *there are* sense data, *there are* modes of givenness, *there are* modes of appearance – we are in less danger of confusing the whole thing. Then we start to enumerate among the things also *Noemata*. When we talk about things we are not talking from the transcendental point of view, we are speaking from the, say, 'natural attitude'. In the 'natural attitude' in some epistemic situations we say 'the thing exists', 'there it is'. *It* is appearing in a certain mode – but we are not talking about the mode of appearing; on the basis of the mode of appearing we say, 'the thing exists,' 'here it is.' We don't have then a double-object theory, which is the sort of thing we get into if we don't remember to keep in mind that this existential operator is used in all sorts of different ways. I am totally in agreement with what Dr. Küng said: We should keep in mind the difference between, as he put it, *noema* and referent – in Frege's terminology the *sense* and the *reference*. In some cases, we have sense but no reference, but these are complications that I think can be taken care of if we remember how we use words.

Jordan: In a way the conclusion of the last paper shows a natural foothold for what I wanted to say which is about that question of words. Oddly enough we seem to have come almost full circle on Husserl's beginning in phenomenology, the *Logical Researches*, which, in the second volume begins specifically with an analysis of language, words, meaning, expression. Basically the problem was whether a scientific philosophy can be formed. Can we possibly, in any way, arrive at the mode of expression such that the things described verbally will be referred to in such a manner that the onus will be on the reader of the description to place himself in such a way, in such a position, as to visualize what if being described? This is a basic position in Husserl both in terms of chronology and, I think, in terms of his whole development. In the *Logical Investigations* he arrives at what turns out to be an eidetic description of the requirements for scientific expression in philosophy. What is at stake is to arrive at a description that it is itself a-temporal, in a way similar to the way the verbal description of mathematical objects is a-temporal. But anyone in any place can be re-read the thing and get a vision of the concepts involved independently of the contingencies of individual conditions of space and time. The crucial next step after this description is to demonstrate that it can indeed be implemented. That such a description is feasible in a way that will coerce even the skeptic into seeing what is put there. This it seems to me is performed in *Ideas I* with the presentation of the reduction and of the transcendental ego. The sensitivity of Husserl to that problem, I think, comes in the notes referred to in the Introduction to the *Fifth Lecture*, where Husserl was despairing of ever reaching this solid foundation that he was seeking. Basically what is achieved there is an elaboration of a tool that is to be used by anyone who wants to contribute to a scientific phiosophy in Husserl's sense. Once, the importance of the transcendental reduction is accepted as absolutely archetypal because without it the demonstration isn't there, the tool does not remain in proper focus; the movement out of contingency into science, out of *doxa*, if you like, into *episteme* is not performed. The balance of Husserl's work consists in an exploration, of this realm that he has uncovered under the special perspective that this tool gives. Hence one should not interpret the results obtained in even *Ideas*, Vol. II and research following *The Formal and Transcendental Logik*, *Experience and Judgment*, and the *Crisis* as a putting in question of the very tool that he is using to accomplish this research. It would be the same thing as accusing an experimental physicist of putting all his science into question every time he devises or describes a new set of data. This may happen on occasion, to be sure, but one should not confuse the tool itself with the result. The descriptions are all *eidetic*. And they are all given at the *eidetic* level except where Husserl specifies it to be otherwise as being back in the naive attitude. As a final note, one thing more should be said: Some of the problems that we have with Husserl's terms such as 'transcendental' or '*epoché*' stem not so much from Husserl's own text and presentation of them as from the fact that they are weighed down with a full historical tradition. Husserl took over those terms, I think, in full knowledge of this fact and with the definite implication that he was thereby settling a philosophic act, giving them once and for all the proper meaning that should be ascribed to them. Hence he was running a risk of confusing associations but he did so because he believed he was establishing philosophy as a rigorous science. Since we were talking about not being a slave to the text I submit that by continuing the use of Husserl's tool we can indeed go on to other and different descriptions in wider fields that may eventually include the kind of ontological worries that some people, such as Sartre and Heidegger, seem to jump into. But it's in the use of the tool in the full spirit of the Husserlian endeavor that we follow Husserl, and not by limiting ourselves to his texts.

Kockelmans: I would like to address myself to Dr. Lingis, but before doing so I would like to make one general remark. Dr. Küng continuously refers to the terms *noema* and *referent*. This is not the phenomenological way of talking. And it causes me a confusion because this term, *referent*, doesn't have any clear meaning to me. It could, on the one hand, never appear in genetic analysis for the simple reason that they belong to a static description of intentionality. The question of hyletic data is brought up for the first time into full-fledged transcendental phenomenology in the first volume of the *Ideas*. There Husserl explicitly says that he is interested in a static description of intentionality and he makes a distinction between what is really contained in the intentional act and what is not contained in it, but to what it refers. What is really contained in the act is on the one hand *hyletic* data and on the other hand, as Gibson translates it 'meaning bestowing acts.' Then he explicitly introduces matter and form in the Greek terminology *Hyle* and *Morphe*. I agree with Gurvitsch that Husserl is using these terms indeed with reference to the Greek conception of hylomorphism. But I also feel that Gurvitsch is right not only in explaining and understanding Husserl this way, but also in pointing to great difficulties Husserl brings himself into by using this language. However, at the moment we are not interested in static description anyhow. We are interested in the genesis of meaning. It is true that the process of genesis of meaning appears to come about in some sense passively. There are certain references which you make on page 98 that seem to indicate that you tried, probably even unconsciously, to make certain relationships between the process of passive genesis, on the one hand, and Merleau-Ponty's doctrine of the body as a source of meaning. You write there: "*Hyletic* data are neither actively produced nor passively received by consciousness; it is rather encumbered with them: it lives them. But they are the very passivity of the consciousness with regard to itself; they are the consciousness laden, affected with itself. To live oneself is not to objectivy oneself; it is to be encumbered with oneself, to be affected with oneself." I believe that these terms could easily be understood as well from Merleau-Ponty's point of view as from Husserl's point of view. My question is: "Do you agree that it would be meaningful to make a distinction between a static approach to the hyletic data and the genetic approach?"

Lingis: Yes, I certainly agree with this distinction and I believe the remarks that you made are valuable. I did not have in mind Merleau-Ponty especially, when I wrote that interpretation of what it means to live. I started with the supposition that for Husserl the sense data are not representational. Their function is not to produce a kind of, as you said, 'second object.' He doesn't say that consciousness is consciousness of sense data or that consciousness perceives them; he says 'it lives them.' It was this thought that I tried to articulate. My reproach to Professor Gurvitsch – he was misled here by *hylomorphic* terminology – is that he says in his book that for Husserl *hyletic* data are pure undifferentiation, pure presence. I think that is wrong. That is what *Hyle* and *Morphe* meant for Aristotle, but not for Husserl.

Spiegelberg: Just two remarks, on the subject of constitution, and relevant to the Półtawski paper and Mr. De Boer's answer. I can agree with Mr. De Boer on the possibility on reconciling the 'principle of all principles' with the ideas of constitution, provided that concerning the idea of the constitution Husserl did not use the term 'creation'. Actually that term, I think '*schöpferisch*', occurs for the first time in Fink's essay to which Husserl subscribed: *Die Konstruktive Phänomenologie*. But what does occur in Husserl is this strange expression 'Leistung', which is strange even in German and we have a difficulty translating it, not knowing whether 'accomplishment' and 'achievement' are adequate. Note that Husserl speaks about the hidden 'Leistungen', 'verborgene Leistungen'. This seems to me to be one of the questions: Can these hidden achievements, constitutions, be laid bare to the principle of all principles – meaning, 'intuiting' or whatever it is. I think it may be possible. One of the things about Mr. Lingis' paper which interested me very much is that he referred to the fact that even Husserl used the word 'interpretation'; and he used the word 'hermeneutics' for the first time. This is perhaps the bridge to Heidegger.

About the hyletic problem, I would like to reminisce first that when I had my doubts about it and approached Pfänder, the Münich leader, he said: "That's exactly what we always objected to Husserl already". In the *Logical Investigations* the term does not occur. What I submit is that one has to look a little more carefully perhaps at the genesis of Husserl phenomenology, at the way in which this term arose and perhaps was dropped. Have you ever seen that in his footnotes to the *Ideas* he uses an almost untranslatable phrase; I put it in German first: "Ich weiss nicht wie ich den Wagen aus dem Dreck ziehen kann". "I can't see how I can pull the wagon out of its ruts." How far he did it when he revived the thing, in the *Formal* and *Transcendental Logic*, I don't know. But to my knowledge the term (*Hyle*) occurs no longer in *Crisis*. It is as if there were perhaps a remnant of, let me call it bluntly, his Humeanism: Hume's 'impressions." He worshipped Hume, continued to worship him, but Hume didn't, of course, discover intentionality. I call for help from the Husserlian experts to show the thing as ontic reality. But Husserl doesn't talk about that relationship between *noema* and ontic reality once the reduction is performed. On the other hand, it could be the relationship between the *noema* and the object as effectively and actively present to me in a certain experience, and the surplus of meaning to which I refer and which I intend and to which Husserl most of the time referred with the Kantian word 'object'. If it is the distinction between the effectively given and the object – the object being the actually given together with its inner horizon and the references to the outer horizon in as far as they play within this totality of meaning – that is meant, I believe it doesn't make much sense to ask what for Husserl the relation between the *noema* and the thing as ontic reality is. Husserl did not say anything about this question. Now we can deplore it or not deplore it, we can claim that it is important or not important, but it is in vain to try to find a solution to that problem in his philosophy except, as Dr. Claesges rightly indicated, in those cryptic passages in which Husserl refers to the possibility of metaphysics. In the book of Iso Kern, *Kant and Husserl*, there is one chapter in which Iso Kern explicitly deals with his question. He compares Husserl's conception of metaphysics with the Kantian and neo-Kantian conception of metaphysics, and be not amazed if, in order to solve this problem, you find there a reference to the existence of God. But my main issue is something different. If I had started to read Dr. Lingis' paper on page 96, then through page 98 I wouldn't have had any problem. I would have known that you were talking about the *hyletic* data as they appear in a constitutive analysis; that is how Husserl talks about it when he deals with active and passive genesis. But on page 99 where you say Husserl lumps together indiscriminately a lot of things, there are also terms which puzzle us how this Aristotelian, *Hyle* concept (Husserl was no Aristotle worshipper) got into Husserl, and possibly how it disappeared.

De Boer: I would just like to underline what Professor Kockelmans said that the distinction between meaning, *noema*, and *referent* is not a phenomenological distinction because it is a distinction within the natural attitude. And that's my main objection to the first paper this morning. On the other hand, in the first volume of *Ideas*, Husserl does make this distinction of *noema* and *Gegenstand*. It is not the same distinction that between the intentional object and the real object, but a distinction within the *noema* itself – between its central point and the properties. I think much confusion about Husserl is due to the fact that this distinction between *noema* and *Gegenstand*, in the sense of central point, is confused with the distinction between intentional object and real object.

NOTE

* The only letter of Ingarden to Husserl which is not lost appears in the Annex to this volume. *The Editor.*

GERD BRAND

THE MATERIAL APRIORI AND THE FOUNDATION FOR ITS ANALYSIS IN HUSSERL

I. PHENOMENOLOGY AS THE PHILOSOPHY OF THE CONCRETE OR MATERIAL APRIORI

Husserl has posed the problem of the life-world, but he has not achieved an analysis of the life-world except in fragments.

If we want to penetrate more profoundly into the dimension which Husserl has opened for philosophy, if we want to travel further on the road which Husserl has indicated, we must first clarify its underlying insight and then retrace Husserl's development of this insight and his corresponding methods. Only thus will we reveal the core of phenomenology.

The return to the life-world is already hinted in the once famous and today almost forgotten motto 'Back to the things themselves' (*zu den Sachen Selbst*). This becomes clear when we interpret the specific phenomenological meaning of this expression.

Three essential characteristics are revealed this way: The decision as to what a thing is cannot be prejudged, but must be decided by the experience of the thing itself and not by something else. Thus phenomenology aims to be a philosophy founded on experience and remaining in experience. This is contrary to philosophies which are concerned solely with forms and functions of knowledge, its conditions, with cognitions which cannot be experienced or with principles of being.

The basic insight of this philosophy is rooted in the unshakeable and all-comprehensive certainty on the foundation and basis on which we move continuously, even if we never consciously thematize it. In fact we usually forget about it because it is so simple and self-understood. Without making any pre-judgment of the things, we can at this point call it: the enigmatic general pre-understanding of all beings in totality. Thus secondly, phenomenology is a philosophy of a pre-understood whole.

If we penetrate this enigma, if we wish to transform the preceding pre-knowledge into knowledge, if we want to bring it to greater determination,

Tymieniecka (ed.), Analecta Husserliana, Vol. II, 128–148. *All Rights Reserved.*

to greater nearness, if the self-understoodness of the pre-understanding is to be understood (*Wenn die Selbstverständlichkeit des Vorverständnisses zur Verständlichkeit gebracht werden soll*, as Husserl expresses it), then we come to the third characteristic: philosophy as phenomenology never totally solves the enigma.

If phenomenology can be distinguished from other philosophies in the way described above, can it not be distinguished by the apriori, which in principle philosophy always seeks?

We define apriori as that which conditions all experience. Through the specification of the concept of apriori we are enabled to define with one stroke the characteristics of phenomenology. Philosophy usually treats the apriori as formal, which means that it cannot be experienced itself, that it is considered as being the not-experienced 'form' of experience, a form, however, which is required by experience itself. In contrast to this, phenomenology considers the apriori as concrete, material. This means that the apriori itself is given, that it is itself experienced as an experience *sui generis* which comprises all singular experiences.

In order to clarify what this means, the phenomenological *apriori* will be distinguished from the formal *apriori*. The formal *apriori* is characterized by at least three essential aspects:

(1) The *apriori* is the condition of experience which is not given in the experience itself,

(2) The *apriori* is immanent in consciousness,

(3) Every certainty is founded on necessity, and this necessity itself is founded on the *apriori*.

Phenomenology refutes this briefly sketched concept. For phenomenologists experience as such is meaningful, sense-generating and unified. To consider the given as a 'chaos of sensations' which has to be 'formed by synthetic functions and forces' is called by Scheler a 'mythological assumption'.[3] Phenomenology must not start from this myth, this concept of experience, but from experience itself.

For the phenomenologist the *apriori* is not immanent in consciousness, but rests in experience itself. Here we must point out that there is no rude confrontation of consciousness and world for the phenomenologist. Consciousness is consciousness-of-, the *Ego* is being-in-the-world, it is world-experiencing life (*welterfahrendes Leben*).

Finally phenomenologists criticize the formal concept of necessity.

Within the formal *apriori*, necessity cannot be evident, it must not be, because it is impossible to experience necessity. What then is the meaning of necessity? Husserl takes this position: If I am at a loss to explain necessity, it does not become more understandable if I push it on another level. Indeed I cannot understand it, even if I call it a transcendental function. The compelling character of the formalistic-*apriori* necessity is proven by Husserl not to be evident, it is blind, incomprehensible.[4] Necessity must and can be made evident in experience, as Scheler demonstrates.[5]

With this we come to a further outline of the concept of experience in phenomenology which is underlying the concrete *apriori*. The concrete *apriori* is givenness of meaning, of sense. Sense, meaning is given in an ultimate, original experience. The ultimate experience is that to which another experience refers to, and can be reduced to, and it is itself such that it does not have to be referred to something else, it is irreducible in its ultimateness. In this sense it is original as the ultimate experience and at the same time originating as giving origin. The givenness of sense shows itself by the fact that a meaningful given cannot be experienced by the intermediary of other givens, only of the experience of itself. 'It says itself', as Wittgenstein would express it. This is not modified by the fact that an ultimate given also refers beyond itself. For example, I can only grasp one thing because it distinguishes itself from others. The concepts of verification and falsification, as they are used by modern Logic and Philosophy of Science, especially by Popper, provides us the opportunity to distinguish clearly the ultimate givenness of sense from other conceptions. In doing this it does not yet become clear how we can unfold with the necessary rigour that meaningfulness which is the pervasive tissue of our life-world. But thus we arrive to pose the very problem. The *formal apriori* can neither be verified nor falsified; its congruity with experience, with reality, of which it is meant to be the condition, is external and postulated. The *concrete apriori* is verified by recognizing with evidence – and not as an external requirement – that it is the fundamental structure of experience which penetrates everyday experience. It is essential, however, – here already appears the problem of a novel relation between *apriori* and *aposteriori* – that it can also be falsified, in the sense that an experience may appear as false, which means that it is replaced by another experience and only by another experience. This is a phenomenological and evident falsification and must not be considered in the terms of the positivistic understanding of falsification which relates to explanatory hypotheses or theories which present formal rules for events in and from experience. The explanatory hypothesis or theory is not itself experience and has no intention to be. For this reason it cannot be verified, but only falsified in the sense that occurring events cannot be explained by the hypothetic rule, thus rendering it false.

When we see something that has sense we just see it in its sense. Seeing the meaningful given is an ultimate. We cannot get around this seeing and approach it as it were from behind. We are always tempted to explain the ultimate given with respect to shadowy frames of references which themselves are not made explicit and are not reduced to their ultimate foundation. These frames of references and this way of explaining coin our habits of thinking without us ever questioning them.

Again and again, over and over, we try to understand understanding once more, to grasp the sense of sense. Not even Sartre can withstand this temptation.[6]

The *apriori* of sense-givenness is a duality, that of the world and of the *Ego* living in it. That there is material, concrete *apriori* means that the world does not present itself to us as a disordered, incoherent chaos, and that any single entity we view in the world has an immediately given sense, which inhabits it.

Two points are certain, however, from the beginning. First, what has sense, when sense is given, what this sense is authentically and how we grasp it, is not clear when we have the overall insight into sense-givenness. With this insight and its implications the fundamental problem only than emerges as a problem in the first place. Second, this problem can be elucidated by subjective activity, whereby the given, the perceived becomes more objective, clearer and more ordered. Never do we have an experienceable consciousness of a 'pure manifold' which would receive its sense only from our activity; but we are conscious that through our activity we disclose sense and thereby in a certain way produce new sense. This activity itself is a given and can be grasped as such. The sense, the meaningful is pre-given, but by our activity it can become more meaningful. This has two important interdependent consequences. The concrete or material *apriori*, when it is thematized, made explicit, is also formalized, without, however, becoming a formal *apriori*. It is then a form of the world and not of thought; it is not a rule of thought but it is prescribed, by the world, for thought.[7]

The rules of thematization and formalization must be taken from experience itself and must not be superimposed upon it apriori, and from without. This means that the material apriori cannot, and must not, be justified by logic, on the contrary, the justification of all logic itself can only be found in the material apriori.[8, 9]

II. THE FOUNDATION OF THE ANALYSIS OF THE CONCRETE APRIORI

The motto 'Back to the things themselves' and the material or concrete *apriori* characterize the fundamental insight of phenomenology. Our task here is to retrace with a few strokes how this insight developed in Husserls work. Our guiding-question is: how can the concrete *apriori* be methodically apprehended?

This insight found its first method in the so-called '*Wesenschau*' (eidetic or essential insight or intuition). The actual basis of the eidetic insight is that meaning, sense, is given in the given, that the 'what' can be seen. This problematic is here assumed to be known; our purpose is solely to make a few critical comments and to clarify the particular significance of the eidetic insight. For Husserl the 'see-ability' of particular, individual entities is the basis of all eidetic insight. The correlation between the apprehension of essences of a higher order and their basis in concrete realities is not clarified by Husserl.

The authentic nature of the apprehension of the un-seeable is not at all explained. Husserl for instance fluctuates between the notion that the perception of a relational context (*Sachverhalt*) is identical with the perception of things and the notion that this perception lies in the process of an intuitive synthesis of the acts in which such rational contexts or categories are constituted.[10]

Finally, Husserl frequently confounds essence with the universal concept. But this is not 'seeable', it cannot be intuited, it is not the essence which Husserl is seeking. The contextual and referential relationships in which all entities, all categories and all realms of being are to be found are confused with conceptual abstractions, precisely because the general essence is confounded with the universal concept. Instead of unfolding this interrelationship Husserl often constitutes general essences as general abstract species.[11] Husserl very soon, however, recognized that the eidetic insight has only preliminary and facilitating functions.

The eidetic insight is not *the* method of phenomenology, it does not place us before the self-given, it is an intermediary stage which can serve to clarify, which presents us with guidelines along which we can proceed to self-givenness.[12] The apprehension of self-givenness which we are seeking cannot be given to us by the eidetic variation which is the most important instrument of the eidetic insight, quite simply because it pre-supposes what it is seeking. Whence should we know what belongs to the eidos, if we do not know it already? However, the eidetic variation can clarify what the pre-given entity, the pre-given theme is, which we want to disclose analytically, revealing the concealed richness of experience that is at its basis. In other words, the eidetic insight does not at all have the function of giving us the essence considered as the basic experience of experience; it is rather intended to help us in a pre-clarification in the search for the self-given by distinguishing the essential from the non-essential, by making the essential more essential.

In the eidetic insight we find one of the gravest misunderstandings of phenomenology, especially if its preliminariness and solely preclarifying character is not seen.

The discovery of the material and concrete *apriori* is the foundation and origin of all phenomenology. Even if the eidetic insight has not yet provided us with the necessary method, we must still adhere to the materiality, the givenness of the *apriori*. Sense and meaning are given to us in the given, they are perceived. Every being has its own perceptibility, and this in a preknown understanding. This perception can be brought to an increasing clarity. It is the source of all legitimate knowledge. Husserl adheres to this insight and claims it to be the unshakeable basis of phenomenology, the principle of all principles.[13] The original, first-hand intuiting is the legitimate source of knowledge.

The so-called phenomenological reduction rests on this principle as well, with its correlate, the *epoché*. This is often misunderstood, even by Husserl himself, as subjectivity being for it-self and valid only as such, as a closed field of consciousness, whose structures must be investigated, while the world or at least its existence is 'put into brackets'. In a preliminary generality we can say that phenomenological reduction means nothing else than the return to self-givenness. Just as any return in general, it has two aspects. Returning to, heading toward, disclosing the things themselves is equally and at the same time bracketing, inhibiting, eliminating what are not the things themselves.[14]

If one takes the principle of all principles as the foundation of philosophizing, and if one asks then, what is to be bracketed, one is immediately confronted with an unavoidable problem. If the material apriori lies in the sense-givenness of things or contextual relationships (*Sachen oder Sachverhalte*), and if it is all-comprising as apriori, if, in other words, every given has sense and thus every given is a legitimate source of knowledge, then why do error and deception exist? It is certain that we err. In fact, is not the problem of error, the problem of truth the very motive of philosophizing?

We can start by giving a general answer to this question. We have first discovered the concrete apiori as the enigmatic precedence of the preknowing apprehension of all being. The enigma of preceding preunderstanding contains already a moment of darkness, hence the possibility of concealment, of error.

But this general answer is not sufficient. What it expounds must be explicited in rigour, since it is a central problem of philosophy.

The question can be put more precisely: if a given has sense and we are deceiving ourselves, the given has not simply disappeared with this deception, but it will show itself 'in truth' as another given. To deceive oneself, to err, can only mean that one has taken one given for another. Thus the question is now: why and how can 'This-thing-there' (*Dies-da*) or a contextual relationship be so given that it is not at all itself? In order to make this question easier to answer we will reverse it, considering it as a point of departure for asking: *How* is it possible for us to bring something to its self-givenness? This question gives rise to a preceding question: *when* is it that a thing is self-given, when is it, in Husserl's vocabulary, evident? Instead of believing that things are themselves just as they are given to us, that we can grasp them immediately as themselves, we have to ask first of all the question of self-givenness itself, and this is the question of truth.

Husserl's answer to the question of truth, if not investigated more closely, seems to be only the phenomenological elaboration of the classical truth concept: truth is adequacy.[15] In a central text of the *Logical Investigations*[16] we find three dominant characteristics of the truth concept which can serve us as a point of departure at an attempt at clarification: positing intention, active thetic statement considered as a synthetizing fulfilling intuition of a congruent nexus, degrees and levels of evidence.

By simply exposing the relationship of these elements, the problem of truth as the question of the how, when and what of self-givenness is not yet solved. On the contrary, misunderstandings and contradictions appear, if, and because, one only considers the first two characteristics, and does not start with the last.

The graduality of evidence develops into the split of two basic concepts of evidence in the later Husserl.

The first concept is that of self-giving itself (*Selbstgebung*). Only because things are given to us and because we consider them as giving themselves, can we ask if they are really themselves. From this first concept of self-giving which pertains to every given, Husserl distinguishes the pregnant concept of evidence, the *pure* self-givenness or self-giving.[17] This self-givenness, this evidence is the one we seek. Because everything is given and, however, does not present itself in pure self-givenness, because the tension, the difference between both modes of evidence manifests itself in modalization, deception, and error, we look for pure self-giving, we inquire into truth.[18]

Misunderstandings and contradictions occur if the intention is misunderstood solely as a positing intention, as an empty meaning, as a pre-intention (*Vorintention*) not yet possessing the reality which it only intends, but which – once it is fulfilled by it – becomes the congruent synthesizing nexus of truth. If intention is considered as pre-intention selecting something out of reality which has to fulfill it, if reality has to yield to intention,[19] if this modality of intentionality is confounded with intentionality as such, if intentionality is taken for 'mere' intention, if consciousness is accepted as empty – to be filled by reality, what then are the consequences? In that case one either comes to a transcendental idealism, in which finally intentionality forms, 'constitutes' reality, then one comes to essential laws of intentionality which rule reality, or one comes to a psychologized phenomenology which still might be science, but no longer philosophy.

What is actually intention and intentionality? Let us raise this question using once again a text from the *Logical Investigations*[20]. This text shows that consciousness as itself is something which it is not itself. It is not primarily and first of all consciousness of itself and simultaneously of its relation to something else, but it is totally consciousness of that of which it is consciousness-of-. The 'content' of consciousness, is not itself consciousness, it is something other than consciousness but of this something other it is consciousness-of-. Also, consciousness *is* not that which it is conscious of, it is precisely consciousness-of-it. Consciousness, because it is conscious, is aware that it is consciousness-of-. Husserl warns that one must not misunderstand consciousness as a sort of knowledge and even less as a concrete knowledge.[21] Consciousness-of- is nothing else but the conscious *non-immanence* of consciousness which can never be 'taken in' by consciousness, which never becomes immanent, it is consciousness-of-transcendence.[22] This consciousness-of- may apprehend and describe itself in a first stage, on a first level of reflection, which is also the source of the misunderstanding of intentionality, in as far as it is considered in isolation.[23] Here we have the basis of the twin concepts 'pre-intention and fulfillment', which is a false foundation because it is considered in isolation. If I consider the objectified act of consciousness and the object which is objectified by it, then I can regard this object as inhabiting the positing of the act, a positing which must be fulfilled by the concrete given object. This ignores that the objectified intention as a positing pre-intention derives solely from its original non-objectified being-with-the object, the given. The seeming fulfillment of an empty intention is in reality nothing else than an un-veiling of the pre-given.[24]

This misleading notion of intention resolves itself in a deeper reflection, in a reflection of a higher order, a reflection which comprises the first one. Therein I discover that I remain permanently original and unreflected consciousness-of-, even throughout an objectifying reflection. Thus I can have various objectified contents of consciousness, in which the *Same* is presented of which I am original consciousness-of-. This is also discovered through reflection.

In this second reflection the consciousness-of- is aware of itself not only as objectifying and reflecting, but simultaneously as continual, unreflected consciousness-of-. It is unreflected in a double way: by continuously being-with-that which it objectifies and by being continuously beyond it.

On this second level, reflection discovers itself in its *non-transparence*, in its consciousness-of- which is aware of its impermeable passivity, of being open to a transcendence, a beyondness, which it can never take in. It is reflection which does not return to its origin, but which recognizes that its origin lies outside itself. In this reflection, the objectification is revealed as a modification of our consciousness, an objectification in which the consciousness-of- mediates for itself its own structure. Simultaneously the reflection shows that this mediation does not preclude the contact with the immediate, but implies this contact. Husserl speaks in this context of a mediating intentionality which goes through something towards something else.[25] The objectifying intentionality whish we grasp at the first level of reflection is a mediated and mediating intentionality, a passageway through which I remain connected with its 'background', the pre-given, pre-present reality.

Having gone through this examination of reflection we can differentiate with Husserl three mutually mediated modes of intention. First, consciousness is always consciousness-of-something, it is always intention as being-with-the-being of which it is conscious, *without*, however, being *directed* toward this being.[26]

From this original non-directedness it is possible to distinguish a first directedness of intention, which is still not an active directing. Consciousness-of- in this modality is not only being-with-being but, as it were, consciousness of this being-with-being, without actively directing itself towards this being. Husserl calls this *passive intention.*[27]

Thirdly, it is possible to differentiate a more active intention from this passive intention. Usually, the active intention alone is considered as intention. This ignores its origin, the affected, passive, non-directed being-with-being. Intention, in its active sense, develops out of the affection preceding it and coming from every being of which we are consciousness-of-. Intention now takes on the character of aiming, it becomes 'intentio', it takes the form of a 'striving' stemming from the affected consciousness-of-, the ego.[28] This striving proceeds from that by which it is affected and goes towards it, to determine it ever more closely. This ever closer determination has the character of unveiling.[29]

In relation to these described characteristics of intention and intentionality the concept of adequacy receives a new content. The term adequate does not signify the adaptation of a pre-intention to reality or vice-versa, but instead the plenitude of self-givenness of reality. The self-given can be more or less self-given, and thereby it can be more or less adequate, i.e. itself. The adequacy of intention now coincides with the adequacy of plenitude, the self-givenness of the perceived. The unveiled, however, and this is essential, remains veiled throughout the unveiling, plenitude remains interwoven with emptiness.[30] All adequacy therefore is in itself inadequacy, which can never be completely overcome, the self-given never appears in its whole self; and therefore affection, turning toward, aiming, unveiling, and being directed never cease.[31]

It has now become clear that intentionality has the permanent form of closer determination and referring-beyond. Thereby we are faced with new enigmas. What is then the 'self-given' which is given to me? Is it 'itself', or is it, at the same time, something other than itself? In connection with this another question arises: if I cannot only deceive myself, in general, but even concerning already established evidence, then how does this affect the principle of all principles, the requirement for original self-givenness? Does evidence as the basic principle of the apprehension of the concrete *apriori* have a meaning which can stand its own test?[32]

The answer to that question lies within the inadequacy of adequacy. Evidence can only be grasped as not totally self-given self-givenness, because it refers to further evidence, and therefore, beyond itself.[33] Each self-giving experience refers beyond itself, is expectation of itself. The possibility of deception is immanent in expectation. Evidences refer beyond themselves, they can be deceived, but – and this is important – they can only be replaced by new evidences.[34] Now a problem is posed. If evidence is never wholly adequate, what certainty has it then? The question of adequacy becomes the question of certainty.[35] It is in this connection that we must see Husserl's concept of apodictic evidence.

The apodicticity of a cognition does not concern its given intuited content, but itself as cognition, and that means as a reflected cognition. Indeed, not being able to imagine that it could be otherwise, which characterizes apodicticity, derives only from reflection. The apodicticity can pertain to a cognition the object of which has a very low degree of adequacy. Already, we can even suppose that the apodictic cognitions will be precisely those with a minimum of adequate givenness. The insight is then apodictic, but its content is not adequate.[37]

The problematic of adequacy, inadequacy, teleology (referring beyond) and apodicticity of evidence, of self-givenness, has brought us to a point from which we must try to fulfill with Husserl in a concrete way the requirements of evidence. If we follow this path, which is not possible in the framework of this study, then we discover that through the horizons of all that is given in our experience, we arrive at the total horizon of the world.

Thereby, the question concerning the concrete *apriori*, the things themselves, the self-given, has lead us via eidetic insight, truth, evidence, intentionality, to the world, to the insoluble, enigmatic, and all preceding interdependence, interwovenness of the *Ego* and the world.

Now we can understand the peculiarity of evidence which, as adequate evidence is never wholly adequate, because no particular being ever exposes itself wholly, and always goes beyond itself. Now we understand that and how evidence has a universal teleological structure, as Husserl once expressed it.[38] This means that the presence of a singular being is never its whole self-givenness, or that it has its self-givenness precisely in referring beyond itself.

Now we understand the double character of intentionality, the already-being-with-being and the being-directed-toward-being, the coming-from-being and the going-toward-it for fulfillment. We realize that no singular being can be recognized apodictically, but that the only apodictic cognition concerns the world-experiencing-life, the Ego-in-the-world.

The solution of the problems of truth and evidence lies in the world.[39] The apodicticity stems from the world, the world-experiencing-life, which experiences itself, and as such. This apodicticity cannot be adequate. No apodicticity of reflected cognitions is in itself apodictic, but only partakes in the one and only apodicticity. The world in its apodicticity is the only thing that is true. From the truth of the world all evidence, all self-givenness and its apprehension receives now its authentic character and its proper name. It is not truth, but truth-in-becoming (Bewahrheitung).[40]

We must stress here that for Husserl the idea of pure self-giving is a construction, and that its apodicticity itself is only relative.

Truth-in-becoming must follow a double path: expansion of self-givenness, since it is always referring beyond itself, and approximation of the already obtained self-givens to their own limits, to the idea of absolute clarity and self-givenness.[41] The idea of pure-self-givenness is therefore a construction. But it is a construction which is the basis for all possible cognitions within the world, it is the idea of truth which I can only attain through truth-in-becoming.[42]

The apodicticity pertaining to becoming-truth is only *relative*, it is related to the apodicticity of the world, of the world-experiencing. Thus

Husserl uses this term which at first seems self-contradictory: 'relative apodicticity'.[43]

Now that the material apriori has been discovered in its authentic extension and in its apodicticity, the question of its systematic apprehension arises anew and on a more profound level. The attempts to answer this question have opened new vistas, but we still must show what replaces the *eidetic insight*, which at first appeared as the method of apprehension of the material apriori. This method is intentional analysis, meaning the unfolding of intentionality. According to the preceding explanations one might think that intentional analysis has to start from the world-experiencing-life. However, the point of departure of intentional analysis in Husserl is not the worl-experiencing-life, the world, it is the particular entity, the singular object in its widest sense: real things, experiences, numbers, contextual relationships, laws, theories, etc.[44]

Husserl quite clearly gives priority to singular reality over the world without further justification. Certainly, the singular experience leads us to the world-experience, but the first has priority over the latter.

The problem is, therefore: how do I apprehend a singular entity in view of its horizon, finally in view of the world? We can find in Husserl five different conceptions of intentional analysis, which he not only does not clearly distinguish but which he also confuses and does not unify.

One conception consists in considering the object, the given, as meant-to-be. Intentional analysis is thus analysis of meanings.[45] Husserl does not use this concept here as 'mere meaning' in the sense of opinion, which simply would have to be exposed, made more graspable, rendered more clear. Meaning is the being-referred-to, which inhabits the object itself, and which can be made explicit. Meaning is that-which-is-there in as far as it always means more than itself, it is the self-being which can be ever closer determined in its being-itself-referred-to. The meant-to-be (*Gemeinte*) is not mere meaning, it is that which is there itself and which is 'meant' in its perception.[46] In Husserl we find a subtle, almost unnoticeable transformation from the 'meant-to-be' into meaning in the more ordinary sense.

We find a second conception of intentional analysis when the meant-to-be is not considered as the same object, having and keeping its self-sameness in its being-referred-to-, but when it is taken as a correlate of consciousness.[47]

A third conception comes about when the given, the meant-to-be itself, is considered as synthesis.[48] This further step is quickly accomplished, almost unnoticed but pregnant with heavy consequences. The meant-to-be dissolves in meaning considered as pure accomplishments of consciousness, the given thereby is considered as synthesis achieved by consciousness.

A fourth conception, intermediate between the analysis of the meant-to-be as meaning and the analyses of syntheses of consciousness, consists in considering the given neither as meant-to-be in its meant sense nor as synthesis of consciousness, but in its subjective givenness The already given is now to be analyzed in its givenness as subjective givenness. Intentional analysis thereby becomes the analysis of the subjective *how* of givens. A text from *Crisis* renders this conception.[49] The horizon here is not any longer horizon of sense-implications, but of cofunctioning subjective modes of appearance. Phenomenology thereby becomes unduly subjectified, psychologized.

The fifth conception, which I regard as the most illuminating and comprehensive, is found in intentional analysis considered as genetic-historical analysis.

We thus arrive at the central method in Husserl's phenomenology: the intentional analysis of truth-in-becoming as its re-enactment, as the re-awakening of its original foundation, as the retrieval of 'institutionalizing' truth in its ongoing institutionalizing.[50] Intentional analytic of truth-in-becoming, as its active re-enactment, remains simultaneously passive truth-in-becoming, which means that the re-awakening of the original institution of truth enters into synthesis with new institutionalizing truth.[51]

We may consider all given 'truths', all purposive constructs stemming from tradition, as pre-meanings, pre-intentions, as positing or void intentions. We have then in the re-awakening of the original evidence an actual fulfillment of a pre-intention. Because there is a continuous synthesis with new institutionalizing truth, intention remains, also in this respect, intention-to-be-fulfilled, and, at the same time, unveiling, truth-in-becoming.[53]

In order to understand correctly the character of intentional analysis, it must be stressed that every unit of knowledge, every fact of experience, every acquisition of truth-in-becoming has its history.[54] In the most simple comprehension of a given of experience, we are co-conscious (*mitbewusst*) that it is also formed through human activity. However hermetic, however merely implicit such a co-consciousness may be, it can always be explicited through an intentional analysis as a *historical* analysis sui generis. Thereby and without any doubt, intentional analysis as analysis of truth-in-becoming takes on the character of an intentional-historical explication.[55]

Finally, there is another element to add which is essential for Husserl. The dimension of the life-world, in which the world-experiencing-life apprehends itself and in which truth-in-becoming takes place, is language. The life-world is a human world, and language pertains to the horizon of humanity. Humanity, for Husserl, is always given either immediately or mediately as a community of language. Language is the authentic medium, the 'between' mediating between the transcendence of the world and the *Ego*. Only in and through language are we related to the world. World and language are unseparably interwoven.[56]

When we try to apprehend the concrete world in which we live, then we try to express it, then we express it in a discourse about the world, and then we have original 'instituting' of truth, truth-in-becoming, truth becoming incarnated in language.[57] Through discourse I apprehend the preceding, pre-given text of the world, which becomes text only through the discourse it instigated, a text which is now an incarnated acquisition. Sense can only become disposable, become a given, when I express it. Intellectual products become lasting language-acquisitions; and through the function of written documentary expressions of language the communalization of humanity, and with that its intellectual and spiritual products, are elevated to a new level.

In this way we see that any form of truth-in-becoming takes place in historical language community. Here the radical elucidation of truth has its deepest foundation. With this Husserl has revealed a new view of this foundation, without penetrating into it. The authentic task of phenomenology is to accomplish the becoming-truth of truth, to lead the concrete apriori to express itself, to shine through in language, to transform the pre-understanding of the world into understanding. Thus, phenomenology is the investigation into the pre-having of myself and the world, a pre-having which is not knowledge and which, remaining encompassing pre-having, is to be ever more known, ever more closely determined.

NOTES

1 "Der Auseinanderfall von Wissenschaft und Leben ist (...) auf seinen Ursprung, die Spannung von Lebenswelt und technischer Welt (Gesellschaft) zurückgeführt. Die objektiven Wissenschaften und das von ihnen beherrschte Universum, die technische Welt, gewinnen nur dann ihre Lebensbedeutsamkeit, wenn sie mit der Lebenswelt (wie auch diese mit ihnen) zur Versöhnung kommen." (Lothar Eley, *Die Krise des Apriori in der transzendentalen Phänomenologie Edmund Husserls*, Martinus Nijhoff, The Hague, 1962, p. 105.)

2 Cf. Eugen Fink, 'Das Problem der Phänomenologie', in *Studien zur Phänomenologie*, Martinus Nijhoff, The Hague, 1966, p. 190.

3 Vgl. Max Scheler, *Der Formalismus in der Ethik und die materiale Wertethik*, Bern, 1966, p. 73.

Cf. Max Scheler, 'Phänomenologie und Erkenntnis', in: *Schriften aus dem Nachlaß*, Vol. I, 2nd Edition, Bern, 1957, p. 383, pp. 397, 433.

4 "(...) angenommen, die synthetischen Urteile a priori seien das Rätsel. Wir urteilen unabhängig von der Erfahrung, wir folgen dem Zug der Notwendigkeit und Allgemeinheit, die doch wohl etwas zum Habitus dieser Urteile Gehöriges ist, wo immer wir sie einsichtig fällen. Die Urteile wollen Ihrem Sinn nach objektiv gelten. Wie können sie das? Die hier eigentlich vorliegende Frage ist nun doch die: Wie ist es zu verstehen, daß ein eigentümlicher Charakter der Notwendigkeit Urteilen von einem gesetzlichen Inhalt wirkliche Geltung verleihen soll, und zwar natürlich Geltung in dem Sinn, den diese Urteile haben, also objektive Geltung –? Ist das nun eine Antwort: es seien diese Gesetze zugehörig zu den formenden Funktionen –?" (Edmund Husserl, *Erste Philosophie*, Part I, Martinus Nijhoff, The Hague, 1956, p. 369).

5 Notwendigkeit ist nach Scheler "ein *negativer* Begriff (...), insofern 'dasjenige notwendig ist, dessen Gegenteil unmöglich ist'". (Max Scheler, *Der Formalismus in der Ethik und die materiale Wertethik*, Bern, 1954, p. 95.)

6 "Les choses ne signifient rien. Pourtant chacune d'elles a un sens. Par *signification* il faut entendre une certaine relation conventionnelle qui fait d'un objet présent le substitut d'un objet absent; par *sens*, j'entends la participation d'une réalité présente, dans son être, à l'être d'autres réalités, présentes ou absentes, visibles ou invisibles, et de proche en proche à l'univers." (Sartre, *Saint Genet, Comédien et Martyr*, Paris, 1952, p. 283.)

7 "*L'a priori* lui-même, s'il est une forme, c'est de l'univers et non de la pensée; il n'est pas une règle, de la pensée, mais une règle *pour* la pensée parce qu'il est imposé par l'univers." (Mikel Dufrenne, *La notion d'Apriori*, Paris, 1959, p. 74f.)

8 "*L'a priori* ne peut pas être soumis aux normes du jugement, ou lui-même jugé selon les caractères du jugement qui l'explicite, parce qu'il est la verité du jugement." (Ebda. p. 87.)

9 "Die phänomenologische Methode zeigt die Vernunft nicht nur als den Inbegriff der Bedingungen der Möglichkeit, ein durch die Sinne gegebenes, in sich sinnloses Material denkend bestimmen zu können, sondern als diejenige, die zuvor schon unmittelbare Erfahrung und Angewiesensein auf ein in ihr unverfügbar gegebenes Wahres ermöglicht. Sie ist im tiefsten Grunde nicht das Vermögen des Entwerfens und Beherrschens, sondern das Vermögen des Vernehmens." (Ludwig Landgrebe, 'Die Bedeutung der Phänomenologie Husserls für die Selbstbesinnung der Gegenwart', in *Husserl und das Denken der Neuzeit*, Martinus Nijhoff, The Hague, 1959, p. 221.)

10 Cf. Edmund Husserl, *Ideen zu einer reinen Phänomenologie und phänomenologischen*

Philosophie, Book 1 (ed. by W. Biemel, *Husserliana*, Vol. III, Martinus Nijhoff, The Hague, 1950, pp. 13, 15, 415, 416 – we quote *Ideen*, Vol. I).

[11] Cf. Edmund Husserl, *Phänomenologische Psychologie*, Martinus Nijhoff, The Hague, 1962, p. 83.

[12] "Fruchtbare Methoden verführen allzu leicht zu unkritischen Übertragungen; es ist eine offene Frage, ob die skizzierte Methode der Ideation in allen Wesensbereichen des Seienden durchführbar ist, ob sich z.B. durch Phantasieabwandlung das Wesen des Menschen fassen ließe. Ja es ist auch eine – von Husserl anfänglich vertretene, später revidierte – Naivität, die im Hinblick auf das gegenständlich Seiende ausgebildete eidetische Methode auf das 'Bewußtsein' und seinen Zusammenhang mit den Gegenständen zu übertragen und so das Problem der Evidenz in einer *eidetisch verfahrenden Korrelativforschung* entwickeln zu wollen. Die Frage nach der Selbstgebung des Seienden, sofern es eine Art hat, kann durch die gegenstandstheoretische Ideation bedeutsame 'Leitfäden' gewinnen; das philosophische Problem aber kann nicht selbst methodisch beherrscht werden durch eine naive 'Selbstgebung' der Wesen. Gerade die Methode der Ideation bildet selber ein Evidenzproblem, in welchem erst das originär gebende Bewußtsein analytisch enthüllt werden und so bestimmt werden soll, was das Wesen 'ist'." (Eugen Fink, 'Das Problem der phänomenologie Edmund Husserls', in Eugen Fink, *Studien zur Phänomenologie* 1930–1939, Martinus Nijhoff, The Hague, 1966, pp. 217–8).

[13] "Am Prinzip aller Prinzipien: daß jede originär gebende Anschauung eine Rechtsquelle der Erkenntnis sei, daß alles, was sich uns in der 'Intuition' originär (sozusagen in seiner leibhaften Wirklichkeit) darbietet, einfach hinzunehmen sei, als was es sich gibt, aber auch nur in den Schranken, in denen es sich da gibt, kann uns keine erdenkliche Theorie irre machen. Sehen wir doch ein, daß eine jede ihre Wahrheit selbst wieder nur aus den originären Gegebenheiten schöpfen könnte." (*Ideen*, Vol I, p. 52.)

[14] Cf. Gerd Brand, *Welt, Ich und Zeit. Nach unveröffentlichten Manuskripten Edmund Husserls*, Martinus Nijhoff, The Hague, 1955, pp. 5 ff., 25ff.

[15] "Danach ist also bewährende Evidenz nichts anderes als das in der Ursprünglichkeit der Selbsthabe vollzogene Bewußtsein der adaequatio rei et intellectus." (Edmund Husserl, *Analysen zur passiven Synthesis* (ed. by M. Fleischer, *Husserliana*, Vol. XI, Martinus Nijhoff, The Hague, 1966, p. 102 – we quote *PS*.)

[16] "Im laxeren Sinne sprechen wir von Evidenz, wo immer eine setzende Intention (zumal eine Behauptung) ihre Bestätigung durch eine korrespondierende und vollangepaßte Wahrnehmung, sei es auch eine passende Synthesis zusammenhängender Einzelwahrnehmungen, findet. *Von Graden und Stufen der Evidenz* zu sprechen, gibt dann einen guten Sinn. Es kommen in dieser Hinsicht die Annäherungen der Wahrnehmung an die objektive Vollständigkeit ihrer gegenständlichen Präsentation in Betracht und dann weiter die Fortschritte zum letzten Vollkommenheitsideal: dem der adäquaten Wahrnehmung, der vollen Selbsterscheinung des Gegenstandes, soweit er irgend in der zu erfüllenden Intention gemeint war. Der *erkenntniskritisch prägnante Sinn* von Evidenz betrifft aber ausschließlich dieses letzte, unüberschreitbare Ziel, den Akt *dieser vollkommensten Erfüllungssynthesis*, welcher der Intention, z.B. der Urteilsintention, die absolute Inhaltsfülle, die des Gegenstandes selbst, gibt. (...)

Die Evidenz selbst ist (...) der Akt jener vollkommensten Deckungssynthesis. (...) Ihr objektives Korrelat heißt *Sein im Sinne der Wahrheit* oder auch *Wahrheit*." (Edmund Husserl, *Logische Untersuchungen*, Vol. II, part 2, Halle, 1921², pp. 121–2 – we quote *LU*.)

[17] "Er spaltet sich jetzt in zwei fundamentale Begriffe. Der *eine* ist der der Selbstgebung überhaupt. Wohlverstanden, reicht er soweit, als überhaupt der Begriff des Bewußtseins reicht als Bewußtsein von etwas." (*PS*, p. 430.)

"Der *zweite* Begriff von Evidenz, der prägnante, ist dann die reine Selbstgebung." (*PS*, p. 431.)

[18] Cf. *PS*, p. 431.

[19] "Wo sich zwei Elemente als Intention und Erfüllung aufeinander beziehen, ist es die Intention, die bestimmt, was ihre mögliche Erfüllung sein kann, und nicht umgekehrt." (Ernst Tugendhat, *Der Wahrheitsbegriff bei Husserl und Heidegger*, Berlin, 1967, p. 51.)

[20] "Es ist jedenfalls sehr bedenklich und oft genug irreführend, davon zu sprechen (...), daß die intentionalen Erlebnisse 'etwas als Objekt in sich enthalten' u.dgl. Derartige Ausdrücke legen zwei Mißdeutungen nahe; erstens, daß es sich um einen realen Vorgang oder ein reales Ich-Beziehen handle, das sich zwischen dem Bewußtsein oder Ich und der 'bewußten' Sache abspiele; zweitens, daß es sich um ein Verhältnis zwischen zwei gleicherweise im Bewußtsein reell zu findenden Sachen, Akt und intentionales Objekt, handle." (*LU*, Vol. II/I, pp. 371–2).

[21] "Beachten Sie dabei diesen bedeutsamen Punkt: Alle Selbstgebungen, die hier in der Erinnerungssphäre, aber ebenso in allen Sphären das Selbst als wahres und identifizierbares, verfügbares Selbst herauszuarbeiten berufen sind, sind in gewissem Sinn 'transzendent' gebende intentionale Erlebnisse; das betrifft alle in eine mögliche aktive Identifikation eintretenden Selbstgebungen. (...) Also jede Selbstgebung hat hier Transzendentes, und jedes Selbst ist transzendent in einem ursprünglichen und guten Sinn – das gibt allerdings eine kuriose Paradoxie. Das erste, urquellenmäßige Transzendente ist der Bewußtseinsstrom und seine immanente Zeit, nämlich er ist das transzendente Selbst, das in der Immanenz der ursprünglich strömenden Gegenwart zur Urstiftung kommt und dann zur frei verfügbaren Selbstgebung und Selbstbewährung. (...) Der Bewußtseinsstrom lebt im Strömen und wird zugleich für sein Ich gegenständlich, objektiv; er wird es als das transzendente Selbst, das in Wiedererinnerungen und Wiedererinnerungssynthesen einer jeweiligen Gegenwart zur unvollständigen und approximativen Selbstgebung kommt." (*PS*, p. 204.)

[22] "Die Beschreibung vollzieht sich auf Grund einer objektivierenden Reflexion; in iht verknüpft sich die Reflexion auf das Ich mit der Reflexion auf das Akterlebnis zu einem beziehenden Akte, in dem das Ich selbst als sich mittels seines Aktes auf dessen Gegenstand beziehendes erscheint. Offenbar hat sich damit eine wesentliche deskriptive Änderung vollzogen. Zumal ist der ursprüngliche Akt nicht mehr bloß einfach da, in ihm leben wir nicht mehr, sondern auf ihn achten und über ihn urteilen wir." (*LU*, Vol. II/1 p. 377.)

[23] "Ja, wir würden überhaupt nicht dazu gelangen, von Leervorstellungen zu sprechen und ihnen gegenständliche Beziehung zuzumuten, wenn es nicht wesensmäßig zu jeder Leervorstellung gehörte, daß sie sozusagen eine Enthüllung, eine Klärung, eine Herausstellung ihres Gegenständlichen zuläßt." (*PS*, p. 71)

[24] Cf. Edmund Husserl, *Phänomenologische Psychologie*, *Husserliana*, Vol. IX, Martinus Nijhoff, The Hague, 1962, p. 428ff. – we quote *Phän. Psych.*

[25] "Ein Bewußtsein von etwas bracht nicht notwendig in sich die ausgezeichnete Form des Gerichtet-seins auf dieses Was, auf seine Gegenständlichkeit zu haben." (*PS*, p. 90.)

[26] "Das Vorstellen ist nicht bloß überhaupt vorstellendes Bewußtsein von seinem Gegenstand, sondern in sich selbst auf seinen Gegenstand gerichtet.

Diese Beschreibung hat nur insofern ihre Gefahr, als es sich nicht um diejenige sehr übliche Wortbedeutung von Meinen, Gerichtet-sein, Intendieren handelt, die sich auf das Ich und seinen Aktus bezieht, wobei das Ich, und in einem total anderen Sinn, Ausstrahlungspunkt einer Richtung, eines Sich-richtens auf den Gegenstand ist. Hier helfen wir uns, in Ermangelung brauchbarer Worte, mit dem Beisatz *passiv*, *passive Intention*." (PS, p. 76 – Accent mine.)

[27] "Die Vorstellung hat nun die Form eines vom Ich ausgehenden Strebens, einer *intentio*, die das wahre Selbst intendiert." (*PS*, p. 85.)

[28] Cf. *PS*, p. 20.

[29] "Der Verhüllungsgrad ist also auch als Grad der relativen Unbestimmtheit, der relativen Armut, Leere zu bezeichnen, die jede Selbstgebung durchsetzt. Sie ist immer unvollkommene Fülle, das ist Fülle, verdünnt durch Leere." (*PS*, p. 206.)

[30] "Darum ist die Rede von Inadäquation, zu deren Sinn der Gedanke eines zufälligen Manko gehört, das ein höherer Intellekt überwinden könnte, eine unpassende, ja völlig verkehrte." (*PS*, p. 19).

[31] "Kann es nicht sein (...), daß (...) die als Norm fungierende Selbstgebung auch wieder sich vernichtet und so das Wirkliche und das Unwirkliche immer nur ein Moment ist, etwas zum zufälligen Erfüllungsprozeß Gehöriges?" (*PS*, p. 109.)

[32] "Nur weil unvollkommene Erfahrung doch Erfahrung, doch Bewußtsein der Selbsthabe ist, kann Erfahrung sich nach Erfahrung richten und durch Erfahrung berichtigen." (Edmund Husserl, *Formale und transzendentale Logik. Versuch einer Kritik der logischen Vernunft*, Saale, Halle, 1929, p. 144 – we quote *FTL*).

[33] Cf. *FTL*, pp. 139–40.

[34] "Wahrheit ist doch Endgültigkeit. Aber Selbsthabe, Erfahrung kann mit Erfahrung in Streit kommen, es kann Modalisierung eintreten. Kann das nicht *in infinitum* fortgehen, also niemals eine Endgültigkeit erreicht werden? Und wenn es eine solche geben sollte, wie davon wissen?" (*PS*, p. 102)

[35] "Die Rede von *absoluter Sicherheit* oder, was gleich gilt, *absoluter* Zweifellosigkeit bedarf der Klärung. (...) Jede Evidenz ist Selbsterfassung eines Seienden oder Soseienden in dem Modus 'es selbst' in völliger Gewißheit dieses Seins, die also jeden Zweifel ausschließt. Nicht schließt sie darum aus die Möglichkeit, daß das Evidente nachher zweifelhaft werden könnte, das Sein als Schein sich herausstellen könnte, wofür ja die sinnliche Erfahrung uns Beispiele liefert. Diese offene Möglichkeit des Zweifelhaftwerdens bezw. des möglichen Nichtseins *trotz der Evidenz* ist durch eine kritische Reflexion auf ihre Leistung auch jederzeit im voraus zu erkennen. Eine apodiktische Evidenz aber hat die ausgezeichnete Eigenheit, daß sie nicht bloß überhaupt Seinsgewißheit der in ihr evidenten Sachen oder Sachverhalte ist, sondern sich durch eine kritische Reflexion zugleich als schlechthinnige Unausdenkbarkeit des Nichtseins derselben enthüllt, daß sie also im voraus jeden vorstellbaren Zweifel als gegenstandslos ausschließt. Dabei ist die Evidenz jener kritischen Reflexion, also auch die vom Sein dieser Unausdenkbarkeit des Nichtseins ⟨des⟩ in evidenter Gewißheit Vorliegenden abermals von dieser apodiktischen Dignität, und so in jeder kritischen Reflexion höherer Stufe." (Edmund Husserl, *Cartesianische Meditationen und Pariser Vorträge*, Martinus Nijhoff, The Hague, 1950, pp. 55–6) – we quote *CM*).

[36] "Erkenntnis geht auf Sein oder Sosein irgendwelcher Erkenntnisgegenstände. Müssen Erkenntnisse apodiktisch sein für Sein und Sosein, damit wir sollen rechtmäßig aussagen dürfen, daß sie sind und so sind? Oder: Müssen alle wahrhaft seienden Gegenstände, alle Gegenstände möglicher Wissenshaft apodiktisch erfahrbar und demnach auch so erkennbar sein? Und nun gar adäquat! Selbst das Ich-denke ist, wenn auch apodiktisch erkennbar, (...) nicht *adäquat* erkennbar." (Edmund Husserl,

Erste Philosophie 1912/24. Part 2, Martinus Nijhoff, The Hague, 1959, pp. 396–7.)
[37] Cf. *FTL*, p. 143.
[38] "'Der' Welt entspricht für uns die universale Synthesis zusammenstimmender intentionaler Synthesen, zu ihr gehörig eine universale Glaubensgewißheit. Aber wie schon erwähnt, es gibt da und dort Brüche, Unstimmigkeiten, mancher Partialglaube wird durchgestrichen, wird zu Unglauben, mancher Zweifel erwächst und bleibt streckenweise ungelöst u. dgl. Aber schließlich gehört zu jedem Unglauben ein positiver Glaube neuen sachlichen Sinnes, zu jedem Zweifel eine sachliche Lösung, und wenn nun die Welt nach manchem einzelnen einen geänderten Sinn erhält, so geht durch die sukzessive Folge der universalen Weltmeinungen trotz solcher Änderungen eine Einheit der Synthese – es ist dieselbe eine fortdauernde Welt, nur, wie wir sagen, im einzelnen korrigiert, von 'falschen Auffassungen', wie es dann heißt, befreit – an sich dieselbe." (*PS*, p. 101.)
[39] "Jede Bewahrheitung ist ein zutage, zur Klarheit der Selbstgebung Bringen eines Verborgenen. Wenn Selbstgebungen ihrerseits wieder Bewahrheiten zulassen und das freie Ich sie tätig fordert und anstrebt, so liegt darin, wie wir wissen, daß auch eine Selbstgebung allgemein ihre Verborgenheiten hat; daß ans Licht der Klarheit, nämlich in Form neuer Selbstgebungen, gebracht werden kann, was in ihr noch in leerer oder verhüllter Weise gelegen ist. Wir wissen ferner, daß die Bewährung sich in zwei wesensmäßig aufeinander bezogenen Typen, einer positiven und negativen, abspielen kann, der einstimmig bewahrheitenden und der entwahrheitenden." (*PS*, p. 201.)
[40] Cf. *PS*, p. 202f.
[41] Cf. *PS*, p. 432.
[42] Cf. *Erste Philosophie*, Vol. II, p. 406.
[43] Cf. CM, p. 95.
[44] Cf. *Phän. Psych.*, pp. 96 and 98.
[45] "Als *Da-Seiendes* ist das Objekt in der Wahrnehmung Gemeintes." (*Phän. Psych.*, p. 184.)
[46] Cf. *CM*, p. 84–5.
[47] Cf. *Phän. Psych.* p. 422.
[48] Cf. Hans-Georg Gadamer, *Wahrheit und Methode*, Tübingen, 1960, p. 246.
[49] "Indem Sinn auf Sinn sich gründet, gibt geltungsmäßig der frühere Sinn etwas an den späteren ab, ja, in gewisser Weise geht er in ihn ein." (Edmund Husserl, *Vom Ursprung der Geometrie.* (Beilage III zur Krisis der europäischen Wissenschafte) ebda., p. 365ff. – we quote *UG*).
[50] Cf. *K*, p. 367.
[51] "Durch das passive Leben gehen also immer neu sich flechtende Synthesen der Erfüllung. Immerfort ein Hinstreben auf Anschauung, (...) immerfort, das Wort drängte sich uns auf, Bewahrheitung." (*PS*, p. 102.)
[53] "Geschichte ist von vornherein nichts anderes als die lebendige Bewegung des Miteinander und Ineinander von ursprünglicher Sinnbildung und Sinnsedimentierung." (*UG*, p. 380.)
[54] "Jede Explikation und jedes von Verdeutlichung in Evidentmachung Übergehen (sei es auch vielleicht allzu früh steckenbleibend) ist nichts anderes als historische Enthüllung; in sich selbst wesensmäßig ist es ein Historisches und trägt als solches wesensnotwendig den Horizont seiner Historie in sich." (*UG*, p. 379.)
[55] Cf. *K*, pp. 369f.
[56] "Redend vollziehen wir fortlaufend ein inneres, sich mit Worten verschmelzendes, sie gleichsam beseelendes Meinen. Der Erfolg dieser Beseelung ist, daß die Worte und

die ganzen Reden in sich eine Meinung gleichsam *verleiblichen* und verleiblicht in sich als Sinn tragen." (*FTL*, p. 20.)

"Die sinnlichen Äußerungen haben in der Welt raumzeitliche Individuation wie alle körperlichen Vorgänge bezw. alles in Körpern Verkörperte als solches; nicht aber die geistige Gestalt selbst, die da 'ideale Gegenständlichkeit' heißt. Indes in gewisser Weise in der Welt objektiv seiend sind sie doch, aber nur vermöge dieser doppelschichtigen (geistigen und sinnlichen) Wiederholungen, und letztlich vermöge der sinnlich verkörpernden." (*UG*, p. 368.)

[57] Cf. *UG*, pp. 372 and 371.

H. L. VAN BREDA

THE ACTUAL STATE OF THE WORK ON HUSSERL'S INEDITA: ACHIEVEMENTS AND PROJECTS[1]

In what follows I shall endeavour to explain the position in which the Husserl Archives in Louvain now are – in the Spring of 1969 – with our efforts to collect the documents concerning Husserl's academic career and more especially his philosophical conceptions and their evolution. Secondly, I shall treat of the stenographic and other originals, which we are now transcribing, and the method used in the elaboration of these transcriptions. Thirdly, I shall mention the texts now in print or almost ready for editing in the series, *Husserliana*. And finally, I shall discuss the projects in which the Archives are engaged in order to further the influence of Husserl's phenomenology on contemporary thought.

I. INTRODUCTORY REMARKS

As is widely known, in the autumn of 1938, some months after Edmund Husserl's death[2] his widow, Malvine Steinschneider, asked us to carry to safety the unedited philosophical manuscripts of her late husband, his personal library, and also a collection of documents concerning his academic career. We readily acquiesced to Mrs. Husserl's request, and, in the months immediately following, brought to Louvain (Belgium) more than 45000 pages of philosophical autographs written in Gabelsberger shorthand by the philosopher himself. We were, moreover, also successful in bringing to Louvain about 12000 pages of transcriptions of stenographs, which had been effected by his former assistants: Edith Stein, Ludwig Landgrebe and Eugen Fink. Finally, in June 1939, Mrs. Malvine Husserl emigrated to Belgium. She most fortunately managed to transfer to Louvain her husband's library, and in addition she brought with her an important set of biographical documents, including such items as letters of academic appointments, correspondence with colleagues and friends, photographs, etc. The library and most of the documents were added to the manuscripts and transcriptions which had previously been removed to Louvain. Thus it was by such rather complicated

Tymieniecka (ed.), Analecta Husserliana, Vol. II, 149–159.

channels that what we may call the *working-capital* of the Husserl Archives flowed into Louvain.

Meanwhile, in Louvain itself, we had not been inactive, and had worked out the basic necessary arrangements for giving a viable status to the new research centre, and with the efficient collaboration of Ludwig Landgrebe and Eugen Fink, the first transcriptions of Husserl's shorthand autographs were made.

II. GATHERING NEW DOCUMENTS

During the years in which Eugen Fink was his personal assistant in Freiburg in B. (1926–1936) Husserl bestowed on him as a gift files of manusscripts (roughly 1200 pages). These manuscripts were for the most part consecrated to the problems of time, and had been written during a fairly long stay in Bernau, a village in the Black Forest near the Feldberg. Last January, Eugen Fink very graciously donated these files to the Archives. As a result of this last and highly important gift, we are certain that we have in our collection, at the present time, all the still existing manuscripts of Husserl's philosophical heritage.

In this connection, we may also mention another important collection of philosophical documents, by whose acquisition, during the first months of 1969, the Archives were considerably enriched. I refer to the Brentano papers. Shortly before his death in Switzerland in January of this year, Professor John Brentano, the son of Husserl's master Franz Brentano, decided to offer us a set of photostats and films of all the unedited manuscripts left by his father at his death in 1917. Accordingly, as a result of this gift, on the 28th ot last February, this second spiritual heritage, which is obviously of incalculable value for our work, arrived at Louvain.

Year after year, since 1950, we have patiently tried to bring together Husserl's correspondence. Our endeavour has not been without fruit and we did succeed regularly in gathering some important pieces of correspondence. However, during the last three years, our harvest has been particularly rich. Thanks primarily to the ceaseless efforts of Professor Herbert Spiegelberg, our collection has grown, continually, so that we are now in the happy position of having in our possession more than 1400 letters from and to Husserl. It will hardly be necessary to add, we think, that the content of this correspondence is either mainly philosophical, or at least indirectly philosophically interesting.

It would also not be out of place here to mention another collection of material we now have in our possession. A voyage we had the good fortune to make to Czechoslovakia in September 1966, gave us the opportunity of uncovering in Prostejov, Husserl's Moravian birthplace, and in Olomouc, where he was a student of the german gymnasium, a lot of useful information about his ancestors and family, about the courses he followed, and about the examination results of his secondary education. Photostats of the original documents which contain these data are, since then, available in the Archives.

III. TRANSCRIPTION OF HUSSERL'S AUTOGRAPHS

In 1939, we began with the systematic typewritten transcription of the multitudinous files containing Husserl's stenographic manuscripts. We straightaway decided to elaborate for each of these files a readable reproduction. This reproduction, moreover, it was considered, should contain – in as far as possible – all the interesting data about its contents. Together with an exact transcription of the master's text (with references to the pages of the autograph), we decided that each transcriber should also furnish them with an introduction containing: (1) the explicit or inferable chronological data about its elaboration; (2) the eventual sub-divisions of its content; (3) the philosophers mentioned in it; (4) a detailed description of the concrete investigations exposed; and finally (5) the impressions of the transcriber himself about the value of these investigations.

We have now reached the following position with our work. About three-fifths of Husserl's autographs (more or less 40000 pages in quarto) are available in transcription in the Archives, and three quarters of these transcriptions can be considered as critically established. As well as those, we are also in a position to state that for all the files not yet deciphered, we possess detailed extracts from each of them. On these extracts we have reproduced all Husserl's chronological notes about the elaboration of the text, besides the titles and sub-titles he added to each part of the text.

Most of these titles of the remaining stenographic texts and all the titles occurring in the elaborated transcriptions have also been noted down on separate cards, while another card-system furnishes the chronological information about the origin of the texts present in the Archives.

If we add one final point – namely that the transcriptions are in principle made in at least 5 copies – I think that we may close this brief exposé about the methods used to make the *inedita* as usable as possible for those wishing to consult them.

The interested scholar will want to know what main topics are more fully elaborated in these transcriptions than in the texts already published.

In the first place, we may state in general that the major part of what the master called *die konkrete Intentional-Analysen* (concrete intentional analyses), can up to the present only be detected and studied in the *inedita*. As all of you know, Husserl gives that collective name to a group of quite divergent evolutive processes, which he uses as his method to rediscover the genesis of the intentional constitution in and through the human consciousness. Most of these analyses were the fruit of long hours of solitary meditation, and they kept consequently the flavour of personal research manuscripts. Writing them down, the master felt himself in the mood of someone being underway – *in via*, as the medieval scholars said – and not yet at the end of his long and difficult road. But, on the other hand, he repeatedly claims that the numerous essays of that type belong to the most important of his philosophical message. We cannot therefore put them aside in our study of his thought.

Let us pass now to the concrete topics treated at great length in those transcriptions which we have elaborated but which are not yet published. The following more or less chronological list may, we think, afford some profitable information.

Quite a number of recently deciphered autographs throw a new light on the origin and the genesis of the *Logical Investigations* (1900–1901) and on the genuine significance of some of its main theses.

For the period 1900 to 1914, all the most important texts on the constitution of physical things and space have been transcribed, and also Husserl's lectures and research-manuscripts on the phenomenological theory of judgment.

We already have, for many years now, a transcription of the lectures on ethics given by Husserl between 1902 and 1914, and during the last few months we have been elaborating a large group of his texts written in the same period on phantasy and its related problems.

We hope that another group of important texts will see the light of day before long. We speak of the manuscripts on the crucial problem of intersubjectivity. These texts were spread over the years 1905–1936. Not only are they already in transcription, but last autumn they were sent to the printer. These texts will take up three volumes of the *Husserliana*, and we hope that they will be published before the end of 1972.

The greatest part of the research for the period 1920–1930 on logical themes is now also in typewritten copies. This will, at least in part, be published in the re-edition of *Formal and Transcendental Logic*, which is being prepared at the moment.

Last - but by no means least - we would like to bring to attention in this not at all complete list, the very interesting texts on *Zeit als lebendige Gegenwart* – time as vivid and ever flowing presence – which Husserl elaborated during the last eight years of his life.

IV. EDITION OF HUSSERL'S WRITINGS

Every scholar interested in phenomenology knows the series *Husserliana*. He will therefore have some knowledge of the method we used in the eleven volumes already published, and will be able to form his own judgment of its worth. Consequently it will not be necessary for us to devote any time to a description of these offspring which have been brought forth by our labours in the Archives since 1950.

It will be unquestionably more interesting and profitable to give some precise information about the content of the new *Husserliana*-volumes which are now in print or which will go to the printer within the next few months, and about others which we hope to publish as soon as possible, but at a date which has yet to be decided.

Some weeks ago we got the last proof–sheets of *Husserliana*, Vol. XII, *Philosophie der Arithmetik* (*Philosophy of Arithmetic*). Its editor, Dr. Lothar Eley, has assembled in this volume a critical edition of the text of the book Husserl published under that title in 1891, and a group of related *inedita*, written between 1890 and 1901. This volume should certainly be published before the autumn of 1970.

The voluminous file with Husserl's texts on the problem of intersubjectivity is now ready, and we will in June or perhaps May of this year receive the first proofs back from the printer. The more than 1500 pages of *inedita* assembled in this file, and which have been prepared for the edition by Dr. Iso Kern, will appear in three volumes of the *Husserliana*

(Vols. XIII, XIV and XV), under the title *Zur Phänomenologie der Intersubjektivität* (*Texts on the Phenomenology of Intersubjectivity*)

In recent weeks Dr. Ulrich Claesges, scientific collaborator at the Cologne Centre of the Archives, handed in a new set of texts (more than 700 pages) which are now also ready for the printer. The main manuscript he used for this sixteenth volume of our edition is the elaborate file of notes which Husserl wrote down for his lectures on the constitution of material things. These lectures were given in Göttingen during the summer of 1907; and he often refers to them in his later writings as *die wichtige Ding-Vorlesungen*, (the important lectures on thing-problems). The texts assembled in this volume will be published under the title *Ding und Raum, Vorlesungen 1907* (*Thing and Space, Lectures 1907*). We think that we are safe in saying that they will be published before the spring of 1972.

Let us now pass to the files not yet ready for the printer, but which are either almost elaborated or whose preparation is far advanced.

Quite recently in Louvain Gerhard Maschke completed the transcription of the manuscripts related to the first volume of Husserl's *Logical Investigations*, which contains his famous criticism of the psychological foundation of logical laws and rules. These manuscripts will, we feel sure, throw a new light on the origin of the *Prolegomena zur reinen Logik* (*Prolegomena to Pure Logic*), as Husserl entitled this first volume.

Also completely transcribed are Husserl's autographs on the theory of judgment (*Urteilstheorie*), written during the years 1905 to 1910. Within the last few months we have been giving them the requisite finishing touches for the edition. The editor of this important file will be Dr. Ursula Panzer, from Cologne, and it is our opinion that this file will offer a lot of elements to bridge the gap between the *Logical Investigations* of 1900–1901 and the *Formal and transcendental Logic* of 1929. We think that this volume could be sent to the printer in 1971.

Apropos the *Formal and Transcendental Logic* of 1929, it should be mentioned that Dr. Paul Janssen, who is connected with the Archives in Cologne, has supplied us with his critical revision of the original version. He has also been collecting a series of unedited materials related to that work, and has reached the final stages of his research. These last materials will indeed be appended to the text of 1929 in the re-edition of that work, which he intends publishing.

Professor Rudolf Boehm is completing the work on his text for a volume containing Husserl's research manuscripts on the phenomenology of time, written during the last years of the master's life. This volume will furnish all the available texts capable of clarifying the above mentioned notion of living present (*lebendige Gegenwart*).

Work is also well under way on a number of other manuscripts, and even though the end is not yet in sight, the elaboration makes steady progress. We cannot, of course, give any idea at this stage of the possible date of publication, but we do not think that it would be out of place here to give you a list of the manuscripts.

First at all, in Cologne, Dr. H. R. Brennecke, who happily has an excellent knowledge of the history of mathematics around the turn of the century, has been working for a couple of years on the various mathematical autographs the younger master elaborated between the years 1887 and 1905. Ultimately, he intends to publish the most valuable of them under the title *Studia Mathematica*, in addition to the text of Husserl's doctoral dissertation, *Beiträge zur Theorie der Variationsrechnung*. This dissertation was written in Berlin under the direction of Carl Weierstrass, and was introduced by Husserl in Vienna in 1882. It is of interest to note that the dissertation was only discovered as late as 1962. In that year Professor Herbert Spiegelberg detected, in the library of the University of Vienna, the handwritten copy of the dissertation which had been presented there to the philosophical faculty. Before Spiegelberg's discovery, no scholar knew of its existence. Professor Spiegelberg very kindly sent the Archives a photocopy of this valuable document.

Karl Schuhmann, who is connected with the Archives in Louvain, has already assembled most of the manuscripts the master wrote for his lectures *Introduction to Philosophy* (*Einleitung in die Philosophie*). These lectures were repeatedly given between the years 1915 and 1924, and on many occasions in the course of them Husserl touches on the same problems which have been partly elucidated in the already edited *Erste Philosophie*, the title of his lessons from 1923 to 1924. Schuhmann's research is progressing quickly, and we may say that his results are really promising.

Finally, and also in Louvain, Edi Marbach is devoting himself to the task of detecting in the *inedita* all the texts related to the somewhat peculiarly sounding topic, 'phantasy'. The gnoseological problems Husserl brought together under that designation were often the object of his attention between 1902 and 1914. The harvest gleaned by our collaborator is quite impressive, and we can be pretty sure that some time in the coming years a volume on *Phantasie* will appear in our edition of Husserl's Collected Works.

V. PROJECTS INITIATED TO FURTHER THE PHENOMENOLOGICAL MOVEMENT

One day in Louvain a gracious visitor said that in his opinion the director of the Archives had to assume more or less the task of being the 'conductor' of what he called the now world-wide phenomenological orchestra. If the director in question has still without any doubt a very long way to go before he can even dream of conducting that distinguished company, we may at least state that the Archives as an institution – apart altogether from the publication of the *Husserliana* – originated some movement with a quite audible international resonance.

We have previously given you a summary description of the transcriptions made of Husserl's autographs and we mentioned the card-systems reproducing the dates and titles occurring in these autographs. Notwithstanding the value these documents may possess for the many scholars interested in phenomenology, it is quite evident that all these cannot be made available in print. In order therefore to facilitate their wider utilisation, we decided to deposit them in different centres, where they could be consulted and studied. Consequent to that decision, the following 'Centres d'Archives-Husserl' were established: Buffalo N.Y. in 1947, Freiburg i.Br. in 1950, Cologne in 1951, Paris (Sorbonne) in 1957, and finally New York (New School for Social Research) in 1966. The academic institutions which agreed to admit such a centre appointed themselves its responsible directors, and furnished suitable rooms in which to locate the documents. In Cologne, moreover, three or four qualified collaborators are regularly attached to the Archives there. We have mentioned many times during the course of this exposé the excellent work they have done and are doing, especially by preparing Husserl-texts for publication.

The directors of the different centres, which I have enumerated previously, form together the editorial board of the collection we started in 1958 under the title *Phaenomenologica*. Since the collection took its official seat, if we may use such an expression, in the Archives at Louvain, the editorial board called upon the director of that centre to act as president of the staff, and committed the task of secretary to Dr. Jacques Taminiaux, who is professor of philosophy at Louvain University.

Anyone who has been following the phenomenological publications which have emanated from our board over the last 10 years will know, we think, the guiding intentions of the responsible editors of the quite extensive series of books we have published in the *Phaenomenologica*. Valuable historical or systematic studies will be accepted only if their subject can be considered as authentically phenomenological; this last term, however, is taken in a relatively broad sense. For administrative and financial reasons we had moreover to decide that only texts in English, French and German would be published in the collection.

From 1958, when we started publication, up to February of this year, twenty-nine volumes have appeared in the *Phaenomenologica*. The editors were particularly happy in recent months that they could publish in their series the interesting letters which Husserl had sent to Roman Ingarden and at the same time to give him a convenient opportunity of recording for us his valuable personal reminiscences of the numerous contacts he had with his former professor.

A few words now on the works being prepared for the series. At present, in March 1969, the manuscripts of seven other volumes are either already set up in type by the printer or awaiting composition. Among them we may note here a voluminous text containing the *Selected Papers* of the late Kurt Goldstein. We owe to Mr. Aron Gurwitsch the privilege of being able to include in our collection a lot of articles of this famous biologist, in whose writings phenomenological subjects were often touched on.

Finally, we would add in this context, that nine more manuscripts are in the secretariat and are being submitted there to a final revision before being sent to the publisher. Among them are two texts introduced by Dr. Dorion Cairns, which will be particularly profitable for every scholar working on Husserl and on phenomenology in general. The first, *Guide for Translating Husserl*, furnishes, in alphabetical order, a translation in English – and partly in French – of the master's technical terminology, and will also provide a useful Husserl-Lexikon. The second contains a detailed, in fact almost verbal, report of the long conversations Cairns had in Freiburg i.Br. with Husserl in 1931 and 1932.

Since 1951, the Archives organised, under the name *Colloque International de Phénoménologie*, several gatherings devoted to Husserl and his impact on contemporary philosophy. We took care also to make the texts and the discussions of these meetings available in print. They were held in Brussels in 1951, in Krefeld (Germany) in 1956, and in Royaumont (near Paris) in 1957.

It can also be recorded here that the direction of the Archives has done a not inconsiderable amount of administrative work in the preparation of the phenomenological sections for the *International Congresses of Philosophy* in Venice-Padua (1958) and Mexico (1963).

Before closing this report we wish gratefully to remember the inestimable thanks to which some of our friends and a group of competent scholars are entitled by their unremitting devotion to Husserl's spiritual heritage.

In this expression of thanks, before all others – but of course still after their deceased mother Mrs. Malvine Husserl – we must recall our debt of gratitude to Professor Gerhart Husserl and Mrs. Elly Rosenberg-Husserl. We will never be able to forget the confidence they so abundantly placed in us in 1938 and 1939, and which they have continued to show us ever since.

Secondly we have to give public expression of our gratitude to the colleagues who direct the different centres of Archives outside of Louvain. Notwithstanding the fact that we meet them only occasionally, we know that their devotion to the one and common cause unites us, and that – *idem volens ac idem nolens* – they think and act always as our personal friends. Our gratitude may help in some small way to balance the kindness and the confidence they granted us in difficult times and under different circumstances.

But, in a very special way, we feel deeply obliged to the already quite extensive group of scientific collaborators and assistants of the Archives. First in that row stand the co-founders of the Archives in 1939, namely, Ludwig Landgrebe and Eugen Fink. As it was primarily their task to be our highly qualified masters and initiators into the arcana of Husserl's autographs, it is not correct to mention them here as our collaborators. However, since 1943, their successors have taken over the burden and are still bearing it. Some of their names figure on the front pages of the *Husserliana*-volumes. There are others whose names have been quoted in this *exposé*, but who are not yet known as editors of Husserl's texts. When we reflect on the mostly anonymous and unseen work they have done and continue to do so efficiently, we can appropriately apply to it the words of the Bible: "Things no eye has seen, no ear has heard, no human heart conceived" (I Cor., II,9). Like so many other devoted scholars, their uninterrupted efforts and efficient labours pass very often unnoticed; they can only be evaluated – and even then but very roughly – if and when their gathered harvest can be and is published. Of each and every one of our collaborators in the Archives we may truly say: *optime meriti sunt de philosophia!*

NOTES

[1] Text of a lecture given on 1 April, 1969 on the occasion of the inauguration of the *Husserl-Archives in Memory of Alfred Schutz* in the *New School for Social Research* (New York). – The same was given in a slightly changed form on 9 April, 1969 for the participants of the *International Phenomenological Conference* (Waterloo, Ont.).

[2] He died in Freiburg in Breisgau on the 27th of April of that year.

DISCUSSION

Spiegelberg: I have a question in connection with the German edition of the *Logische Untersuchungen.* Will it give a comparative text of Husserl's first and second editions of that work? Only then will it be possible to compare the significant changes in these two editions.

Van Breda: To answer your question, let me first make a comment about our critical edition of *Ideen I.* In the Archives we possess three copies of that work annotated by Husserl himself, along with some other comments. After long discussions we decided to incorporate these notes and comments into the text of our 1950 edition. Shortly afterwards we realised that we should have given them as chronological footnotes. In this way the student could follow on every page Husserl re-reading his own work in 1913, in 1916, in 1922, and in 1925. In the present 1950 edition this is only possible by referring constantly to the critical apparatus. – For our edition of the *Logische Untersuchungen* we won't make that mistake. We will follow instead the example of the critical edition of Kant's *Kritik der reinen Vernunft,* which reproduces simultaneously, as you know, the first and second editions. We will do the same for Husserl's two editions, if necessary by using two different print-letters. Furthermore, we will add as appendices lots of additional notes and comments written by Husserl but never published. We may add here a special remark concerning the sixth *Investigation.* We have in the Archives the texts of eight attempts at a new elaboration of it. He even sent two of them to the printer, got them back as proof-sheets, but he then refused to allow their publication. We have in mind to publish in our critical edition, along with the already known texts of the first two editions, these eight mentioned attempts – or at least a large portion of them – as appendices. Needless to say this will be a huge textual undertaking, but we all know that this work is crucial for a proper understanding of Husserl's epistemology.

DISCUSSION

Questioner: Dr. Brand, is your interpretation of Husserl's phenomenology or of phenomenology in general basically the same as Cassirer's notion of philosophy in *The Philosophy of Symbolic Forms*?

Brand: I'm not interpreting Cassirer; I'm interpreting Husserl. You find in the by line to the *Crisis*, I think in the appendix, the text on which my interpretation is based. Also you find several passages in *The Formal and Transcendental Logic* where language becomes for Husserl incarnated and thus mediates our discourse.

Questioner: Just one other point on this theme: There are a few letters of Husserl to Cassirer which were published in which Husserl criticizes Cassirer's project as not being radical enough, not ultimately getting back to the transcendental origin. I believe the implication is that according to Husserl, Cassirer falls ultimately into a kind of historical relativism.

Brand: I'm not sure, but I think that this only became clear to Husserl at the end of his life. You can read, for instance in Heidegger's letter to Father Richardson (in Richardson's book), that as of 2 or 3 years ago he still considered Husserl's phenomenology as purely static, as a-historic. Heidegger seems not to have read this, but nevertheless this is finally Husserl and you can find many passages to this effect.

Questioner: In your discussion of the community of language, on page 142, it sounded as if you were interpreting Husserl very much like Wittgenstein. I wonder if you would make the claim that language is the limit of reality (reality taken in the Husserlian sense), in view of your statement that "world and language are inseparably interwoven."

Brand: No, reality is not limited to language. Language does not limit reality. But there is an interwovenness between them and I see a similarity between Husserl and Wittgenstein, Wittgenstein alone, and not any other analytic philosopher, when Wittgenstein says that you cannot explain the world with words. What we want to say is always beyond words.

It is a strange relationship that the pre-understanding which we have can only become understanding when we express it through language.

Tymieniecka (ed.), Analecta Husserliana, Vol. II, 160–169. *All Rights Reserved.*

This is why I have said that a text is pre-given and only becomes a text through interpreting it. I can recognize it as a text in its pregivenness only by interpreting it. That's a difficult situation to describe and will always make phenomenology difficult. One has the difference between the world and language.

When I see and express something, or even without expressing it, as it were, I try to double it. I try to present it to myself, to express it to myself, but it's still the thing which stands in front of me. This strange doubling of the world through my seeing it in language is very difficult to grasp but it also gives us the opportunity to find the origins of the presentation itself, of the concept.

Sokolowski: I have some brief questions. First: you distinguished three levels of awareness, a type of being with the object without being concious...

Brand: No, I don't, Husserl does. He does so very clearly in the passive synthesis.

Sokolowski: Do you think that this first level is an expression of hyletic awareness in a later form?

Brand: I wouldn't call it '*hyletic* awareness'. I would call it 'affective awareness', because in the passive synthesis, Husserl speaks several times of 'affection'.

Sokolowski: Do you think that the word '*hyletic*' fits at all in the three levels?

Brand: No, I wouldn't use the word '*hyletic*'.

Sokolowski: The third level is a kind of spontaneous move of consciousness which would be an interesting correlation with the *Auffassung* in the *Ideas*.

Brand: No, it's much simpler than that. When you're lying on the beach, the sun is shining, the waves are rolling, you are not thinking about anything special: then you are with 'it', without being directed to 'it'. Then you might grasp 'it' – *that* is the passive *intention*. But this is still not real directedness towards that which you are doing; from then on you may go further to say to yourself: "I'm lying here, seeing the sea, seeing the beach, now I want to know what this is all about". Then you aim at your activity.

Sokolowski: I'm contrasting one Husserl with the other and he should be simple in both cases. But I find those problems analogous; the description you gave of the first two levels is very much like the *hyletic* appreciation of something which then has a *foundation*.

Brand: I'd agree with you. Don't let's refine about terms. If you want to call it *hyletic* – as long as 'hyletic' does not mean 'sense data' which have only a function but which cannot be experienced – then I would agree with you. Affection can be experienced. That is something which can be experienced and phenomenology has only to do with things which one can experience... in one way or another. There are several ways of experiencing.

Sokolowski: Secondly: you said that the only apodictic speech possible is about the ego and the world.

Brand: That is the only apodicticity at all.

Sokolowski: Why should that have apodicticity, and not other experiences?

Brand: When one examines the problem of evidences, as I try to do it, then one finally (and Husserl at some place) asks: How can we go about the problem of evidence? Must all evidence be adequate and then apodictic? How can one examine the problem of evidence, of selfgivenness? One can only try in a concrete way to show *what* is evident, *what* is self given. When one follows this way, then one arrives finally at the total horizon of the world. Here we have switched full circle back to the things themselves where we had an enigmatic pre-understanding of all beings in totality: It's the world; the world is the bottom. The only thing that is really apodictically true (let me remind you that apodicticity concerns cognition, not content) which one is *absolutely* sure of, is that one lives together with others in a world. Therefore self-givenness is a construction, since a particular cognition is only apodictic because it is within the world; one experience can only be replaced by another on the foundation of the world. So you have only the 'Bewahrheit'.

There's one truth that is about the ego and the world and it is *Bewahrheit*. You always go on in a dynamic way.

Sokolowski: In that excellent paragraph on page 133 you work out two concepts: The 'underivable' and the 'original'. You accept certain statements as derivable from others and some that are not derivable; you explain that they are not derivable because they are 'originary'. I think Husserl would say so too. But in fact you have *two* concepts: 'The underivable' and 'the originary'. I wonder if we can consider them as identical?

In other words, is the non-derivability of the final context of the world and of the *ego* also a guarantee that it is originary with the type of self-givenness, with an intuition of this own?

Brand: You touch a very difficult problem which Husserl, I think, didn't solve.

Sokolowski: That seems to be Derrida's problem in interpreting Husserl. He seems to believe that there is a linguistic tendency to go ever farther in one's inquiry, but then suddenly one simply stops at a given point because of compulsions of language. There seems to be no originary givenness operating there. The givenness would be a myth.

Brand: There is a certain difficulty in Husserl and in other phenomenologists which consists in this. The only truth is the truth about the ego and the world. Is this, however, the only thing that is given? There are other things that are given and are ultimate. There are singular entities that are given, and I think that the method to arrive at the underivable and the originary is simply to follow the circle of definition. If, in order to define something one must use that which one wants to define, then one stops and has an ultimately given; there seems no arbitrariness in it.

Sokolowski: No, just because it is underivable in the language doesn't mean it is arbitrary certainly. A final question: You talk about a constant mixture of givenness and absence in intentionality. Is that your criticism of Tugendthat?

Brand: No, I criticize Tugend on one account, that he considers Husserl's notion of the concept as the concept of intention to be fulfilled by a reality.

Sokolowski: Yes, and therefore he takes intentionality as a whole to be only this part, or rather he takes the part for the whole. He takes the empty intention for the intention itself. Now I would agree with your position in terms of *schlichte Wahrnemung* but when you talk about propositional intention, intention which is not perceptual, couldn't that be a totally empty perception?

Brand: No. I indicated at the end of my paper that when we come to the true method, the central method of Husserl – intentional historical analyses – then we can consider the constructions we find in our daily life, which come from tradition, which you were talking about as pre-intentions, as an empty meaning. Then the re-enactment of the first truth is the fulfillment of the intention yet at the same time always unveiling.

Van Breda: I had an occasional meeting with Jacobson who said to me that the manuscripts of Husserl are the only valuable source for settling the philosophy of language. He said that *The Logical Investigations* are his classic. He said he knew very precisely the evolution between *The Logical Investigations* and *The Formal and Transcendental Logic.* He said in the first period Husserl was putting the level of language and the level of being in the world too much one outside of the other. I would like to add that the concept of *Idealism* in Husserl comes not at all from the classic concept of Idealism. The first time Husserl uses the term 'idealism' it is to oppose the heresy of his youth, the 'realism', as it was represented by Locke, what he later on will identify as the implicit philosophy of the natural standpoint. He wants always to prove that although passive synthesis plays an important role in institutions there is also an activity of consciousness. 'Idealism' means for him 'intervention': *Leistung der Subjectivität des Bewusstseins.* When he identifies himself as an idealist, that means first of all a defense against the kind of realism he doesn't like in Avenarius' *Empirio Kritizismus* what he learned for instance, in the school of Brentano. 'Idealism', when he uses the term, never means directly a world stemming *out* of consciousness but before all an attack *towards* a world and knowledge *about* the world without consciousness, or without its activity. And to stress the intervention of *Leisten,* he uses the word 'idealism'.

Strauss: I don't want to argue about the interpretation of Husserl but about the results of his thoughts. Is it not the case that through language and speech we first of all belong to the life world? In talking we certainly are part of the life-world. So much so that we have to learn it 'locally'. That we can see here: not everybody can understand German or French or so. These are local, historical, geographical events.

Furthermore, there is the difference which the French make, I think, between *le mot* and *la parole.* If I were to say that tomorrow night we will have dinner in the church colleges it would not be right because it would not confirm this statement with the situation in which it is spoken. So that there are two aspects of language: the 'saying' (if I may use that word) and the 'meaning' pronounced. Although tomorrow we will understand the meaning of those words, 'we will have supper in the church colleges', it would not fit the concrete situation in which we talk with each other today. Although the meaning of today remains unchanged, it will signify tomorrow that day of tomorrow. It has quite a different meaning everyday. And the same is true of course for the verb, for the past tense and the future. In our language when we use the past tense (and this is I think the other aspect of language) we go beyond the life-world, we exalt the power in the life-world, which is not in the things themselves and ...

Brand: When you say 'life-world' do you mean 'The natural world'? Or do you mean the *life-world* with these four elements I have tried to expose which is itself already a purposive construction made by philosophy?

Strauss: My *life-world* is the world of the everyday man. It's not reflective but the world as we live it.

Brand: It's the invariant structure underlying your everyday life.

Strauss: Is the *life-world* as, I think, you have in your mind the one already conceived and interpreted by philosophy?

Brand: No, not really. It has aspects of being concrete and at the same time being interpreted but finally it is an invariant structure. It is not a pure structure but it can exist only in and through its variations. We can thus find it.

Strauss: But although we speak in time and use the temporal expressions, we transcend the moments of the past and future in statements which are not bound to time. Even if I were to say such a simple sentence as, "This room has a square shape", I say something which transcends the individual planks of this room. To speak about this room means to apprehend (I'm not strict now about words) but to apprehend or comprehend a unity in nature (and I would doubt that it is really a part of the physical world), because this plank here has really nothing to do directly with that wall over there. We unite these four walls which we cannot even see with one glance into the conceptual unity of one room and I think in all this we transcend the *life-world.*

Brand: But that IS the *life-world.* You're just making a contribution to the analysis of the *life-world.* What you expose *is* the *life-world.* It is what Husserl considers the *life-world.* Husserl and you are in complete agreement here.

Strauss: I agree with you.

The same is true, and I can give another example when we speak, as we always do, about 'today'. Nobody has ever experienced the day as such. Whenever we talk about 'today' we mean it from a given point at this moment, in the morning, at noon or at night. But we are able to conceive this extension of sunrise and sunset of the day as one unity and through language we are able even to grasp it, grasp something, comprehend something which (perhaps not only in the *life-world* but in nature) is far beyond, transcending the moment. So that I think I agree with the point that the *life-world* is already transgressing nature.

Brand: Absolutely.

Kockelmans: I do not at all disagree in any sense with the paper of Dr. Brand. But I have a doubt about the first sentence of the paper: "The *life-world* is the central theme in Husserl's thought". According to Dr. Brand I will admit immediately that it is true. In his book, *Ich, Zeit und Welt*, he has put forth a similar view. But what bothers me is this: If we take the *Ideas* or *What is Philosophy*, *The Cartesian Meditations* and even *Experience and Judgment* I believe that Husserl was always very proud of the other invention or perhaps 'discovery' he had made, that wonder of all wonders: the transcendental subjectivity. The subjectivity seems to be thought of in these works much more than the life-world. I do not deny that, especially in the later works, from the later period of Husserl's life, the life-world was a very important idea. But, if this sentence of Brand's is true, in the way it stands here, then not only do the other works of Husserl suggest something which after all would not appear to be the case but I'm also concerned that from the beginning, great commentators like Fink, Landgrebe, Ricoeur and many other people who have written about Husserl would have completely and wholeheartedly misunderstood it.

Finally one wonders what would be the point that people like Heidegger, Merleau-Ponty and Sartre are trying to make when they start talking about being-in-the-world instead of the *transcendental subjectivity*? These problems are somehow connected.

Brand: The *life-world* is the central theme of Husserl's thought from the beginning on *in a teleogical way*. I try to show how in a certain way it is present in the motto, 'zu den Sachen selbst'. Finally it comes out very clear – this does not alter at all Husserl's other writings or his commentators because one may encompass with this concept the totality of Husserl's work. It's one way of interpreting it. There are other ways of course. I don't see any difficulty there. But Merleau-Ponty, I would say, sees this question the way I see it because, for instance, he very clearly says in his preface to the *Phenomenology of Perception* that the whole of *Being and Time* is only an explicitation of Husserl's *Lebenswelt*. When you read his books, especially *The Visible and the Invisible* it is absolutely clear that he sees it that way.

I do not deny that Husserl can be misunderstood. Ricoeur once said something to this effect: Husserl was a man who in a masterful way did certain things and then in an equally masterful way denied them and crossed them out; the history of phenomenology is a history of heresies from Husserl.

Kockelmans: I personally found it was a very strange statement that the *life-world* is the central theme of Husserl's thought and that probably it is to be understood teleologically. For after having read the *Crisis*, I read much later the *Phenomenological Psychology* in which to my knowledge in his published works....

Brand: The word '*life-world*' appears for the first time in 1913.

Kockelmans: I am not talking about the appearance of the word, but Husserl attempted in the *Phenomenological Psychology* to indicate a very brief outline of what an *ontology* of the *life-world* could mean, what it could consist in. I would really recommend you to read this if only to see how difficult it was for Husserl to find his way in this matter because that conception of *life-world* is in my view certainly unacceptable. In *Crisis*, in many instances, this very primitive idea of the *life-world* is highly transcended. If a great philosopher around 1925 still is struggling with the very same problem of the constitution of the world and is able to say only that space and time and causality are the basic formal structures of *such* a world, and does not really transcend this picture in any great sense, then I still feel somehow reluctant even to admit the teleological idea.

Gadamer: I have three small remarks. One is an addition to your dialogue with Dr. Kockelmans. I think we can give the answer to his question by saying that only by starting with the problematic of the *life-world* could Husserl claim that his transcendental *ego* was a real universal starting point. I think it is quite clear that although the term 'life-world' came late into his thinking, Husserl's claim nevertheless is that he did not limit his own thinking by the sciences. Whereas in the Kantian approach the world was only the object of scientific cognition which was founded by the transcendental ego and by the categories, Husserl's great innovation was to go behind these limits; therefore the discovery of the *life-world* gives the possibility to make, or to form, or to show how the transcendental ego itself is universally constituted. Then some remarks to your own expose. I found it very helpful, but I think, first, you must be a little more cautious or you must concede to me that apodicticity is a predicate of evidence. It is an element of Husserl's theory of evidence. Then, I have a doubt about your claim that the only apodicticity of cognition that exists is that concerning the *ego* in the world. I think it would be right to see the *ego* as a primordial factor constituting the world.

Brand: Where I said 'the world-experience-life-knowing-itself-as-such', I would agree with you. The apodicticity belongs to that *ego* reflecting upon itself.

Gadamer: Of course, it is the functioning *ego*; his own power to function forms the content of this apodicticity. Therefore I think that that point was always ambiguous and equivocal in Husserl's thought. In our discussion of the morning 'self-givenness' was only another expression for 'constituted self-givenness'. Self-givenness is not a 'natural' feature of what there is that could be found without any preparation. Constitution is the inverse form of reduction and this means that by constitution what is intended receives the character of self-givenness. As you remember, that is Husserl's concept of phenomena.

Brand: If we consider constitution not as a construction, then of course, we do agree. I try to smuggle in, I confess, the concept of 'institution' because when we have first acquired a certain truth then it's more an 'institution' than a constitution. But that is a question of terms.

Gadamer: I think that the only important point is that in the discussion of Ingarden I find that self-givenness and constitution cannot be opposed. The only aim of constitution is to have the result of the evidence of self-givenness, to avoid all the possible mistakes and smuggling in of none-vident presuppositions. Going back to the self-constitution of the *ego* as constituting, we must admit that in posing the ego as a constituting agency Husserl has opened the great realm of evidences which is transcendental phenomenology. A second question – and here I am somewhat hesitant. It is a problem of language. On the last page of your paper you mention that it is selfevident that the position of the *life-world* is not separable from the linguistic approach to the world. This point is undoubtedly right. It is remarkable that Husserl himself does not develop it. He describes the subjectively related *life-worlds*. I think all our speaking about the *life-world* misses the point if we don't realize that for Husserl it is the relativity of *a manifold of life-worlds* and not the existence of a common and universal horizon of life-world which is the problem. The relativity of the life-world is the theme of the *Crisis*. Why is Husserl so reserved in the point of linguistic approach? My question is mainly philosophical and not one of the interpretation of Husserl. It may well be that the real and authentic importance of language, as a basic attitude and form of experience, depends upon renouncing the attempt to found all of our knowledge in the transcendental *ego*. Could it be that only in going beyond this concept of a transcendentally constituting *ego* are we aware of the primordial and basic function of language?

Brand: I agree with you. Husserl always wavered. He was in '*Schwebezustand*' and he even often mentions that he is in '*Schwebe*'. I think that is the main reason. There is another reason too. Husserl was conscious of starting something and he wanted to leave things to be done by others.

PART II

PHENOMENOLOGY AND HERMENEUTICS

H. G. GADAMER

THE SCIENCE OF THE LIFE-WORLD

Although the science of the life-world is the most discussed part of the doctrine of the late Husserl, there seems to be a permanent necessity to examine what is novel in this doctrine. Does it open new paths of investigation or is it only a new and clearer outline of the programmatic intentions phenomenology had from the beginning? It is a specific character of the Husserlian style of thinking to correct himself and to repeat himself in such a manner that they are indistinguishable. Therefore the introduction of the concept of *Lebenswelt* wavers between a mere description of the authentic approach which Husserl chose for his phenomenological investigation, and which distinguished him and his philosophical interest from the dominant Neo-Kantian and positivistic scientism, and a new self-criticism which may not attain the great goal for which Husserl longed throughout all his work, namely to found philosophy as a rigorous science, but which would make this goal appear attainable. To this we may add that the self-interpretation of Husserl is by no means a trustworthy canon for the understanding of his meaning, because his self-interpretation oscillates between continually renewed self-criticism and teleological self-interpretation, to such a degree that, for example, Husserl pretends that his own psychological approach in the *Philosophy of Arithmetic* would be the prefiguration of phenomenological investigation of what he calls constitution. Further, regarding the critical objections raised by Scheler and Heidegger against his foundation of phenomenology, he repeated with perseverance that they did not understand the authentic meaning of transcendental reduction. That is the only fixed point in his self-interpretation. Therefore, it seems to me that it is a methodical demand to interpret the doctrine of the *Lebenswelt* against the background of this permanent complaint which he raises against his philosophical contemporaries.

On the other hand, one must concede that the new word, *Lebenswelt* (life-world), one of the very few new words proposed by a philosopher which has had a success of its own in ordinary language, has a very

Tymieniecka (ed.), Analecta Husserliana, Vol. II, 173–185.

broad meaning which expresses very well the specific character of Husserlian thinking over against the widespread philosophies of Neo-Kantianism and positivism. It does not restrict the task of philosophy to the foundations of science, but extends it to the wide field of experience of daily life. It is quite understandable that this wider sense of the concept of life-world proposed in Husserl's later work would be accepted and acclaimed by many scholars who by no means intended to follow him on his path toward transcendental reduction. No, in perfect opposition to it, they used the slogan of life-world in the sense of turning away from Cartesianism, or at least they wanted to legitimate their own investigations as independent analyses of the social and historical world in the context of a phenomenological anthropology. This is not unjustified, in so far as Husserl himself acknowledged that it is a genuine task to work out an ontology of the life-world. To do so does not necessarily require that one follow Husserl's own way of transcendental phenomenology and transcendental reduction. As a matter of fact, it is not only the intention of Husserl's phenomenology to go back behind the whole of scientific experience to the simple phenomenological data, like sense-perception, or practical experience, and to legitimate claims of validity over against the sciences – it is a point of very fundamental interest that the life-world claims its own phenomenal legitimacy by going back to the constitutive intentionalities: for, this field of themes represents a mode of givenness, or, better, a realm of original modes of givenness. It was a great insight of Husserl to see that it was meaningless to reduce these modes of givenness to a system of scientific truth lying behind them, like the ontology of the objective world. It was one of the first steps which Husserl made, already in his *Philosophy of Arithmetic*, to recognize in the example of the symbolic number that there exist no monolithic and dogmatic concepts of givenness at all. What could the concept of givenness mean, for example, in the case of infinite numbers, which nevertheless have a well determined mathematical meaning? In the same sense, the modes of givenness of the life-world represent in themselves objects of intentionality and must be founded in their character of phenomenon by constitutive analysis. Such an analysis which follows the correlation between intentional object and intentional act, and determines the meaning of the intended by the intention, includes necessarily that the stream of consciousness containing all the manifold of intentional acts, willed as well as anonymous, constitutes also the modes of givenness of the life-world and not only the objectivity of scientific experience. It is thus important to observe that even the first step of Husserl's investigations went beyond the objects of experience in the sense of the science of facts.

Only the analysis of this correlation between intentional act and intentional object is able to disclose the naiveté of an ontology of the world based on the objectivism of mathematical natural science. Such an ontology of the world misses the decisive question of the idealizations involved in its method and therefore floats in the air, Husserl says. But on the other side, the intentions within the life-world-horizon also involve idealizations and include therefore intentional acts cooperating in their building-up. It was already the program of the correlative constitutional analysis in the *Logical Investigations*, not only to thematize the constitutive intentionalities correlated to the objects but also to work out the basic structure of consciousness as a whole, which by its own streaming temporality builds up all validity of objects. In this sense, the thematic of the life-world is not completely new, but is manifestly in view where Husserl investigates the deepest layer of phenomenological constitution research, the self-constitution of the inner consciousness of temporality – investigations going back to a time before the first programmatization of the idea of transcendental phenomenology.

But after, when the philosophical project of transcendental reduction integrated all the Husserlian investigations in one systematic philosophical framework, and when philosophy was programmatized as a rigorous science by starting from a last-founding (ultimate) apodictic evidence, it is a new problem to examine the procedure of transcendental reduction with regard to its rigorousness and its unambiguity, and at this moment the problem of life-world will be important – and at this moment the expression 'life-world' was created. It is well known that Husserl in the *Ideas* interpreted the new style of scientific philosophy founded in the analysis of transcendental subjectivity in the sense that now for the first time an idealism of a really transcendental character would be achieved. To demonstrate it, he followed the Cartesian way of a universal methodical doubt resulting in the apodictic evidence of the *ego cogito*. Certainly, this Husserlian Cartesianism was very far from the authentic motives of Descartes, and Husserl was fully conscious of it. In the first place, the universal doubt of the world would not claim in the eyes of

Descartes that it resulted in the foundation of all knowledge in a new philosophical evidence, but that it served to legitimate the mathematical natural sciences as authentic knowledge of the objective world. Therefore the ego, that *fundamentus inconcussum* which resists the most universal doubt, was by no means the transcendental ego, by which Husserl wanted to build up the order of evidences and to found philosophy as pure phenomenology. The Cartesian ego was only a little piece of the world which remained and by which in quite different theological ways the cognition of the world could be legitimated. Nevertheless Husserl recognized by the example of Descartes' consideration of doubt the universality of suspending all validity of the world-belief. Analogically, he developed his own doctrine of the transcendental *epoche*, which brackets and suspends all positing of reality and also the validity of the eidetic sciences, like mathematics. Undisturbed by the claims of modern science the phenomena, the modes of givenness as such, should be demonstrated and acknowledged in their ontological status by investigating their phenomenological constitution.

The enormous field of investigations outlined in its full methodical autonomy in the *Ideas* claimed to be all-embracing. With the principle of transcendental reduction Husserl was going back behind all the usual philosophical debates about standpoints and worldviews. He felt misunderstood when people expected from his own analysis of transcendental subjectivity a decision about standpoint-questions or even imagined to find realistic elements, for example, in his doctrine of the hyletic data grounding sense-perception. I think he was correct. With the same radicality he combated the confusion between phenomenology and psychology because all sciences of facts were excluded from the inner field of transcendental phenomenology. Only, starting from this basis of transcendental phenomenology they regained a new legitimacy in the form of eidetic sciences of a new style. Founded on transcendental subjectivity, philosophy had obtained a basis broad enough to resolve in principle all imaginable philosophical problems. That was the claim Husserl made in the *Ideas* and retained and repeated in his last work, the so-called *Crisis*.

Certainly Husserl recognized the enormous difficulty of retaining the transcendental attitude firmly and unerringly. He not only objected to his adversaries that they were falling back into an uncritical natural attitude or that they didn't understand the radicality of the transcendental reduction, but he also acknowledged that the danger of a falling back into the natural attitude is always present for everyone. Therefore he discussed untiringly in many of his phenomenological investigations the problem of intersubjectivity: how is it imaginable to grasp the constitution of an alter ego by the transcendental ego, when this alter ego has in itself the same character as the transcendental ego? But his solution of the difficulties is ultimately the same and an unambiguous one: only on the basic ground of transcendental subjectivity, only in the radical solitude of the transcendental ego, only from the standpoint of a transcendental solipsism can one legitimate the concept of we" that is the experience of other subjects each of which is in itself and from its own '*ego*'. To be sure, the problem of intersubjectivity was considered not only by Husserl but also by his school as a crucial one, and many scholars treating this problem postulated a break-through and a rejection of the transcendental attitude as a whole. Some of them even pretended that Husserl himself had already done so. But Husserl was right in contending that this is a wrong appearance. It is only from the side of the transcendental ego and its constitutive efforts that the problem of intersubjectivity can be resolved, that the 'like *ego*', the intentionality of the so-called transcendental *Einfühlung*, can be interpreted – it is curious enough that only such an *Auffassungsform* should authorize the mere perceptual object, an extended corporeal thing, to be an alter ego. Husserl conceded expressly that the problem of intersubjectivity was not respected enough in his own following of the Cartesian way of doubting. He therefore dedicated numerous papers to invalidating all of the objections against his theory of transcendental reduction following from this point. But that this was no real danger to his own foundation of philosophy as a rigorous science seemed to him absolutely certain.

It is the same with the problem of the life-world, but nevertheless this problem, even less than the problem of intersubjectivity, loses nothing of its permanent actuality in the later works of Husserl. What was the actual problem? Husserl recognized this problem in a double form: once in the form of self-criticism directed against his own description of the transcendental reduction in the *Ideas*, and also in another form, that the problem of the life-world has special importance in the transcendental foundation of philosophy which remains always actual. Certainly he will contend at the end that the entanglement between life-world and transcendental reduction can be resolved and will only appear as irresolvable as long as the way of reduction is not worked out exactly, but he recognized more and more that for the working out of the reduction the problem of the life-world involves special difficulties and paradoxes.

On this point the rigorous consideration of the historical world was unavoidable for him. He repeated, what in the beginning, in the outline of his philosophical program, the *Logos*-article of 1911, was one of his main points. There he recognized besides the danger of a naive and unreflective naturalism the other danger of the *Weltanschauungsphilosophie*, the danger of impatient desire for hasty philosophical anticipations and as a consequence scepsis and tiredness. The heavy shattering of the First World War – it meant for Husserl not only the loss of one of his sons but also the separation of his home country, Maehren, from the older Habsburgian empire – had a personal impact on him, and other culture-wide events, for example, the breakdown of popular idealism the rise of dialectical theology, and the new thinking of Scheler, Jaspers, and Heidegger, claimed his attention. The deep earnestness grounding his own simple and innocent personality engaged him fully, then and forever, with the unique question: how can I become a sincere philosopher? A philosopher for him was a self-thinker, a man who wanted to give an ultimate account for all his thoughts and convictions, beginning with the basic problems of science (Husserl was a mathematician) but expanding over all the problem of human life, and for whom every uncontrolled and unproven conviction must appear as a loss of his own inner confidence in himself. It was in the context of this lifelong questioning for ultimate self-legitimacy that the demons of historicism and the scepsis involved in it continued to disquiet him. He hoped to find by the clarification of the life-world a new and ultimate clarity and the beginning of a new epoch of sincerity and reasonability transforming all coming generations of human beings.

He acknowledged his own mistake, which he had committed in the construction of philosophy as rigorous science, that is, as the coherent theory of the transcendental reduction, and he felt that it was a consequence of his own mistake that the demands of *Weltanschauung* overflowed the dams of scientific sobriety of thinking. In his *Ideas* he believed that he arrived on the field of pure subjectivity and apodictic evidence by the bracketing of all posited reality, of all objects of science. He didn't realize that after the suspension of all objects in the world by the suspension of the general thesis of reality, the belief in the world as such, in the horizon-intentionality of the world anteceding every positing of entities, was not suspended also – and that meant that in the investigation of constitution, to claim to build up every objective validity by starting from the transcendental subjectivity might permit uncontrolled prejudices to slip in. It was not really a pedantic desire for absolute precision and rigorousness which controlled the way of reduction and denounced the mistake – this mistake might be a fatal one. For, the horizon of the life-world into which life goes with full self-confidence and which is never an object by itself, represents a cardinal problem for the philosopher as such. Clearly, he himself as the one who reflects, in the transcendental sense of the word, is surrounded by this horizon of world with complete self-evidence. One look at the fields of investigation in ethnology or history informs us that spaces and times bear highly different life-worlds in which highly different things pass as undoubted self-evidence.

Of course, it seems to be the way of science to recognize the objective facts and the objective laws and to make them controllable and at the disposal of everyone. There alone seems to be truth. But the way of scientific investigation follows quite different aims posed by willed decisions which go beyond the natural self-givenness of the life-world and involve a specific idealization or mathematical description of the world. In this respect the bracketing of the scientific cognition of facts by the *epoche* presupposes the validity of the life-world dimension of pure self-givenness; but "now we are embarrassed as to what else could be claimed scientifically as established for everyone and for all time," writes Husserl in the *Crisis*. The word 'now', of course, means here not 'now after the first *epoche*', but 'now after we have recognized the manifold and the relativity of the life-worlds and their precedence for every experience of objects'. In this sense, the thematic insight in the *Bodengeltung* of the life-world, which as such was not perfectly new, involves nevertheless the rise of new problems. Certainly the subjective relativity of the worlds of life may be analyzed in its universal structure and the *a priori* of a life-world, a universal *a priori*, can be disclosed as one which by no means is the objective *a priori* of the traditional metaphysics or the sciences but which grounds all the sciences because it precedes as *Bodengeltung*, as basic validity, every science, including logic. Is it not a new Fundament for all truth and does it not displace the transcendental *Ego*?

Reading the description which Husserl gives of the new role of transcendental reduction in his *Crisis*-article, the only explicit summary of it, one is astonished that the old well-known problems and insights of the earlier program return although in changed dress. The analysis of the *a priori* of the life-world and its methodical function describes it as a change of attitude which is exactly the change described as the transcendental *epoche* in the *Ideas*. (In passing it should be noted that the title of paragraph 38, whoever may be the author, is erroneous. There are not two fundamental modes of thematizing the life-world but the unthematic validity of the horizon of life-world as opposed to the thematization of it by a universal turn of interest, *Interessenwendung*.) What is new in the new description and differs from the older Cartesian way of graded doubting or graded reduction is that the turn of attitude is a total and unique act. Every graded bracketing of values would occupy the universal ground only by other positions but could not suspend it.

Therefore it is quite understandable that to thematize or to bracket the basic validity of life-world as a new moment in the transcendental investigation of the autonomous realm of intentionalities, the very universal structure of the manifold worlds of life with their changing horizons of the world or, as we might call it, the *eidos* life-world. In this sense it is a new way of reduction, since the transcendental ego in which the reduction results in this form will be able to resolve complications which appeared insolvable before. It is quite right that the universal horizon of life-world embraces by itself the transcendental subjectivity when the horizon articulates itself in its own subjective relative structure by going-into the world. The world is the unique world, an essential singular, yet really a highly variable and a manifold of modes of givenness. *The* world is never self-given, it is only the pole of objectivity. In going-in further and further, and disclosing the field of life-world experience, the direction towards this pole is maintained and hence the pole remains functional. The *epoche*, that is, the willed thematization of it, discovers the transcendental subject-object correlation in its purity. But what is involved here with all its mysterious implications? What can be the world constitutive subjectivity which is itself a part of this world? I think it is we, the men for whom this world is valid. We are many egos among whom I am one ego. It is necessary to analyze this dimension of intersubjective experience of the world by constitutive clarification. One can investigate all these modes of givenness; for example, what is an acquaintance? What is there involved and presupposed for a common horizon of life-world? Like: present for others and others present for him, a circle of acquaintances with the open possibility to expand this circle, the internal levels of nearer and more distant acquaintances, of friends and enemies and foreigners, and further the anonymous horizon of the society with all its patterns and rules where one is acquainted and which is nevertheless quite a different thing from the circle of our acquaintances. One sees that it is the world itself which is concretized by such intersubjective experiences: that, and not an objective world of mathematically describable *a priori*, is *the* world. Nevertheless I may concede all that and realize that every ego, like myself, possesses the same capacity by freely willing to change his attitude and to work out the *epoche* and to investigate the transcendental a priori of correlations – I may concede that the transcendental subjectivity allows and postulates even a transcendental community – still we cannot avoid the paradox that the world-constitutive subjectivity, though it may be a manifold of such constitutive subjectivities, is a part of the world constituted by these subjectivities and therefore brings into play all the specific subjective relative characters of the personal horizon which distinguishes for example the negroes of the Congo or the Chinese farmers from Professor Husserl. How, considering the unsuspendable specification of the pre-given horizons of life-world, is science possible at all?

Following the text of the Crisis-article, the solution of the problem explained there seems to be neither new nor problematic – it is the old answer that the transcendental functioning subjectivity of the ego, by which belief in the world is constituted, may not be confused with the ego which is a part of the world and is constituted with all its experiences of the world. That which constitutes all the forms of world, for example, the world of dreams, the world of children, the world of beasts, the historical world, the problems of birth and death, the problem of the sexes, is not one ego beside others and is not in the world as one of its parts but functions only as the ultimate functioning ego in all its absolute apodicticity but also in its unique philosophical solitude.

But it is not the text of the article alone which is here the source – and this source is unclear enough, in so far as the discussion of newly arising difficulties always interrupts the stream of thinking. Besides this text (which was never completed) there exists a lot of tractates and notes of the same epoch which give the right idea of what is the point for Husserl and what stimulates him further and further. But the composition of the Crisis-article itself confirms it also. Husserl speaks of historical investigations which became unavoidable because of the great variety of subjective-relative world-horizons, involved in the *a priori* of the life-world. The historical investigation which he himself undertakes concerns the origins of scientific objectivism in the physics of Galileo, who had full consciousness of the specific problematic and idealization grounding natural sciences. Husserl's analysis is a genetic ideal-typical construction and treats Descartes, Hume, and Kant under the norm of definitive transcendental phenomenology and especially of transcendental life-world. It may be that the overlooking of the life-world and the radicality missing in Kant's transcendentalism and in the thinking of the Neo-Kantians comes from the narrowness of the concept of scientific experience basic for them, but does it follow from this that the new thematic of the life-world cannot be investigated apart from historical investigations? This is by no means the case when the *eidos* life-world is described as a philosophical task. Husserl speaks there about the totality of all imaginable possibilities which are included in the horizon or into which the horizon can be resolved by explaining it. Clearly, that is the problem of the idealization involved in the world of life. One may concede that one cannot enter into the manifold of life-world horizons without beginning with the present life-world horizon in which one lives and that one must interpret the world in its concrete infinite historicity. But when one reads, "All possible worlds are variants of the world which is valid for us," or that the world is built up only as a "perpetually streaming" constituted something in the sense of an infinite idea, this means certainly that the world of life is not the idea of the world itself which constructs itself in a continually streaming change by a continuous series of corrections, nor can it ever be the object of a science in the traditional sense of objective science. It is precisely this unexamined acknowledgement of objective science, which was a mistake of former philosophers, that is understandable only in light of the concept of life-world which grounds this idea of science. The Lebenswelt is in principle an intuitively given world, perhaps in the flowing and fluctuating of its streaming horizon, while the world of science has rather a symbolic givenness of a logical substruction which can no more be given by itself than the infinite series of numbers.

The life-world therefore, has a universal structure of a finite subjective relative world with indeterminate open horizons. Starting from one's own finite life-world and the memory of the well-determined variations of our history since the Greek epoch, the validity of this life-world is disclosed by limiting the objective world – *a priori* of the sciences. Certainly. But can it be doubted that the *a priori* of the life-world is available also in the old phenomenological fashion by varying and trans-stylizing and describing its *eidos*? The very self-interpretation of the historical way which Husserl uses in his analysis of Galileo, written by Husserl in 1936, confirms that beginning with one's own life-world represents the only access to the *aeterna veritas*. To disentangle the paradoxes coming from the continuing validity of the ground of *Lebenswelt* for everyone and at every time, it is necessary only to retain firmly the transcendental meaning of the *ego* as a pure working ego. The text of the article clearly proves it.

Nevertheless, the opposition to objective science which determines the way of human civilization from the time of the Greeks is not arbitrary, but present in our own world-horizon. In this horizon alone, the naive dogmatism of science can be understood by historical exploration of its origin and its limits of validity, and the prejudices involved in this form of science can be overcome. Rigorous science in the sense which the young Husserl professes and never revokes may be science, but science of a new style, a universal account and self-examination based on a change of attitude, and certainly not derived from the idea of objective science but from the situational cognition involved in the direct form of life-interests. This is nothing new. But one must concede that in a certain sense Husserl's own lifelong task changed at the moment when he entered the way of historical self-clarification, thematizing the personal life-world presuppositions of philosophizing. That is the way of the *Crisis*-article. Nevertheless, without any doubt this new way leads to the old end of transcendental phenomenology, which is based in the transcendental ego (leaving aside the self-constitution of this ego). This alone is rigorous science, clearly not in the traditional sence of science or in the sense of traditional philosophy, but in the sense of a new will to live, "to become acquainted with oneself in one's former and predetermined future being." But that is the old aim of ultimate absolute self-cognition. One hears there the old tones. The world of life in all its flexibility and relativity can be the theme of a universal science.

Regarding all that – and in the appendices and in the planned preface for the *Crisis*-book it is well-confirmed – I contend that there were only external, contemporary reasons why Husserl, after the general discovery of the *a priori* of the life-world and its historicity, joined this new way of transcendental reduction with historical considerations. They are means to the goal of opposing the spirit of the time (*Zeitgeist*) and the historical scepsis entailed by it. The names of Scheler and of Heidegger are mentioned and many reflections are dedicated to the right use of historical studies. The result, however, cannot be different from what is reached by other ways of Husserlian thinking: the transcendental phenomenology, and the transformation of philosophy into phenomenology as the only meaning of the history of philosophy.

To summarize all the texts published in the sixth volume, Husserl would agree with Oskar Becker, who formulated it long ago, namely, that the contribution of *Being and Time* to the problematic was restricted to the 'fixation of horizons' of historical existence left open by Husserl himself. And the claim of rigorous science remains untouched. Self-consideration achieves itself in a knowledge for everyone, and Husserl adds proudly: in a universal praxis of humanity which is ready to be led consciously by phenomenology.

Really? Is this the solution for the problems of modern life and modern civilization, which demand the bridging of the increasing gap between practical and political prudence and the anonymous validity of science? Can phenomenology really prescribe – or recommend at least – the ways of men in the life-world by simply the example of the philospher? He finds his legitimation through the insight into the complicated relations of practical knowledge and the ambitious enterprise of science. It was the ultimate aim of the life-long thinking of Husserl to become a sincere philosopher. Perhaps he succeeded personally in overcoming the alternate complications of the basic reality of life-world and the speculative principle of the transcendental ego. But this knowledge will be a science, although in the new style of transcendental phenomenology. Nobody can doubt that here the tension between the running flux of time and life and the philosophical claim of eternal truth remains.

It was the ontological inversion which Heidegger proposed, which lent a new prestige to the existential process and delivered philosophical investigation from the vain pretension of going beyond scientific knowledge by an *a priori*-science of essences. Philosophical thinking is not science at all. It clarifies not only the conceptuality of traditional philosophy and its basic function for all the sciences, but opens the mind for the inner mystery of language and the world view involved by it. Thematizing our fundamental concepts and experiences, like being, time and space, life and death, mind and body, man and God, nature and truth, it discloses the scientific approach to the world in its own limited validity and denounces the tyranny of willing by the means of science and technology as the clash of reason with itself. There is no claim of definitive knowledge, with the exception of one: the acknowledgement of the finitude of human being in itself. Grasping the chances involved in it for infinite self-correction, humanity goes further in the permanent dialogue of one with the other even about the life-world.

K. KUYPERS

THE SCIENCES OF MAN AND THE THEORY OF HUSSERL'S TWO ATTITUDES*

The relationship between Husserl and Dilthey underwent a serious change during the later period of Husserl's life. Whereas he opposed Dilthey's historicism at first in his *Logos* article in a trenchant requisitory, he later fully ackowledged the significance of historicity in culture and human existence. But he also recognised the significance of Dilthey's theory of the humanities, albeit with certain restrictions and criticisms on the lack of precise formulations and methodically certain solutions.

This acknowledgment of Dilthey's merits is attended by the acknowledgment that the development of a general science of the mind and the acquiring of insight into the relationship of phenomenology to this science of the mind would mean nothing less than a new access into phenomenology, which, in his opinion, would be more meaningful and far reaching than from psychology.

As phenomenology based its existence on the descriptive psychology of Brentano, and as Husserl has retained his faith in this origin despite everything, this acknowledgment of Dilthey's endeavour is the more remarkable. However, it gives rise to the question how meaningful this new way really is.

It is closely bound up with the conception which is based on a typically German tradition: a dualism in the theory of science, viz. the distinction between the natural sciences and the humanities, which finds its origins in Hegel, but which, especially by Dilthey's contributions, albeit with some hesitation on his part, particularly with regard to the nomenclature, has led to a fixed terminology, which by the way is difficult to represent in other languages. It must be conceded that it belongs to the French and Anglo-Saxon tradition to make a distinction between *science* on the one hand, and *arts* and *humanities* on the other. This does not create any dualism in the theory of science if the arts and humanities are simply excluded by virtue of the notion of science operated with. Yet, endeavours are made to include psychology and the social sciences both in the theory of science and in the practice of its cultivation, and to make them meet the

Tymieniecka (ed.), Analecta Husserliana, Vol. II, 186–195.

recognised norms and methodical standards. However, in the German theory of science a fundamental distinction is made between the natural sciences on the one hand, and all the other sciences, covering human consciousness, behaviour and achievements on the other. These are summarised by the term '*Geisteswissenschaften*'. The strongest opposition against this dualism comes from the *Unity of Science Movement* and finds its most precise formulation in the physicalism of Carnap. Phenomenology in general and Husserl in particular have sided with this dualism. It can even be said that Husserl has tried to produce a theoretically stronger and philosophically corrected version of Dithey's theory by stating phenomenology as the universal science of the mind (person) or, more general, of subjectivity.

Next to the acknowledgment of the significance of historicity as being of essential moment, there is also the adoption and continuation of the task to give a philosophical basis to what in the German theory of science is usually called '*Geisteswissenschaften*'. In this respect the criticism exerted by Husserl on Dilthey (such as the manner in which Dilthey has provided this basis by distinguishing between description and explanation, justifiedly rejected by Husserl) is of minor importance. To us the most important aspect seems to be that with this acquirement a relatively important shift in the conception and objective of phenomenology takes place. This shift consists of the acquirement of the idea of a general science of the mind as a new, better and further reaching way in phenomenology. Only via the theory of reductions had the idea of phenomenology, originated from the psychology of Brentano, been able to develop into a universal, viz. transcendental philosophy. We cannot expound on the difficulties embodied in this and shall only point out that while Husserl radically rejected and resisted with Frege the psychologism of his time, he was nevertheless accused, among others by Carnap, of never having grown out of psychologism himself. In the new conception of phenomenology, in which it is accommodated in a universal science of the mind, it is advisable to distinguish three moments. Firstly the anthropological distinction between body (*Körper*, *Leib*), psyche and spirit (person), then the theory of two different attitudes that can be assumed, leading to a two-fold sort of '*Gegenstände*' and sciences, and in the third place the metaphysical background and framework in which the theory of the sciences of the mind can be placed with regard to the philosophy of the mind or the absolute consciousness of the mind.

Husserl says frankly in his introduction to the *Konstitution der geistigen Welt* (*Ideas*, Vol. II, *Husserliana*, Vol. IV, p. 172) that he is concerned with metaphysical and theoretical scientific distinctions, particularly with those between psyche (*Seele*) and spirit (*Geist*) and that the elucidation of these distinctions decides our entire philosophy of life basically.

This shows very clearly that not only theoretical scientific considerations are to be dealt with, but also that the primeval metaphysical polarity of nature and spirit constitute the factual theme and background of the theoretical science controversy. The classification of natural sciences and humanities is finally governed by a philosophy of the spirit. The fact that this has assumed such an important role in the German theory of science and has become a relatively stable tradition may be ascribed to the influence of Hegel. The renewed great importance of Husserl's anthropological tripartition in body (*Körper*, *Leib*), psyche (*Seele*) and spirit (*Geist*), this in contrast with the Anglo-Saxon and Cartesian bipartition, must not be considered simply as a return to Aristotle and Hegel, although Hegel's metaphysical tradition makes itself undeniably felt in Husserl's writings. This is rather based on an attempt to attribute to the psyche (*Seele*) as a reality a mediatory role and function in the diametrical antithesis between nature (body, *Körper*) and spirit (*Geist*).

Although the three moments mentioned in the constitution theory of Husserl are closely connected and final evaluation can only be stated in this framework, we should like to restrict ourselves here to Husserl's theory of the two attitudes (*Einstellungen*), the naturalist and the personalist or humanitarian one, which is of extreme importance for the hermeneutics and theory of science of our time, because of its connection with the increasing tendency to arrive at objectivation and formalisation and finally even at mathematisation in the humanities as well. This is conceived as a way not only leading to preciseness, the ideal of all sciences, but also to genuine science and thus to a unification of all scientific methods and techniques. This advance in objectivation and formalisation in psychology, sociology, linguistics and economics is considered as a step forward in its elimination of less exact methods and ways of thinking.

The position of the theory of the humanities is generally defensive and conservative as regards the current prosecution of certain sciences, especially the historical ones, whereas the opponent, known in its extreme form as the protagonist of physicalism, whether or not connected with the name of Carnap, leads the offensive and in its claim to unity of science aims to be an anticipation of a possible future development of all empirical sciences towards the model of natural science.

What does Husserl mean with the two 'attitudes'? Shortly it amounts to this. His basis is a natural attitude (*Einstellung*) towards the world as it is given before, which continually manifests itself as '*Umwelt*'. This term is not identical with social environment, but the personal, subjective way the world is given to each of us can be conceived differently for anyone concerned. This is the '*life-world*', i.e. the world in which we live, conceiving nature as it appears to us. This gives rise to a two-fold scientific attitude, the natural scientific one and the personalist or humanitarian one. The first attitude is concerned with the examination and definition of the physical nature '*an sich*' regarding physical nature. This occurs by idealisation i.e. by designing models and formalisations on a mathematical basis. But the entire world can also be conceived as nature. In that case the conception of nature is enlarged thus covering what the Greeks comprised under *physis*: the entirety of what exists, the '*All-Natur*'. Later this was limited to what Husserl calls: physical nature. This is correlated to a universal inductive experience and the claim of a universal natural science, in which all phenomena are examined without exception on the basis of inductive experience. This also includes the acknowlegdment of the right of unlimited application of the principle of objectivation and formalisation respectively idealisation, also with regard to the spiritual atmosphere and the spiritual relations of the subjects. Thus naturalist prejudice, as Husserl phrases it, is linked to this, the prejudice being that nature is also the essential essence of the subject. Naturalism considers man exclusively as being only completely filled extension and therefore the world as such as no more than expanded nature. In the attitude opposed to this, the humanitarian or personalist one, just these spiritual relations demarcate the field of examination. Man remains the point of departure as a person, such as it is the case in the natural attitude, and another is nature such as it appears (*erscheint*) within this attitude. This is the crucial difference between this kind of sciences and the natural sciences. The *life-world* of persons eludes the naturalist attitude in spite of the fact that it purports to examine the totality of realities, and even the most subtle natural science theory is not able to grasp this. The critical question can now be asked which sciences are supposed to be comprised under the title 'humanities'. Generally there is a tendency to equate them with the science of man as a human being (human consciousness, behaviour, achievements, attitudes, culture). Both Dilthey and Husserl are especially concerned with the position of psychology within the humanities. In accordance with opinions of the end of the 19th and the beginning of the 20th century, psychology was considered a basic science for all sciences about man, and a science was postulated to fulfill this demand and to function suitably as the basis for the humanities, indicated by Husserl as descriptive concrete cultural sciences. This gave rise to the recognition within psychology of two possibilities leading to two modes of research and method: a natural scientific and a humanitarian mode. This recognition of two entirely different methods and directions within one single science is remarkable as such, containing as it does the recognition of a dualism of a principal nature, not only regarding sciences in general but also within a certain science, in this case psychology.

The right of existence of both attitudes can be found back in Husserl's theory of the two attitudes in *Ideas*, Vol. II and it is especially developed in the 'Annexes' 'Annex' (from before 1930) in Volume VI of the *Husserliana* (*The Crisis*) containing the most mature phrasing and the most suggestive perspective. While basing phenomenology itself on a certain attitude and developing this idea further in his doctrine on the reductions, Husserl develops similar ideas with regard to the sciences of man.

The attitude decides what is to be the point of departure as the object, and what method is applied to it. In case the naturalistic attitude is applicable, mind and spirit are subordinated to what is considered to be part of nature. If the humanitarian scientific attitude is applicable, man or his fellow-being function as data to whom the world appears circumbiout world and who acts, thinks and evaluates in terms of the world as a natural starting point. The naturalistic attitude shares with the naturel scientific one the intention not to restrict itself to the manner of appearance, but to focus on the 'an sich' and therefore on the truths 'an sich'.

Remarkably enough Husserl in principle recognises the right of both attitudes. This is in particular valid for psycho-physics, which deals with man as a real unity with a double layer in an objective nature and which examines the relations of dependence of the psyche and spirit to the body.

This recognisance, however, is in certain ways restricted. The spirit can indeed be defined as dependent on nature, and even be naturalised itself, but this can only be done to a certain degree, since, when exceeding this degree, that which provides meaning to nature, would be lacking (*Husserliana*, Vol IV, p. 297). Nature is a field of transient relativities. In the end all of them are relative to an absolute which supports all relativities, i.e. the spirit.

It should be added that this theoretical acknowledgment of the right of a universal world or natural science, in which the universe is conceived as in one universal science deducible from axioms following the example of physics, has a drawback which decides in favour of a humanitarian attitude.

This is because Husserl connects the idea of phenomenology itself with the idea of the foundation of the humanities within a general science of the spirit. The backing of the method of the humanities as being focussed on concrete intuition of the world as it directly appears to us, the life-world as the world we live in as human beings, and thus as the basis for the humanities, receives confirmation and appreciation from Husserl. The reason for this is that phenomenology itself has adopted the description of this *life-world* as the basis for aiming at exactness, objectivity and the ability of understanding and explaining the '*An sich*' of the natural sciences.

With this defense of the humanities as pertaining to concrete intuition, phenomenology defends itself and its standpoint as taken with regard to the original right of the data of intuition, the original intuition and what is given in this intuition as the source of all validity and evidence.

This will even lead to a priority being granted to the humanities above the natural sciences, since they only can lead to philosophical depths whereas the natural sciences cannot, the depths being those of the last being (*Husserliana*, Vol. IV, p. 366). They are according to Husserl the sciences of subjectivity. Whereas the subjective aspects are regarded to be of no importance in the natural sciences, and therefore eliminated wherever possible, this subjective aspect receives from Husserl the mark of being the aspect of truth in the better sense of the word. In their original form, concentrating on perception, the humanities already constitute truth and must retain this mode of intuition in their further development. Husserl even proceeds to proclaim the proposition that the natural science can only be accommodated within the realm of philosophic truth if it is interpreted as being '*entquellend*' and deprived of the absolute meaning it pretends to have, but does not really in fact possess, according to Husserl. Its pretention to gain knowledge about the 'in itself' of nature would be withdrawn in a transcendental interpretation and finally also be understood subjectively in the transcendental sense. Natural sciences too are a community enterprise by scientists and can as such and in its aspirations towards exactness and objectivity be explained and understood only from the historical attitude ('*Wir, die Naturforschenden*').

Dilthey tried to find a descriptive psychology next to and in contrast to the natural scientific psychology as a basis for the humanities. Husserl adopts this idea and modifies his phenomenological foundations of psychology into a psychology of the person, a psychology of the personalist attitude, i.e. a universal psychology of the subjectivity leading to a transcendental phenomenology. Within this framework the objectivity of the natural scientific attitude, directed on an investigation of the '*An sich*' of nature and the world, is revoked and reinterpreted as issuing from and based on the subjectivity of the investigators. This is accompanied by the appropriation of Dilthey's view that man's world is a socio-historical world. Therefore the theme of historicity becomes an integral part of the phenomenological analysis. But what the humanities precisely are, and what method should be most applicable remain questions.

The personalist attitude is the guiding view of the universal psychology, providing the basis of all *Geisteswissenschaften*.

However, this attitude can be connected closely with the historical attitude in the present case. According to Husserl we are concerned with descriptive-intuitive sciences and as such with historical sciences too. Therefore the personalist attitude here coincides with the historical one, such as it has been developed over a period of time by the philological-historical sciences. This has nothing new to offer with regard to Dilthey. This philosopher too has identified the humanities with historical cultural sciences. They bear for him the mark of carrying through the historical method. But Husserl is more positive in the handling of the teleological view, especially in his historico-philosophical concept of the situation and meaning of European culture.

One factor which was a flaw in Dilthey's theory has not been amended by Husserl but has grown into a problem and a matter of uncertainty. Dilthey did recognise the possibility of systematic humanities next to and in contrast to the historical ones, but they were of no particular concern to him. Husserl recognises the possibility of an objective science of humanities as opposed to the historical ones, and on the other hand of an a priori phenomenological science of being, but this occurs in the form of a surprising and intriguing question, rather than in the form of a clear answer. The objective '*Geisteswissenschaft*' is based on the view that the entire concrete factual world of experience can be idealised and can therefore be constructed mathematically. But this would by consequence demand the availability of a world-embracing form of mathematics. On the other hand he asks himself if it really goes without saying to base an investigation into the factual spirits in their historicity on an essential knowledge of the spirit and of the community they live in, and he questions the self-evidence to extract 'exact notions' and exact, unconditionally valid truths from that community as ideal poles for all spiritual factuality (*Husserliana* Vol. VI, p. 306). In a footnote he counters himself with the question: "but what does essential knowledge of histoical knowledge look like in this respect?"

All this touches the problem of generality. For all typology here bears the mark of historicity, not of the strict, possibly mathematical generality. It is therefore impossible to evade the impression that both for Dilthey and for Husserl the sciences of man are identical with the historical humanities of cultural sciences. They find their basis in a general science of the mind. It is true that Husserl recognises psycho-physics and even an objective form of humanities, operating purely inductively with physical and mathematical methods, but yet the humanities find their true fundaments in a science of subjectivity, since they are empirical sciences of historical subjectivity themselves.

With regard to this many problems are prominent, only a few of which connected with each other, we are going to mention.

(1) It may be asked if it is possible to call the sciences of man only and exclusively historical. Does this tally with the increasing tendency in these sciences towards the formation of theories and in connection with this towards objectivation and formalisation? Is this pursuit identical with naturalisation or is it rather a pursuit of a higher form of abstraction pointing towards structural analysis? The equation of the sciences of man

with the historical sciences or humanities clearly shows the mark of the 19th century and the beginning of the 20th century. Thus linguistics was primarily considered as a historical science: 'linguistics is history of language' (H. Paul). With the rise of a general linguistic theory and science, matters have changed completely. The same is true for social, political and economic sciences. The interpretation of the science of art as being the history of art, is no longer truly valid either.

Only law and theology were excepted from it in the 19th century, being dogmatic sciences. They were only regarded as sciences in so far as they consisted of the history of law and the history of religion.

This gives rise to the question what is really the task of a theory or philosophy of science. Is it purely descriptive as to the methods followed and aims pursued, is it standardising or rather providing ideas for guidance?

Husserl leaves us no doubts. It can hardly be said that he gives a description of the method actually used. He rather provides a philosophy of the spirit and of the person in connection with the conception of nature and the world.

This agrees with the tradition defensively and metaphysically. The dualism it defends is ultimately the dualism of nature and spirit. With regard to the sciences it leads to a juxtaposition between the science of objectivity and exactness on the basis of a narrower and a broader conception of nature on the one hand, and a science of subjectivity based on a conception of the spirit and person on the other.

(2) The idea of an objective mind, for Dilthey an attempt to evade the influence of Husserl's criticism, deserves more attention in the historical cultural sciences and humanities than Husserl himself thought fitting to display. It must be admitted that Husserl is not alien to the concept of the objective mind, but he also awards a certain place to the achievement (*Leistung*), to products of the human mind and culture, but in his case personalism in the humanities dominates to such an extent, that the idea of an objective mind (works, works of art, institutions, political systems, forms of language) are not very prominent. This can be understood in terms of the fact that he regards the humanities as sciences of subjectivity. The pursuit of objectivity and factual accounts in linguistics, philology, literary and historical criticism is especially directed towards this aspect (forms of language, the work, the work of art). In the history of philosophy

itself it is also advisable not to regard this history as merely a collection of personal views and personal world pictures. This is of great importance for the hermeneutics of these sciences[1].

(3) Husserl's theory is much more than a mere description of the scientific methods since it recognises the function of ideas and norms exceeding positive science. On the one hand the theory does not transcend this science, but indicates a direction it is already moving towards, such as the idea of the infinite and true world as the correlate of the idea of a science of the world ('*Weltwissenschaft*') on the other hand the theory does transcend by the putting forward of ethical ideas such as the idea of a true society and of true humanity characterised by the idea of a true and genuinely personal individual life (*Einzelleben*).

In the end it revokes the idea of a universal science as being a natural science and gives priority to a universal, personal science, and thus to the humanities. This is based on the consideration that all theoretical questions are ultimately based on man as a living person, and that all science is an undertaking of humanity.

Ultimately all this is put forward as the absolute ground of the transcendental subjectivity, which is the theme of the transcendental attitude and of phenomenology as a universal absolute '*Geisteswissenschaft*', in which the world is not accepted as given before, but is revealed in its mode of appearance and as 'in itself', containing truths 'in itself' pertaining to it, as a world constituted in and by the transcendental or pure consciousness.

NOTES

* The German version of this paper has appeared in volume I of the *Analecta*.

[1] See K. Kuypers, 'Die Geschichtlichkeit der Sprache und die Aufgabe der Philosophie', in *Das Problem der Sprache* (ed. by H. G. Gadamer), Wilh. Fink Verlag, München, 1967, pp. 387–407.

T. KISIEL

REPETITION IN GADAMER'S HERMENEUTICS

It was Heidegger who made 'repetition' into a fundamental category of the hermeneutical process. It refers to *Dasein*'s explicit transmission of the possibilities that are as having been. However, it is not simply a matter of reiterating the tradition, but of 'destroying' or transmuting it in order to expose the possibilities relevant to the historical situation in which *Dasein* finds itself. Repetition thus understood is not a static reproduction of a meaning fixed once and for all, but the revival of a meaning subject to continual review and revision. Repetition in this sense keeps the tradition alive by constantly exposing the new and hitherto hidden possibilities of meaning that it offers.

The problem of the repetition of the tradition is found in Husserl as well. When he confronts the task of reactivating the origins of geometry[1], Husserl explores the conditions that make the transmission of any tradition possible. First and foremost is the repeatability of ideal objects. On the one hand, the Pythagorean theorem is intentionally achieved in a series of transient subjective acts in the individual geometer's mind, which as such are restricted to a certain time and place. On the other hand, the Pythagorean theorem as a sense formation has a unique objective existence which makes it accessible to all men at any time as identically the same sense, no matter how often it is spoken or written, or in whatever language it is reproduced. Not the repeatability of actually multiplied copies of, say, Euclid's *Geometry* or Goethe's *Faust*, but their non-localizable identity as one and the same sense for all men in all ages. Such an omnitemporal objectivity can only be 'irreal' or ideal, inasmuch as anything real is necessarily individuated in time and space.

But it is through language that the theorem in the mind of the individual inventor acquires an existence in the world and in fact becomes accessible not only to contemporaries but also and more significantly, through the written word, to later generations. Objectification through the written word foregoes the need for the continual presence of active subjects by preserving idealities for the historical tradition in the form of a virtual

Tymieniecka (ed.), Analecta Husserliana, Vol. II, 196–203.

communication. But the virtuality of language not only preserves meaning but also 'seduces' [2] by promoting a passivity toward these virtual meanings which induces a forgetting of original meaning that may eventually lead to a crisis in understanding. The loss of the lifeworld meanings in the current formulations of mathematical physics is a case in point. Rooted in the natural linguistic tendency toward *Sinnentleerung* reinforced by the one-sided technical rationality of present-day science, the crisis is at bottom a 'disease of language'.[3] Its cure is in the reactivation of the most original meanings that have been sedimented in the words and then submerged under further deposits of meaning.

Thus the repetition or reactivation that particularly interests Husserl is the excavation and recovery of the primordial meanings that lie at the root of a tradition. Likewise, Heidegger's repetition is directed toward the most incipient beginning of Western thought, in order to transmute it into a radically new beginning.[4] By comparison, Gadamer's hermeneutic [5] has more moderate aims. Instead of Husserl's archeological reactivation and Heidegger's future-oriented repetition, Gadamer considers man's response to his tradition simply in terms of a "participation in a present meaning" (p. 370). The transcendentalism of a transworldly subjectivity in its objectivating function is explicitly abandoned in favor of a starting point in the facticity of a current situation in which, e.g., an interpreter understands a text is such a way that both are inextricably intertwined in the spontaneous movement of tradition. No method of understanding and interpetation is offered. The aim is simply to describe how understanding actually happens, where understanding is not so much a method, not "so much an action of subjectivity, but an entering into a happening of tradition, an occurrence of transmission in which past and present are constantly being mediated" (p. 275).

Now, to belong to a tradition means to inherit a network of presuppositions which serve as precedents that naturally provide the anticipations of meaning that guide our understanding of anything whatsoever. Without these prejudgments, in particular those of our tradition of language, we would in fact not understand anything at all. The issue for finite understanding is therefore not a matter of discarding prejudices in order to begin absolutely, but to determine what distinguishes the legitimate prejudices from those which obstruct understanding. The opportunity for this occurs when the message transmitted by the tradition strikes us as strange and thus poses a question to us, in effect challenging the disposition of our prejudgments. The temporal distance that is evoked between us and the transmitted message should not be considered as an obstacle to be overcome, in the hopes of perhaps returning to the text in its "original first" (p. 370), but as a distance to be maintained and cultivated for its productive possibilities of understanding. For temporal distance is not an empty void, a "yawning abyss" (p. 281), but a fullness of the continuity of transmitted articulations mediating the present with the past, a mediation which in fact has already occurred precisely in and through language. Temporal distance not only serves to filter out the sources of error that accrue from being too close to, e.g. a mode of art or a historical event, but also to release out of its fullness new sources of understanding manifesting unsuspected relations of meaning. It thus suppresses prejudices of a particular nature and permits prejudgments that are more comprehensive in scope to come forth. It is temporal distance that serves to distinguish the viable prejudgments from the obstructive prejudices. Here we have tradition at work, and it is not too much to speak of a genuine productivity of the historical process. It is thus that tradition is a happening, an occurrence that retains its currency by constantly drawing on us to evoke the new and timely possibilities of interpretation that it offers, continually renewing itself in accordance with the changing times.

The 'hold' that tradition has on the interpreter, for all of its finitude and its situated character, now appears as something positive. The time of tradition and its medium of language are no longer obstacles to be surmounted, as Husserl's idea-centered account tends to suggest, but positive and creative possibilities for *relevant* interpretation. And it is no surprise that a situation-centered hermeneutic should find its model in concrete juridical application rather than, say, the abstract mathematical example for which Husserl had such a propensity. In a situation that is historical through and through, the law, the text, the artwork is never a constancy of a meaning fixed once and for all by its author, but the continuity of a meaning which presents itself through an unbroken series of perspectives mediating present with past. It is because of this state of affairs that the reading is never a simple "repetition of something past, but rather a participation in a present meaning" (p. 370).

The nature of the transmission of the written word particularly brings all this home. For with the written word, language is alienated from its empirical execution and becomes virtually "contemporary with every present" (p. 367). Detached from its authors as well as its intended reader, i.e. from all things psychological, it is elevated into the sphere of meaning pure and simple. Finally detached from the contingency of its origin, what is put down in writing is freed for new relations of meaning exceeding that which was intended by the author. It enters into its history of meaning (*Sinnsgeschichte*), in which new meanings emerge in accordance with the horizon of transmitted presuppositions in which it is read. The apparent deficiency of the written word, the lack of such helpful factors as intonation and attitude through which the spoken word really interprets itself, now appears as an advantage. Without these aids, the interpreter of the script must concentrate on the essential, namely, to give full weight to what the text means. It is especially true here that the 'thing' (*die Sache*), the topic or subject matter, is the central concern of the hermeneutic consciousness.

But if this is not to be equated with what the author intended, then what exactly is repeated in the interpretive situation? Conscious of the productivity of history, interpretive understanding is suspended in the tension between the sameness of the transmitted text and the peculiar difference of the current hermeneutical situation, in which the text has fallen out of its customary familiarity and demands reinterpretation. The first and foremost task of hermeneutic is really to appropriate this single and same message to the times, and in this sense to amalgamate the horizons of present and past. Since this application of the text to the times necessarily takes place in and through language, the task is one of administering the translation that the concepts of the past demand. This translation is achieved through the mediation of certain interpretive concepts which do not become thematic in the understanding of the text, but which themselves disappear into what they bring to language, namely, the topic itself in its present sense (p. 375). Hence the interpretation is not evoked in place of the text, or as something superadded to the text, but is in fact the text itself in terms of the meaning it presents in the current situation. Accordingly, the interpreting concepts "belong to the inner articulation of the topic (which is meaning)" (p. 449), while at the same time they are a "new creation of understanding" (p. 448) through which the transmitted sense finds its concretion.

A similar unity is to be found in the reproduction of an artwork. When, for example, a symphony or a play is performed, the reproduction at its best does not become thematic as a reproduction. Instead, through it and in it the work comes to be represented and evoked in a new immediacy. The authentic experience of the work is precisely this non-distinction between itself and its mediation. The creative element which the reproductive artist brings to the work, here the musician or the actor, disappears into the epiphany of the work itself. The reproduction re-presents the work itself, makes it present again, no matter how far removed in time it originally may have been, and in fact in such a way that the work undergoes an enhancement in Being (p. 133). And the work depends on the reproduction in order to represent itself, through which it obtains its authentic Being (p. 112).

The problem of repetition in Gadamer's hermeneutic thus issues into comprehending the "paradox" (p. 448) of how one and the same message transmitted by the tradition is always understood differently, according to the situation of the receiver of this message. The metaphysical solution of a meaning 'in itself' to which interpretations are superadded, like accidents to a substance, is out of the question for a 'hermeneutic of facticity.' For one thing, we do not first know a text 'in itself' like an abstract universal, from which we then proceed to determine what significance is applicable to it in one or another concrete situation. Rather, the text is a universal which must first be understood in the concrete situation in which we encounter it, if it is to be understood at all. Hence the concretion of meaning derived from mediating the text to the hermerneutic situation in which one finds oneself is the very universal that understanding seeks. Interpretive understanding does not respond to a text as to something *bloss vorhanden* in a supratemporal state of static sameness, but instead to its perpetual and inexhaustible giving of meaning (*Sinngebung*) in the dynamic continuity of a history. And yet it is the same work that remains and perdures in its varying interpretations through changes of time and circumstances. It is there in all of its changing aspects. It is contemporaneous with all of these and they with it (p. 115).

When the unity between the textual meaning and its varying aspects is now called a 'speculative' unity, this conscious borrowing of a Hegelian expression is to be understood in terms of a 'speculative dialectic' that occurs within the "living metaphoric" (p. 409) of language itself, in its "inner dimension of multiplication" (p. 434). Speculation etymologically refers to the self-reconciling speculation of mirroring. Just as the mirror image has no existence of its own but only serves to highlight the spectacle that offers itself in it, so the interpreting concepts do not assert their own existence but serve to articulate the subject matter itself. The same evanescent existence belongs to language generally, which at its best effaces itself in favor of its subject matter. Basically, words have no *raison d'être* other than to reveal their topic in its topology. And the *topos* of language is always the world, the total context of meaningfulness, 'the whole of Being,' to which every word ultimately bears witness. The speculative structure of language is such that every word mirrors the totality of meaning, as 'though a glass, darkly,' and not simply a fixed and given being. Finite human discourse is such that it "brings into play a totality of meaning without being able to say it totally" (p. 434). It is this inexhaustible depth to language that 'speculative' discourse at its best, e.g. the texts that over the years have become 'classic,' has supremely succeeded in bringing into play. To speak speculatively is to speak pregnantly, by holding together an infinity of the unsaid with what is actually said (p.444) Because each word lets the whole of language appear as its suggestive unsaid, it has infinite possibilities of finding ever new answering 'anti-words'. The ensuing dialectic is an inexhaustible 'play on words', a self-propelled language game, the 'mirror play' of language itself, or better, of its topic, Being itself. For the authentic speculative movement is the 'doing of the topic' that presents itself in language. And the ultimate speculative unity is between the language and the topic, in which Being presents itself in language and language simply serves to present Being, so that the distinction between Being and self-presentation is not a distinction after all. "Being is *language, i.e. self-presentation*" (p. 461).

At this ultimate level of non-distinction or immediacy, we experience language in an intensely intimate familiarity that simply cannot be explained by any sign theory of language, in which words are designative tools used by a thought somehow separate from them. To intimately experience a word is to live in it in such a way that there is no distance between it and what it intends. Here a word 'is' what it means, and does not simply 'stand for' some 'speechless pregiven' that is on hand. It is thus that we are seized by words and led by them to make ourselves at home in the world. And after all, there is no difference between living in a language and living in the world. Our most immediate experience is linguistic through and through, and this linguistic world experience provides the meaningful context that guides all of our understanding. It is thus the 'element of thinking,' the immediate medium that permits us to function as men, just as water permits a fish to be a fish.

How this medium mediates is the crucial problem in understanding the peculiar temporality that constitutes the repetition of a tradition. For it is under the aegis of the life of language that a transmitted text or an artwork lives a life or its own and 'speaks for itself' out of a surplus of sense, reflecting new facets in new situations in a variegated display of emerging senses. What has been called the timelessness of art is more precisely this capacity to be "contemporary for every present" (p. 115). The great artwork (or the classic text) is like a recurring feast whose very Being demands that it be regularly celebrated. And a true celebration is never a mere imitation but a *mimesis* in its original sense, which always involves a transmutation and an intensification in order to re-present the work in a way never before seen, a way that appropriates the work to the present and makes it 'topical.' It is this apparently contradictory movement, to be at once idiomatic and abiding, that gives the repetition of a tradition viewed from the radically finite situation of the interpreter such an "extremely puzzling time structure" (p. 117). For "it has its Being only in becoming and recurrence" (p. 117) and "a being which only is, in always being another, is in a radical sense temporal" (p. 118n). It is this creative iteration of time that invites and in fact demands creative repetition on the part of man. Gadamer describes it as the speculative movement of Being articulating itself, dividing itself from itself, past from present, old from new, the said from the unsaid, horizon from horizon, only to recollect itself in its speculative unity. Heidegger speaks of the temporalizing of a contemporaneity in which the three dimensions of time are mutually implicated in a reciprocal play in which they pass into each other and at once hold each other apart, thus yielding an opening, a clearing, the contemporary world. And Heidegger himself has not been able to say just how to think this remarkable realm of identity and difference.

NOTES

[1] Edmund Husserl, *Die Krisis der Europäischen Wissenschaften und die Transzendentale Phänomenologie*, Husserliana, Vol. VI (ed. by Walter Biemel), Martinus Nijhoff, The Hague, [2]1962, pp. 365ff.
[2] *Ibid.*, p. 372.
[3] Jacques Derrida's introduction to Husserl's *L'Origine de la géométrie*, Presses Universitaires de France, Paris, 1962, p. 91n.
[4] Martin Heidegger, *Einführung in die Metaphysik*, Niemeyer, Tübingen, [2]1958, p. 29.
[5] Hans-Georg Gadamer, *Wahrheit und Methode: Grundzüge einer philosophischen Hermeneutik*, 2nd edition, expanded with a supplement Mohr, Tübingen, 1965. Page references to this text will be incorporated into the body of the paper.

GUIDO KÜNG

INGARDEN ON LANGUAGE AND ONTOLOGY

(A Comparison with some Trends in Analytic Philosophy)

This paper has two aims. First it wants to draw attention to the works of Roman Ingarden, which contain to-day's most elaborate phenomenological descriptions in the fields of language and ontology (i.e. concerning the acts and products of linguistic activities, and concerning the formal structures and existential dependencies of beings). Unfortunately many of the more fashionable contemporary writings on these subjects in the phenomenological-existentialist tradition are well below the level of Ingarden's precision, and the discussions could profit greatly if Ingarden's works became required reading.

On the other hand the tradition of what can be called 'analytic philosophy' (in a very broad and vague sense, including what I call 'logistic philosophy' and 'ordinary language philosophy') has developed sophistications of its own. But I believe that despite their differences the 'analytic' and the phenomenological approach are not as incompatible as it seems at first sight, and that on the contrary an encounter can be mutually enriching. To help the cause of such a meeting of minds is the second aim of my paper. In particular I want to point out how both Ingarden and the 'analytic' philosophers have discovered that ontology is a prerequisite of metaphysics.

I. PHONEMIC STRUCTURES, MEANINGS, PURELY INTENTIONAL OBJECTS AND IDEAS

Ingarden's account of the phenomena of language represents a further development of the investigations of Husserl, and as such it is at the antipodes of most 'economy-minded' systems in the 'analytic' tradition. I do not want to give here an exhaustive account and justification of Ingarden's distinctions. Let me only recall some of them which document his profound disagreement with any kind of positivism, physicalism, behaviorism, psychologism, etc.

(a) Ingarden agrees with those who see even in the sounds of language

Tymieniecka (ed.), Analecta Husserliana, Vol. II, 204–217.

something which goes beyond the merely physical. Already in 1931 he gave a rich phenomenological description of the distinction between the physical sound (*Lautmaterial*) and what linguists call to-day the phonemic structure of a word (*Wortlaut*). In logic we have become used to a similar distinction between the expression tokens and the expression type, where the latter is usually defined as the class of the corresponding expression tokens. It has to be noticed however that Ingarden – as it is becoming for a phenomenologist – pays close attention to the exact shades of the differences, and that for him a word type is a certain '*Gestalt*', i.e. something which is not quite the same as a class and which is also not the same as a species.[1]

(b) Ingarden was also one of the first philosophers who refuted the dogma of the Vienna Circle according to which the meaning of a sentence was its verifiability. In a paper for the 1934 International Congress in Prague he pointed out, for instance, that there are sentences which have such a meaning that they are not verifiable. Meaning must therefore be something more basic than verifiability. In a similar vein he stated in a rather condensed paper for an international meeting in 1962 in Oxford, that both the word tokens occurring in concrete statements and the word type by itself had each their intuitive meaning, and insisted that even the meaning of the word type had to be distinguished from a rule for the proper use of the corresponding word tokens. Unfortunately Ingarden did not elaborate, but his main argument would probably consist in a careful description of what is a rule, and how learning a rule presupposes some understanding, some grasping of meaning.[2]

(c) Furthermore Ingarden rejects like Frege and Husserl, the psychologistic view that meanings are mental and strictly private. Meanings of words are intersubjectively established ways of focussing on things which are intersubjectively present. Notice that if I want to understand what somebody tells me, then I do not concentrate my attention on the mental acts of that person, but on the subject matter which we are both talking about.[3]

But if we say that meanings of words are intersubjectively established ways of focussing on the subject matter (i.e. that 'intentions' are ways of intending), does this not lead us back to the view rejected above, namely that meanings are rules governing the use of words? This is not the case. To find the right focus, the right tuning in on the subject matter, is an experience of the kind: "Aha, now I have got what you mean, now I see it also". Once I have found the understanding, I can formulate rules. In order to make use of the rule that the expression 'red' applies properly to situations where something red is present, I must learn to see red, to see the difference between red and not-red. Wittgenstein is of course right when he insists that learning to see a difference and learning to apply an expression to the difference which I see, goes normally hand in hand. But this does not mean that seeing the difference and applying the word are the same.

(d) Unlike Frege and the early Husserl, Ingarden rejects also the view that meanings are ideal and timeless. He does accept the existence of eternal ideas, but in his view the meanings of words are not such ideas. The meanings of words do have a history: they are established at a certain time, they can become modified, they may be lost again. They are objective[4] and can even outlive the individual mental act with which they came into being, but they are nevertheless purely intentional, existentially dependent (heteronomous) with respect to human thinking.

(e) According to Ingarden meanings, these intuitive ways of focussing, are not the only entities depending on human thinking. Even what the speaker is focussing upon, the *object* of his talk, may sometimes be something purely intentional. For instance in a novel the author talks about persons, states of affairs, events, etc., which do not exist in reality, but exist only as creations of the author's mind.[5] Here too, as in the case of the meanings, Ingarden fights against psychologism. If the persons and events described in a novel were considered to have an absolutely private residence in the mental acts of the artist, then it would remain a mystery how the reader can gain access to them. Even the fact that a writer can create a certain fictional character on one day and can come back to it on another day without remembering his concrete mental acts, seems to have no satisfactory explanation in a psychologistic framework.[6] The acceptance of purely intentional objects is thus not gratuitious indulgence in multiplying entities beyond necessity, but serves a concrete explanatory function.

The realm of the purely intentional objects, which are neither real nor ideal has been of special concern to Ingarden, since his basic interest lies with "The Controversy About the Existence of the World",[7] and epistemological idealists conceive what is ordinarily called the real world as precisely such a purely intentional object, i.e. for them real objects and fictional objects are two (different) kinds of purely intentional objects. (I will say more about this shortly.)

II. TWO DIFFERENT SEMANTICAL FRAMEWORKS

After this glimpse at some of the rich distinction made by Ingarden, which are quite in contrast with the austerity usually found in 'analytic' philosophers, I will now try to minimize somewhat this contrast.

The first thing to remember is that the dogmas of the Vienna Circle and the skeptical views of the later Wittgenstein are *not* representative for analytic philosophy as a whole (in the broad sense in which this term is understood here). For instance G. E. Moore, the father of ordinary language philosophy, is in many of his views in agreement with phenomenology in the broad sense; and the work of many representatives of contemporary conceptual analysis is essentially comparable with that of the phenomenologists.[8]

But in logistic philosophy, too, there are philosophers whose views are quite compatible with phenomenology. According to Frege, there exists an intermediate level of senses or meanings between the expressions and their referents. Carnap himself has outgrown many of his former prejudices and, progressing from syntax to semantics, has made room in his ontology not only for classes, but also for intensions, i.e. objective concepts, properties, propositions, etc. Carnap's basic difference from Frege is merely that the intensions are not shelved on a special intermediary level of senses, but occur in the domain of reference. Actually I believe that this difference of semantical frameworks is one of the main difficulties for a mutual understanding between logistic philosophy and phenomenology. Let me therefore insist on it:

The traditional 3-leveled semantical framework (Phenomenology, Frege, A. Church):	The 2-leveled semantical framework (Meinong, Russell, Carnap etc.):
signs	signs
|	|
sense	|
|	|
referents	referents, resp. designata

It is interesting to note that the innovation of the 2-leveled semantical framework is not only due to economy-minded logistic philosphers, but has its root also in Meinong. It is well known that Meinong held the curious view that there are not only existing and subsisting objects, but that there must even be objects which have no kind of existence (*Dasein*) at all. He was driven to this opinion because of his referential conception of truth and because he thought that certain true statements were talking *about* square circles or *about* the non-existing (and non-subsisting) golden mountain.

Russell agreed with Meinong on the fundamental importance of the conception of truth as a correspondence, as an isomorphism between the words of a statement and what the statement is *about*, i.e. the referents or designata of the words. This correspondence theory of truth was the cornerstone of Russell's refutation of Bradley's monistic idealism: if to the multiplicity of components of a true statement there must correspond a multiplicity of elements in reality, then monism must be wrong. It was the great achievement of his theory of descriptions that it defended the correspondence theory of truth by showing how the statements which are seemingly about non-existing objects can be transformed into statements which are clearly no longer *about* these mysterious non-entities. [9]

It would lead too far to explain here in detail how in the two-leveled semantical framework of Russell and Carnap ideal or purely intentional entities appear among the designata. I have dealt with this question elsewhere.[10] Instead I will now point out an interesting development which has brought the two frameworks somewhat closer to one another.

III. ONTOLOGY AND METAPHYSICS IN COMTEMPORARY LOGISTIC PHILOSOPHY

It is typical for the development of logistic philosophy from Russell to the generation of Carnap, Quine, Goodman, etc., that the straightforward realism of Russell has given way to a more Kantian position: the universe of discourse, i.e. the universe of the designata, is no longer simply identified with reality as it is in itself. Instead contemporary logistic philosophers have found that reality, 'the world', can be described in different systems, whose universes of discourse are articulated differently. This means that the two-leveled semantical framework has been supple-meted by a third level:

signs
|
designata = universe of discourse
|
reality in itself = *the* world

The basic categories of the designata of a system are to-day called the *ontology* of this system. Each semantically developed logistic systems has to specify its *ontological commitment*. Actually two tasks can now be distinguished: (a) what I call the *ontological* task of describing different ontological commitments and (b) what I call the *metaphysical* task of deciding which ontological commitment to adopt, i.e. the metaphysical question which ontological commitment fits reality best. (It may well be that alternative universes of discourse fit reality equally well). Carnap distinguishes in this vein *internal questions* within a systematic framework from *external questions* concerning the choice of a systematic framework (of a universe of discourse, i.e. of an ontology, and of specific rules of inference, i.e. of a logic).[11]

There are different views concerning the nature of metaphysics. According to Carnap and Quine the metaphysical or external question is a purely pragmatical question. Any ontology can be tolerated provided that 'it works'. Any further going metaphysical question with respect to the nature of reality is for them meaningless.

N. Goodman's position is different. He has only a limited tolerance: a philosophically acceptable ontology must first of all be understandable; it has to satisfy a criterion of intrinsic intelligibility, not just anything goes.[12] Secondly he believes that many different systems *do fit* reality. Some systems capture the world in a finer network than others, but it seems that for Goodman it can even happen that two alternative systems give an equally detailed map of the world. Goodman concludes from this that the world *is many ways*. He emphatically rejects the view that there is *one* way the world is, but that this way is hidden from us.[13] (Actually Goodman agrees with Carnap and Quine that to ask for *the* structure of reality is meaningless. But Carnap and Quine merely express their skepticism with respect to metaphysical answers, while Goodman explains metaphysically why a certain metaphysical question is meaningless.)

Other 'analytic' philosophers have on the contrary the metaphysical conviction that there *is* a unique way the world is. For instance Wilfrid Sellars assumes that reality is what the ultimate scientific image (if it ever could be arrived at) would depict.[14]

IV. ONTOLOGY AND METAPHYSICS IN PHENOMENOLOGY

The questions concerning the world as we picture it and the world as it is in itself are of course familiar to phenomenologists. But let me explain how these questions fit into the phenomenological semantical framework:

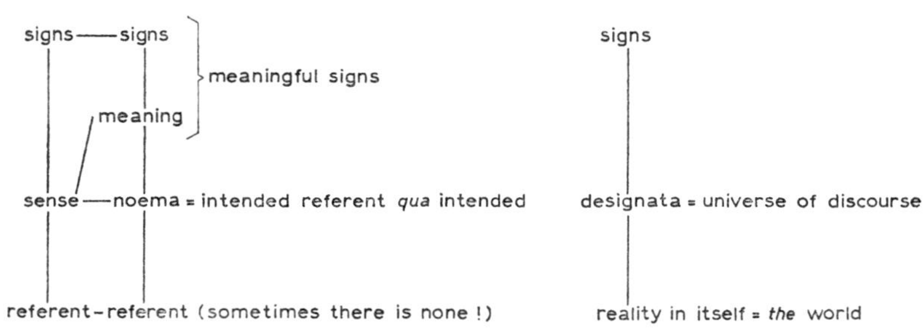

The logicians' universe of discourse has its counterpart in the noematic world of the phenomenologists, while reality in itself is the domain of reference of the phenomenological system. Notice especially that Husserl's *noema* belongs like the meanings to the level of sense and *not* to the level of reference. Dagfinn Føllesdal has documented this crucial point very clearly from Husserl's own texts.[15] The *noema* is only the *intended* referent *qua intended*, it is not the *actual* referent (*qua referent*).

It can happen that there is an intended referent but no actual referent. For instance, if a certain Mr. X would believe that there are centaurs grazing on the lawns of the University of Waterloo, then the belief of Mr. X would have centaurs among its intended referents. But there are, I guess, no actual centaurs corresponding to it. Meinong might have said that this belief was concerned with non-existing centaurs. But Husserl, I trust, would have agreed with me that this belief cannot be *about* non-existing centaurs. There simply is no such thing as a non-existing centaur. Furthermore our Mr. X does not claim that *non-existing* centaurs are grazing on the lawns of the university of Waterloo, he claims that *existing* centaurs are grazing on the University of Waterloo.

Notice also that Mr. X is not making a claim *about noemata*. Rather his belief is *through noemata about physical reality*. (The *noema* is so to speak the tip of the arrow of sense which points to a certain 'point' in physical reality, but nothing can be found at this 'point'.)

Actually the *noemata* themselves *do* exist (though not on the same level as their referents), they are not non-existing entities. They are purely intentional entities, and Ingarden has shown that purely intentional entities always have an ontological double structure. *Qua* purely intentional entities they always have a certain mode of heteronomous existence, while in their 'content' (*Gehalt*) any mode of existence can occur: thus the centaurs of Mr. X are depicted as having real physical existence.[16]

The case of a contemporary poet writing about grazing centaurs is different from that of our Mr. X. The poet knows that his centaurs are only fictions and an astude reader can usually find out from the style of such a text that it is a work of fiction.[17] The schema is then the following:[18]

signs
|
noema, in the content of which a fictitious world is depicted
|
referent: this fictitious world

Like any noema, the noema *qua noema* has a certain mode of heteronomous existence. In the *content* of the *noema* of the poet a fictitious world is depicted. This fictitious world is itself a purely intentional object which involves two further modes of existence: the fictitious existence of the fictitious world *qua fictitious*, and a mode of existence depicted in the *content* of the fiction (e.g. the real existence fictitiously attributed to the centaurs). Notice that while the belief of Mr. X has *no* referent, the text of our poet *does* have a referent.

I think that a similar subtility is operative in the controversy between realism and idealism. The following seems to be the schema of Husserl's position when he starts transcendental phenomenology:

signs (resp. *noesis*)
|
noema of the autonomously existing real world
|
(?)

Husserl starts out (like everybody else) as a realist, though in transcendental phenomenology he brackets the question whether there is actually an autonomously existing real world or whether there is no such referent.

With Husserl's conversion to transcendental idealism the situation changes (though Husserl himself seems not clearly aware of this change):

signs (resp. *noesis*)
|
noema of the *so-called* real autonomous world
|
?

Husserl sees now the world differently; his noesis and therewith also his noema have undergone a change. He assumes now that the autonomy of the world is fictitious.[19] The question which before was bracketed, is now answered: in the negative. [20]

If idealism is right, then there is no referent fulfilling the realist's noema, but there is a heteronomously existing referent fulfilling the idealist's noema. If realism is right then there is a referent fulfilling the realist's noema, but there is no referent fulfilling the idealist's *noema*. The metaphysical question of idealism vs. realism can be decided if it is possible to show that either the *noema* of the realist or the noema of the idealist contains in its content an impossibility, i.e. if it can be shown that either the notion of an autonomously existing world or the notion that the autonomy of *the* world is fictitious is impossible. This I think, is roughly the conception which underlies Ingarden's great work.[21]

As in logistic philosophy we can distinguish two tasks: the ontological task of describing alternative ontologies, i.e. the task of describing alternatives frameworks operative in the noematic content of our thought about the world, and the metaphysical task of deciding which noematic content is fulfilled by reality, i.e. the task of actually committing ourselves to certain existential assumptions.

This procedure has some similarity with the Leibnizian startegy of surveying all possible worlds and then picking out the actual one. But Leibniz, unlike the contemporary philosophers, did not consider alternative proposals for the 'analytical' truth concerning the world, he only considered alternative possibilities of factual truth within a pre-given semantical (logical and ontological) framework.

Ingarden defines ontology as "the a priori analysis of the content of ideas".[22] This does not say exactly what I have been saying, namely that ontology is the description of the possible contents of *noemata.* But the connection the two formulations becomes obvious, when we remember that for Ingarden the ideas are the measure of what is a priori possible. The content of a *noema* depicts a possibility, i.e. contains no *apriori* impossibility, if and only if there is an idea of which, what is depicted by the *noema*, would be an instantiation.[23]

Actually ontological descriptions and metaphysical decisions go hand in hand. If more elaborate ontological analyses show that what initially looked like a possibility is *apriori* impossible, then this entails an automatic metaphysical decision that no actual referent fulfills this noema.

But not all metaphysical questions seem to have an *apriori* solution. If beings are *apriori* possible whose mode of existence is contingent existence, then only actual experience can tell whether such beings actually exist. It seems, e.g., that there might have been no experience of a world at all. Only actual experience tells us that we do in fact live through such an experience. Is it a *noesis* with a realistic or with an idealistic *noema*? Ontology can describe the difference between the two, but again only actual experience can tell which kind of noesis and noema somebody has in fact at a given time. Does the noema have a referent? If the ontological description of a noema shows that its content involves an impossibility, then the metaphysical decision that no such referent exists is apriori justified. But if the object depicted in the content of the *noema* (a) involves no impossibility and (b) has a mode of contingent existence, then the affirmation that it has a referent must again be based on the actual experience *qua* actual.[24]

V. PHENOMENOLOGICAL ONTOLOGY AND LOGISTIC ONTOLOGY

I do not claim that doing phenomenological ontology and doing logistic ontology is the same. The aims are clearly different: the phenomenologist is primarily interested in the richness and the subtle nuances in all what he finds intuitively given. Therefore his great principle is the principle of non-miserliness.[25] The logistic philosopher, on the other hand, aims at the construction of formal systems where everything follows strictly from a certain basis of primitive terms and axioms: his first principle is a principle of economy.

Nevertheless phenomenological ontology and logistic ontology are essentially comparable and complement each other. The formal systems of the logistic philosopher are not merely games of uninterpreted symbols. They want to tell some of the ways 'the world' is. Therefore phenomenological ontology could be of interest to a logistic philosopher in two ways: (a) the phenomenological descriptions could enrich his understanding of the ontologies he has committed himself to in his systems. (b) The descriptions of phenomenological ontology could serve as inspirations for the further progress of his system-building; there are new logics to be developed, new features to be mapped in formal systems, and this presupposes some intuitive inspirations.[26]

The phenomenologist, on the other hand, will find that the formal systems are tools which can help to strengthen his intuition, to sharpen his vision in certain respects. Remember how difficult it would be to develop our mathematical intuitions without the help of formulas and formal systems. How primitive would geography be if it refused to draw maps. Of course geographers should not abandon the study of the earth and restrict themselves to the study of maps, but wherever we intuitively grasp a structure, there it makes sense to develop maps, to construct formal systems.[27]

NOTES

[1] R. Ingarden, *Das literarische Kunstwerk. Eine Untersuchung aus dem Grenzgebiet der Ontologie, Logik und Literaturwissenschaft*, 1st edition Halle, 1931, 3rd edition Tübingen, 1965, pp. 30–61. R. Ingarden, *Vom Erkennen des literarischen Kunstwerks*, Tübingen, 1968 (1st Polish edition 1937), pp. 16–21.

[2] R. Ingarden, 'Der logistische Versuch einer Neugestaltung der Philosophie', *Actes du huitième Congrès International de Philosophie à Prague 1934*, Prague, 1936. R. Ingarden, 'L'essai logistique d'une refonte de la philosophie', *Revue Philosophique de la France et de l'Etranger* **60** (1935), 137–59.

[3] *Vom Erkennen ...*, p. 21f., p. 364f.

[4] Cf. R. Ingarden, 'Betrachtungen zum Problem der Objektivität', *Zeitschrift für philosophische Forschung* **21** (1967), 31–46 and 242–60.

[5] Another important kind of purely intentional objects are social institutions like Universities, States, Churches, etc. Notice also that the meanings themselves can become what our talk is *about*, can become the objects referred to by a meta-discourse.
[6] *Vom Erkennen...*, p. 362f.
[7] *Der Streit um die Existenz der Welt*. This is the title of the multi-volume "Summa" of Ingarden's work, which is still unfinished. The three volumes which have been completed up to now have the subtitles: Vol. 1 *Existentialontologie*, Tübingen, 1964; Vol. 2, Part 1 *Formalontologie, 1. Teil Form und Wesen*, Tübingen, 1965; Vol. 2, Part 2 *Formalontologie, 2. Teil, Welt und Bewusstsein*, Tübingen, 1965. (1st Polish edition in two volumes 1947–1948). Substantial parts of Vol. 1 have been published in English translation by H. R. Michejda as *Time and Modes of Being*, Springfield, Ill., 1964.
[8] Cf. G. Küng, 'Language analysis and phenomenological analysis', *Akten des 14. Internationalen Kongresses für Philosophie, Wien 1968*, **2** (1968) 247–53. G. Küng, 'The role of language in phenomenological analysis', *American Philosophical Quarterly* **6** (1969), 330–34.
[9] Furthermore the logician Russell was happy that the transformed statements did no longer violate the logical laws of non-contradiction and of the excluded middle. – I do not know whether Meinong was acquainted with Russell's paper ('On denoting', *Mind* **14** (1905) 479–93), but J. N. Findlay (*Meinong's Theory of Objects and Values*, Oxford, 1963, p. 53f.) mentions that Meinong rejected a suggestion similar to Russell's theory of descriptions. – Most logicians after Russell have accepted the two-leveled view of semantics. Only A. Church has further developed the 3-leveled semantical system of Frege (A. Church, 'A Formulation of the Logic of Sense and Denotation', in *Structure Meaning and Method. Essays in Honor of Henry M. Sheffer*, New York, 1951, pp. 3–24). – A further very interesting alternative view is the one of S. Lesniewski where the universe of reference contains only concrete individuals, but where (extensional) meanings make up the ranges of quantification (the latter are in Leśniewski's system *not* the same as the universe of reference). I am preparing in collaboration with Dr. J. Canty (Notre Dame) an article on this original semantical framework.
[10] Cf. my book *Ontology and the Logistic Analysis of Language. An Enquiry into the Contemporary Views on Universals*, Dordrecht, 1967.
[11] R. Carnap, 'Empiricism, semantics and ontology', *Revue Internationale de Philosophie* **4** (1950), 20–40.
[12] Actually even the tolerance of Carnap and Quine has a limit: an acceptable ontology must be free from contradictions. The extreme platonism of Cantor's unrestricted set theory gave rise to antinomies and has therefore been rejected.
[13] N. Goodman, 'The way the world is', *The Review of Metaphysics* **14** (1960), 48–56. N. Goodman, 'A world of individuals', in *The Problem of Universals. A Symposium*, Notre Dame, Ind., 1956, pp. 15–31.

Notice how Goodman's view agrees with the phenomenological view that the notion of an in principle unknowable *Ding an sich* is nonsensical and with the Husserlian conception that the experience of an evident truth is the experience of a *fulfillment* of a meaning intention and not the experience of a *correspondence* between two things.
[14] He calls his conception "a Peircean conception of truth as 'the ideal outcome of scientific inquiry'". Cf. W. Sellars, *Science and Metaphysics. Variations on Kantion Themes*, New York, 1968, p. 50. – Sellars insists like Husserl on a realistic interpretation of science and rejects instrumentalism. Man in everyday life and the scientist are talking about the *same* reality. Cf. E. Husserl, *Ideen zu einer reinen Phänomenologie und phänomenologischen Philosophie*, Vol. 1, § 52.

[15] 'Husserl's notion of noema', a paper presented at the meeting of the American Philosophical Association (Western Division) in St. Louis, Mo., 1968.

[16] The ideas, too, have a similar ontological double structure: *qua ideas* they have a kind of autonomous existence, while in their *content* any mode of existence whatsoever can occur: there are ideas of real things, of purely intentional objects, of ideas. Cf. Ingarden, *Der Streit*... Vol. 2, Part 1, where also interesting differences between the ontological double structures of the purely intentional objects and the ideas are pointed out.

[17] Cf. Ingarden, *Vom Erkennen*..., §13.

[18] Ingarden, as far as I know, does not discuss the distinction between a purely intentional object as noema and a purely intentional object as referent, and I do not know whether he will accept my distinction. Ingarden *does* distinguish between the 'transparent' objectivity (*'transparente' Gegenständlichkeit*) of a work of science and the non-transparent objectivity of a work of fiction (*Vom Erkennen*..., p. 35, p. 154), but I do not know how he would relate this to the noema-referent distinction. Does he mean that in the referent of a work of science there is more than what is depicted in the content of the noema (that the noema contains variables which are 'filled in' by properties in the referent), while in the referent of a work of fiction there is only as much as is depicted in the content of the noema (that the referent contains exactly as many variables as the noema)? Or does he mean that only works of science but not works of fiction do have a referent?

[19] His position is closely analogous to the case of the poet. – I wonder, however, whether the idealist could not say that the world is a purely intentional object *sui generis* which, unlike the purely intentional object of the poet, does *not* have an ontological double structure, since it does not imitate another reality. It would then only have one mode of existence, namely a mode of heteronomous existence. Ingarden, (*Der Streit*..., Vol. 2 part 1, p. 211) says that *every* purely intentional object has an ontological double structure; but why must that be? – I also neglect the question of the transcendental egos, the actual existence of which Husserl still assumes, and to which some appearances in the world 'refer'.

[20] The discussions about transcendental phenomenology are often vitiated by a serious ambiguity as to what is meant by the step from the natural attitude to the transcendental attitude. Does this step merely consist in the provisional bracketing of the question of the actual existence of the referent of the natural attitude and in the focussing on the noema and the noesis of this attitude? Or does one mean by it the conversion from the natural to an idealistic attitude? Whereas transcendental phenomenology in the first sense may well be accepted into a realistic epistemology, transcendental phenomenology in the second sense is for a realist clearly inacceptable.

[21] Of course more than one kind of realism and of idealism has to be considered. Ingarden distinguishes (in *Der Streit*..., Vol. I, p. 188) not less than 64 cases!

[22] 'Die ontologische Betrachtung besteht in der apriorischen Analyse der Ideengehalte', *Der Streit*..., Vol. 1, p. 33.

[23] There are, e.g., noemata of square circles, but there is no idea square circle; this means that square circles are apriori impossible. Notice also that in virtue of the ontological double structure pointed out by Ingarden the noema *qua noema* is neither square nor circular, the incompatible properties occur only in its *content*. Unlike the non-existing objects of Meinong, the noemata do as such not violate the laws of logic. – The 'if-and-only-if' statement made in the text indicates that there is a material equivalence between ontological statements and certain metaphysical statements about

ideas. But formally only the analysis of the content of ideas, and not the existence statement, belongs to ontology.

[24] Ingarden's distinction between ontology and metaphysics allows also for a rigorous formulation of the puzzles concerning the ontological argument for the existence of God. If the ontological description of the object depicted in the content of a noema could show with certainty a) that this object involves no impossibility and b) that it had a mode of necessary existence, would this be enough to justify the metaphysical affirmation that this object does actually exist? The question just stated seems to be itself an ontological question, but I do not know how it has to be answered.

[25] The term is due to K. Menger, but he does not use it with respect to phenomenology. J. M. Bocheński used to enounce a similar basic philosophical principle: "Why should it be simple if it can be complicated".

[26] Cf. the contemporary development of modal logics, tense logics, deontic logics, logics of belief, logics of questions, etc.; the building of systems whose individuals are events; the task of defining formally the different kinds of organic unity, functional dependence, etc.; ...

[27] Notice that even a false map can be useful. It directs us what to look for at a certain place, and even in the case where we find something else than what the map predicts, the map has been instrumental in furthering our knowledge.

DISCUSSION

Küng: I have one question. You mentioned that Husserl himself said that the building up of the *life-world* can take on different forms. The Negroes in Africa have a different *life-world* from the Chinese and from Mr. Husserl. In this respect, of course, there is the question: "Does not science in return influence our life-world, so that actually the life-world is not only something more basic and better than the scientific image but the scientific image can also contribute to improving the manifest image?" I'm using here the terminology of Wilfrid Sellars and I think that it's important to see how analytic philosophers like Wilfrid Sellars have exactly the same problems as those discussed here. Therefore this is a very basic question: "Cannot the scientific image influence in return the manifest image of the *life-world*? And influence not only in a bad sense?" Of course, maybe Husserl would have stressed the negative influence of the scientific image which comes out in the manifest image of positivists who see the world in a somewhat wrong way. On the other hand, scientific progress seems to teach us also to see the world in a better way than we did before and so that our Western manifest image may be in some ways better than the manifest image of the Negroes in the Congo or that of the Chinese farmers, at least before Mao Tse Tung. Therefore science has also a positive function in the formation of the *life-world.* This connects up with the problem of idealism and realism, because what does 'a better manifest image' mean? It means a manifest image which fits reality as it is better. Only if one accepts a realistic position, *that there is a reality as it is*, can you talk about this manifest image as better or less good. So here I think we come to an extremely basic and contemporary philosophical problem on which I would like to hear your comment.

Gadamer: Science has certainly influenced our *life-world.* In fact, Husserl attempts to follow up this influence in an historical analysis, showing that at the beginning of the 17th century there was still present a full awareness that the view of the world was being approached in a different way. In the full perspective such approaches are evolving,

Tymieniecka (ed.), Analecta Husserliana, Vol. II, 218–241. *All Rights Reserved.*

relative, whereas with the development of the natural science this view of the world became dogmatically fidgeted. What Husserl opposes is the dogmatic attitude of any one approach towards the *life-world*. The life-world cannot be identified with, reduced to, any scientific view, since a scientist is not exclusively the man who thinks in some way or other but is first of all a person who lives.

On the other hand, natural science is only one of the elements which influence our civilization. This influence has not only negative but also very positive effects upon human life. To mention only one: none of us would be here at this conference without the modern technic and civilization that make it possible. The philosophical problem lies not in making an inventory of bad and good influences of science upon life, but to follow up and to observe the dangerous consequences of *reason in evolution*, if reason, as it seems to be the case, has completely taken over in forging the world view of today.

Laskey: May I say, sir, how much I appreciate your article and it makes me think that perhaps Ingarden and Husserl aren't that far apart after all. I'm recalling from memory and I may be wrong, but I think in letter 46 from Ingarden to Husserl Ingarden says that he admits, of course, the significance of the whole transcendental stage but he asserts again the priority of his own ontological investigation. My question concerns a way of humble reconcilation. Could I repeat the question that Ingarden raises in his own paper: "What is it that forms the guiding thread of the constitutional inquiry from the deepest layer of the originally constituted consciousness of time up to constituted units of sense through which the things of the world and other objects are given?" Now, I'd like to put a corollary to it, namely is it or is it not the case that the *epoché* does eliminate some elements of major importance? Surely the facts of science are not altogether hypothetical constructs by any means! They are different, of course, from what is intuited, but I'm wondering if there aren't some elements which are perhaps left out; with a reconciliation, perhaps, both of these might be put in. If one were to make a case somewhat different for the necessity of a prior scrutiny before the transcendental analyses.

Gadamer: We must first of all distinguish between facts and essences. It is an old Husserlian distinction accepted also by Ingarden. The problem, which you raise, does not concern the progress of the cognition of facts but rather our attitude which we as human beings may adopt towards life, the world, science; as such it is in general the problem of the life-world.

You pointed out yesterday what you believe to be a paradoxical situation, arising from the claim that phenomenological reduction is supposed to proceed forward in evident steps. I do not agree with you and would reverse your statement by saying that it is only by way of constitution that we may have cognitive evidence. It may be observed in Husserl how the procedures of reduction are meant to track and to clear the way towards a genuine selfgivenness from all the intrusion of dogmatism.

We may distinguish several steps of achieving the selfgivenness of phenomena. The fundamental step is to establish a hierarchy of evidences within the inner awareness of the streaming self-presence of temporality. Once this is established, all further steps of our constitutive analysis will be evident. In general, evidence resembles belief. The apodictic evidence of the self-awareness of the inner stream of temporality is of a different nature; it ressembles the Cartesian ego-cogito but within a context different from the Cartesian; its context accounts for its apodictic certainty.

While describing all the ways and levels of constituting the objective reality, Husserl gives also an excellent description of how the constition of the external world relies upon the constitution of the alter ego and of the human community. I repeat: constitution and self-givenness is the same problem.

Sokolowski: My question is addressed to Professor Gadamer concerning his own thinking. You speak at the end of the lecture about an infinite self-correction of man in history. I believe you talk about a *Wirkungsgeschichte*, which in your book is constantly being reinterpreted. Yesterday Dr. Brandt spoke about anticipations of meaning which I think fit very nicely into this schema. There are two questions then: First of all is talking about history not also transcendental, so that you wind up again with the same tension that Husserl had? Secondly, on a broader scale, isn't this another form of "Western intellectual imperialism?" I mean, a way in which one world view becomes imposed upon others? Or at least a claim is made that one world view should be assumed by all other civilizations.

Gadamer: These are two great problems; however, I will attempt to answer them. I will begin with the second one. The occidental way of life is, of course, the presupposition of my own personal and spiritual existence. But in feeling and observing the tensions modern civilization manifests, I believe that we are now more open than ever to admit that there are quite other experiences of the world than those occidental culture has developed. Heidegger, for example, makes a fiction of a conversation with a Japanese scholar, and I myself am very interested in the possibility of forms of speaking other than the subject-predicate grammatical structure that the Aryan languages allow. It is not only a question of different races or of different languages but also of different ways of using language. For example, occidental and oriental ways of describing things, which makes such great difficulties now in the negotiations between the American delegation and the Vietnam delegation. Also when one is at the beginning level of studying a language, e.g. English, one doesn't speak in the same direct and idiomatic way as in one's own. One circumscribes, one mirrors, one reflects, one gives hints which others cannot grasp. I think therefore that it is exactly due to the crisis in which our civilization is that we are more open to other cultural traditions than before.

The first point: is universal statement about history not also a transcendental statement? Of course, transcendentality cannot be avoided when we reflect. Reflection is not the same as the direct approach to things. If one asks what the basis of reflection is, if one asks how one can attain validity for one's own views and statements, I must confess that within the transcendental perspective it cannot be answered. Pragmatism is wiser than transcendentalism on this point.

Strauss: In Descartes and in Husserl doubt plays an important role. And is doubting not a specifically human attitude? One cannot doubt existence. All doubt concerns propositions which say that something is so and so and in doubting one wonders whether he should say it. One wonders whether it is or could be otherwise. And therefore doubt also refers to the concept of the *factum*, to the facts. Even the word that expresses that the *factum* as a past tense form, is something definite. But the *factum* that this door is closed refers to the human situation of considering the possibility that it could be open. All our observations require that we go beyond that which is immediately accessible to our considerations. And all this seems to me to be involved more with the human situation and our placement in this *world* than with a transcendental *ego*. We don't meet the other *ego*, we meet the other fellow. And I think this is a quite different situation. When I read the passage here on page 12:

> What constitutes all the forms of world, for example, the world of dreams, the world of children, the world of beasts, the historical world, the problems of bith and death, the problem of the sexes, is not one ego beside others and is not in the world as one of its parts but functions only as the ultimate functioning ego in all its absolute apodicticity but also in its unique philosophical solitude.

I wonder: Is the transcendental *ego* creating? Does 'constituting' mean 'creating'? I see you wouldn't accept that. And yet it seems to me that this ultimate functioning *ego* has a deistic character. It almost acts like a *deus ex machina* or *loco transcendentalae* and it seems to involve the problem of our embodiment, or to start it from the other way around, that we are living human beings.

Gadamer: I agree perfectly with your remarks on several points. Of course the way of doubting described by Descartes or by Husserl is not a real doubting, it is a methodical doubting. This makes a great difference and any Cartesian knows that this doubting is an artificial operation. Descartes had no idea of the transcendental *ego*; it is Husssrl's own discovery. Kant made a good advance towards the discovery of the transcendental *ego* by the synthesis of apperception but only in respect to the constitution of the natural sciences. According to Husserl that is not enough. Concerning your remark about the deistic background of the transcendental *ego*, I think it is the last form, I will not say of the deistic approach, but of the *intellectus agens* and the *noesis noeseos*. It is a problem of metaphysics, which presupposes an infinite, presenting, acting center in order to account for an infinite advancement of experience and for its general framework. It would fall into the scope of the old metaphysical tradition. This point I cannot follow. On my view the finitude of the human being is so basic a feature of man that we must find new specific approaches to account for the specifically human experience instead of claiming to have a divine insight or a divine possession of truth.

Questioner: I have a question about the transcendental *ego*. If we interpret the egological project of Husserl in the sense of the late Heidegger and in the sense of your own work, can we not see that egology, not as the study of the self-constitution of *ego* but as *Sprache des Selbsts*, has its foundation in '*das Selbst der Sprache*' (which was indicated, for example, by Heidegger in *Unterwegs zu Sprachen*, in the later part of it)? Can we not see, then, that the problem of the transcendental *ego* is finally the problem of the language of the self which is founded in the self of the language?

Gadamer: Yes, that is quite a problem. The problem is whether we are willing to accept the concept of a last self-identity of the *ego* as a basic problem of philosophical science or whether we base ourselves on living speech for the progress of human dialogue. In the second case the self would be not the self of one or other person but the self of language. I think however that in analysing our linguistic experiences we have always an encounter with ourselves. To use a word, to find a word, to be satisfied with an expression is like a self-confirmation. In this way we distinguish for linguistically expressed experience two opposite points of view: to have a self and to find one self confirmed by the self of the language. The identity or the inner fusion of these two forms of experience is the very life of human experience in all its linguistical media. Perhaps it is not only the language as a formal construct but also all the different and expanding content of experience we encounter in speaking with one another that gives us the experience of selfconfirmation or of self-refutation.

Tymieniecka: I would like to make two comments. They are in fact rhetorical questions, since as I have already pointed out several times, there are certain aspects of the theory of the *life-world* in Husserl as well as in his followers that seems to raise real and quite fundamental difficulties. Both of these points refer to the fundamental claim of Husserl and of all who followed his theory of the transcendental constitution. The constitution of the life-world is supposed by Husserl to account for all philosophical problems that could be raised. One of the problems which seems to arise is the following one: If the life-world is emerging spontaneously within the strictly individual subjectivity of every *ego* there arises not only the question of intersubjectivity in terms of understanding in human communication (which Husserl is indeed trying to solve and I myself would think maybe in an adequate way) but also the question of intersubjectivity in terms of interaction which goes beyond understanding and psychological communication as such; interaction of the monads among themselves, as Leibniz would say. Husserl, in the *Cartesian Meditations*, has related his theory of the *ego* very radically to Leibniz, presenting it really as a monadology. The *ego* is supposed to be a monad in the Leibnizian sense. Leibniz, in order to account for the interaction of all the individual monads within one world, had to rise above the realm of this world. Let us say, to use Husserlian terms for Leibniz, that he had to leave the order of the *life-world*, which each monad is projecting by itself, he had to seek the principle of the coordination of these individual *life-worlds* in the 'constituve system of the universe', in the universal harmony. The universal harmony is not intrinsic to what we could call the Leibnizian *lived-world*. The question is whether Husserlian theory, or Husserlian monadology if you want to say so, can solve this question of the ordering of the world-order, within the theory of *life-world* alone.

There is a second point, which seems to me equally important; it puts again in doubt the claim of self-sufficiency of the life-world's constitution; its claim to account for all of what there is. Although the language of the constitutive analyses is rich, aesthetic, full of emotive overtones which have an artistic impact upon us, the way in which Husserl analyzed its modes and processes is strictly rational. Constitutive analysis proceeds in strictly rational and universal terms. Of course, it is due to this fact that, as Professor Gadamer has justly said, we cannot despair that the theory of the *life-world* can contribute to the universal science of phenomenology. However, grasping the life of consciousness in this absolutely universal system of the cognitive procedures we have left out something radically different, namely the life of the soul or the empirical natural *ego*. This remains completely suspended, and of course, here the question has to be raised: It is all very well to reach the originary constitution – which means again nothing else but formal universal structures – but what about the spontaneous stream of psychic life which is carrying on this whole process? There would be no constitution of the *life-world* had it not been carried by experience which proceeds by various types of emotional life and which we cannot grasp in universal structures as I have pointed out in my lecture. This experience is what we call the life of the soul in Pascalian terms and is the stream that is being constituted into meaningful structures. Unless there would be found a coordination between the strictly universal aspect of the constitution and this pulp of its life, this really palpitating stream which goes on, we cannot really account for the *life* of consciousness as such.

Gadamer: I think the second question has been overlooked by Husserl in the *Crisis*. Concerning the first problem, we are much more modest. Leibniz had an enormous pretension in extrapolating the problem of the harmony of human reason in different monads. Husserl has no answer to this problem. He limits himself to describing only the basic meaning of the being of many subjectivities within the same world. He has no 'over-science'. Therefore you are perfectly right when you say that all problems were not answered by Husserl. Your claim was that he pretends to answer all problems. But you must recognize that there are problems that cannot resonably be raised or answered. With this Husserl would have agreed. Consider the problem of *realism and idealism*. On this point I find our discussion of the last few days too peaceful. All agree, of cource, that the existence of the external world is a great problem. But it is not the problem that concerns Husserl. When we follow what Husserl accomplished, we cannot maintain that the question of the existence of the external world is a reasonable one, and I myself as a hermeneutic being cannot answer a question that does not arise within my world-horizon. When I have no reason to doubt I have no basis on which to invalidate doubt. I think Husserl is quite right in asserting that the world-belief is never put in doubt by living beings.

Concerning the second question, I think that Husserl is really suffering from this problem. "How can I combine the necessarily contingent world view I have by my own traditions of historical and personal experiences with the claim of a universal science?" All his reflection (the appendices of the six volumes), a tortuous repetition of the same self-examination, stems from this point. I think his answer was that the starting points are contingent but we did learn (that was my description in my paper) how to discern eidetic structures among the contingent. We can proceed scrutinizing by imagination individual experience and find out the common structure.

Kockelmans: I have two questions: One has to do with Professor Gadamer's text. It's on page eight. Because of the conciseness of the text there seems to be an ambiguity in the last lines:

He didn't realize that after the suspension of all objects in the world by the suspension of the general thesis of reality, the belief in the world as such... was not suspended also...

This sentence at first sight seems to be in contradiction with the explicit statements of Husserl since *Ideas*, because I believe that when he talks about the reduction, he explicitly says that what he is suspending is the *belief in the world*. Professor Gadamer has tried to avoid the confusion by putting it between brackets, that is to say, in the horizontal intentionality of the world antecedent to the very positing of entities. Does this statement mean that Husserl at the moment when he introduced for the first time the reduction did not sufficiently realize that the totality of meaning, which as meaning is left after the reduction, cannot be taken as an absolute but is relative at least in so far as it is only a totality of meaning which has its origin in a particular life-world and not in another? Is that what you mean?

My second question has to do with your discussion with Professor Sokolowski and the remark made by Professor Brand. Am I wrong in assuming that our Western Civilization has indeed all reason to be proud of itself in regard to other types of civilization because of the fact that it produced science? I do not want to express any depreciation of any other forms of civilization and culture but is science not an enormous good for mankind? Something to be seriously proud about. And when we talk about trying to overcome the crisis created by science, I fear that many phenomenologists go astray, claiming that to be a 'genuine phenomenologist' means to disavow science. *If there is anyone and anything to be blamed is it not much more the philospher who, since the 16th century, never has been able to explain the genuine meaning of science and, even more important, has been unable to show its genuine function and meaning within the world in which we live?*

Gadamer: To the first question: Yes, my lines are almost a literal translation of the remarks of Husserl (I think in the second volume of *Ideas*). To reconcile this apparent contradiction I have examined the texts and I think that you are right in supposing that there is a manifold of the life world to be assumed. It is enough recognized in the *Ideas*, but one can explain this omission by the fact that the horizon character of the *life-wold* was not sufficiently acknowledged. The second question: Yes, I agree perfectly but with one addition. That is that the greatest part of philosophers and scientists who are believed to be philosophers also have reason to accept the blame.

Klibansky: It is difficult to do justice to such a rich historical theme in a few minutes. I will try to fulfill this part of Professor Gadamer's desire that our proceedings should be less peaceful. I shall concentrate on some of the main points. He showed us that Husserl maintained even after he recognized the central importance of the problem of the *life-world* that philosophy should retain its character of a rigorous science. This he maintained to the end. And in the conclusive paragraph of his paper, at the very end, Mr. Gadamer adds, "Philosophical thinking is not a science at all", it is something different. If I may comment very briefly on these two sides. First, as far as Husserl is concerned, in the sixth book, (I am not a Husserl scholar but I've read a little and I know friends of Husserl) we have this statement:

> Philosophy as science, as serious, rigorous, nay apodictically rigorous science, this dream can no longer be dreamt, is dreamt out.

We have, of course, to look at the context in which this remark was made. And I am not sure whether I graps the context. But I do remember one thing in connection with Husserl's recognition, or even emphasis, that history is necessary for radical philosophical thinking. Historical genesis is a further dimension in Husserl's thought, which, although not altogether absent in the earlier Husserl, gained a new emphasis here. I wonder if it is correct to say, as Mr. Gadamer tells us, that the reasons Husserl adopted this new way of the transcendental reduction with historical considerations concerned only external, contemporary conditions. Is there not to be found something totally new in these later writings, in which phenomenology is represented as the secret longing of the whole development of European thought? Phenomenology appears here as a *telos* of European science and philosophy. This is shown in some detail for instance in Husserl's presentation of Locke, where he sees the first idea of the method of reduction; in Berkeley he sees the prediscovery of the problem of constitution; in Hume the first draft of the pure phenomenolog Then, as to the metaphysical consequences of phenomenology, who, according to Husserl, showed us the way, though he was far to the left? It was Leibniz. Husserl points out again *auf die von Leibniz in genialem "aperçu" autizipierte Monadologie*. In Leibniz Husserl sought the first sketch of the world in which the activity of perception is comprehended as reality and the reality of the world is comprehended as an outcome of perception. It seems then that in the later writings we find a new approach to the foundation of the transcendental reduction of phenomenology. The statement of Husserl about the lost hope of philosophy as rigorous science is a small point of *Ideas*. But we have to interpret this statement. Was it just a passing doubt? Or does it not express a doubt at all? I grant Mr. Gadamer that it was Husserl's intention to maintain phenomenology as a rigorous science in spite of these, perhaps passing, inclinations.

But Mr. Gadamer himself asserts that philosophy is not a science. I would like to ask him: "Should we not as philosophers, when we speak about science, first state what we mean by science?" I am not a specialist in the philosophy of science and yet here again and again I find in discusssions between philosophers and scientists a fundamental misunderstanding. Science is viewed by philosophers as a fixed body of timelessly valid truth. However, no reasonable scientist, no scientist who reflects upon the foundations of his science, would accept such a statement. The scientist who reflects upon the foundations of his science recognizes the *historical* character of all foundation. All science is historical. It might be said that philosophy itself is, of course, historical too. It is not the historical character of philosophy and the historical character of science which constitutes a fundamental difference. It might be said that science claims no apodicticity save in a very limited realm, that is, the recognition of finitude, which Mr. Gadamer points out very well. But if science is not a fixed body of timelessly valid truth what is it? Here we find that for the scientist, science is in fact an ideal; an ideal in a threefold sense. It is an idea in the Kantian sense, as a regulative principle by which we organize and order all our cognitions. It is an idea in the Stoic sense of its own historic consciousness. Finally and mainly it is an idea in the Platonic sense as a norm in regard to which we order all of our empirical cognitions whether they satisfy this norm or not. This norm, of course, is strict but our reaching this norm the way in which scientists do, all scientists up to our time and scientists of the future, is of course conditioned by language, by his position here and now. Here I contend that science as well as philosophy is the attempt at reaching knowledge which transcends the 'here and now' from which it is gained, to reach knowledge which is supposed to be timelessly valid, though our claim for this timeless validity, we know very well, is always conditioned by our position 'here and now' But is it not a problem both for science and for philosophy that we want to transcend time and to transcend subjectivity, that knowledge which the scientist and the philosopher aim at is to be valid not only for him but for others as well? It should be a coherent and consistent body of truth. It should be an attempt, and it is an attempt, at formulating in a coherent and consistent way the body of our experiences. But the experiences with which we are concerned in philosophy are different from the experiences with which we are concerned in what is normally called 'science'. There are also different standards of verification and falsification for both. Can we say however for that reason that philosophy is not science at all? Undoubtedly it aims at being scientific. It is not a rhapsody of unrelated insights into the mysteries of language, laying bare the world view implied in it. Mr. Gadamer's own work aims at a standard of rigor which I consider to be scientific.

The last question: is Husserl really an academic philosopher? Have we all this time dealt only with problems of academic philosophy? In that case we would have done what Marx and the Marxists reproach us with. They claim that so far philosophers have only interpreted the world but the time has come to change it. Is there an attempt in Husserl to change the world? When we speak about the *Crisis*, of course this crisis starts with the *Grundlagen Krisis der Wissenschaft*, then it goes on to the *Zusammenbruch des europäischen Menschentums* and means a breakdown not only of the sciences but of the European humanity. And it raises the problem of, and here I think I do justice to Husserl's work: *das Problem der Erneuerung des europäischen Menschentums*, the problem of the renovation of European humanity. What does he offer to this effect? Is it only sincerity – or is it more?

Gadamer: That is an issue which merits considerable attention. Unfortunately, it cannot be discussed now to the full extent it deserves. What is at stake is the relation of philosophy to science. You raise three problems. The first concerns the interpretation of Husserl's statement: "dieser Traum ist ausgetraümt". It is the nature of historical investigation, to which Husserl at his late period related the transcendental phenomenology, that seems to compromise the ideal of philosophy as a rigorous science and the possibility of achieving it. We both agree upon this interpretation, however you go further saying that turning to historical investigations Husserl discovers the continuity of the European culture as a whole, beginning with the Greeks and culminating in the stage of the transcendental phenomenology. That is, he gives to the transcendental phenomenology a historical foundation. It is very moving, in fact, that a thinker like Husserl, who was not historically minded, began in his later days to inquire deeper and deeper into history. And yet, and this is crucial, in the midst of the contingent nature of history he established the transcendental phenomenology in its own, absolute right, claiming for it an autonomy with regard to all contingency. I wanted precisely to stress this point that may be the *aeterna veritas* and must be approached through ways of contingency. Concerning your second point, I agree with you that the idea of science should never be dogmatic, in the sense of the medieval science, and that therefore the claim to a definitive possession of truth can never be realized. All that can be claimed legitimately in science is that the direction of investigations is legitimate. In this respect, I also agree with you that philosophy has no different scholarly status than science does. But scholarly work is always imperfect and, in the first place, the scholarship side of philosophy is not the point of its highest claims. Consequently, one should be careful, despite all the affinities between philosophy and science, not to identify them completely. In this respect there is still the problem of verification to be considered. I am the last to say that philosophy is an arbitrary venture. A philosopher must, in his work, exercise a great self-control and criticism. It belongs to the meaning of science to have some means to verify its body of cognition and so it would seem that whenever we have some form of verification of cognition, we have to do with science. However, no matter to what degree philosophers accept and satisfy this claim of science, I would again point out that philosophy goes beyond this claim, but not in the sense that philospohy has a higher claim. We cannot make the distinction between science and philosophy along the line of a supposedly different object of inquiry. Yet despite these common features I said that philosophy is not at all a science. Then of course the expression "at all" has to be qualified. In the first place it is an emphasis to distinguish between two very similar things despite their similarities. The third question, however, will throw a light on this issue, namely, what are the reason for which I reject the idea of philosophy as a rigorous science. Indeed, Husserl was a profound human being who not only felt deeply the historical and political situation in which he lived but tried by his own effort of philosophical reflection to answer its challenge. His throught was not limited to theoretical problems but like a missionary, he tried to throw it into the life struggle of humanity. In the papers of the sixth volume of the *Husserliana* there is a considerable number of appeals made to human responsibility in which he urges humanity to strive towards a higher ideal, which would be proposed by the insights of the transcendental phenomenology. Yet I do not believe that it is a sufficient solution to the cultural crisis of today.

Questioner: In regard to *Ideas*, Vol. II, Husserl is concerned among other things with explaining the possibility and limits of an objective science of the spirit, which would be the science of the soul, or psychophysics. There's a passage right near the end of the main body of the *Crisis*, a text in which Husserl says something like: "here is no such a thing as a science of the spirit within the natural attitude, that is a science of subjectivity meant either as a *Natur-Wissenschaft* or as a *Geisteswissenschaft*." He adds, that this must be corrected. His attempt to correct it has never been very clear and I was wondering whether you see a tension or an incompatibility between the final outcome of the analyses in the *Crisis* with Husserl's position in *Ideas II*, about the possibility of an objective science of the soul or even of the spirit.

Kuypers: It is one of the principal points of my paper that Husserl distinguishes *Naturwissenschaft*, natural science, and *Geisteswissenschaft*, the sciences of the humanities, but ultimately he doesn't recognize a juxtaposition between them but gives priority to the sciences of the humanities. He admits thus the possibility of a *universal science* including the human soul in psychophysics (with all the methods and techniques of natural science) but from the transcendental point of view he assumes this juxtaposition and this intention of the natural science. He brings out the ultimate reference of natural science to human sciences as dealing with persons and the human community to which the scientist himself belongs. Scientists themselves are a community of people who are exploring nature and utilizing exact methods. He who intends to see what nature is in itself will at the end find that he will not find nature in itself (it is the illusion of the scientific point of view), but, (and this is the transcendental point of view) in a correlation of nature and consciousness. Consciousness must be examined however from the transcendental point of view. Husserl agrees with the intention of the natural science movement with the use of methods of mathematics as far as it goes, but he revokes the conclusive interpretation that 'scientism' has given of these activities.

De Boer: I agree with Professor Kuypers that Husserl at the end of *Ideas II* revokes the juxtaposition of the two attitudes but that nevertheless the natural attitude is not illegitimate in Husserl's view. It is a possible method of inquiry to develop a natural science of mind. I would ask you whether it is not on this point that lies the difference between Dilthey and Husserl. I think that Husserl is more liberal with natural science. He attempts to reinterpret the results of the natural science of mind, whereas Dilthey seems to propose that an *alternative* to the natural sciences of the mind be developed. Husserl's intention to reinterpret the results of science in a phenomenological framework seems more fruitful than the approach of Dilthey.

Kuypers: I agree with you that the difference between Dilthey and Husserl is that for Husserl it is a question of attitude and not of method. Dilthey, in contradistinction, was of the opinion that the difference between natural science and humanities is that whereas the latter gives a description, natural science gives an explanation. For Husserl one can use this or that method but the result must be explained ultimately from the transcendental point of view. It is not a juxtaposition as for Dilthey but it is a question of attitude. With regard to the human being and human culture, he recognized, on the one side, the rights of the natural attitude, the mathematical method as applicable to all kinds of objects, and, on the other side, he gives a deeper foundation for the phenomeno-logico-historical disciplines in the theory of the *life-world* considering always human beings as persons. The historicity of culture is recognized by Husserl after his criticism of the relativism of Dilthey.

Gadamer: I agree with you. I have only a remark to make about Dilthey's relationship to science. You underestimate how much Dilthey was in his own attitude a universal scientist. You take into consideration only the last period of Dilthey's thinking in which he was so under the impact of the new hermeneutical approach of Husserl that his own approach was already modified. In a new publication, Peter Krause has published for the first time the basic papers of the Breslau teaching of Dilthey, the period in which he prepared and developed a full integration of physiology, of biology, of anthropology in the naturalistic sense of the word. And therefore I think Kuypers is right in saying that the primacy of the humanities depends upon what I call 'the universal *life-world* horizon'.

The community of scientists is seen as the last foundation for the possibility of the scientific approach.

Questioner: If with some phenomenologists we accept the statement that a phenomenologist is not supposed to ask the 'why' question or a causal question then the criticism against Husserl as far as the spontaneity of the *ego* is concerned may be met by saying that the question itself is a question of 'why'. In other words, the question raised previously of the spontaneous dynamic aspect of the personality may be dismissed by saying that it is a 'why question'. But if it is true that a phenomenologist is not supposed to ask a 'why' question then how can he develop an objective science of the cultural human world since that science would involve a series of statements as well as predictions, and both causal statements and predictions would involve hypothetical descriptions of something not given to us. If we accept the first statement as true then there are contradictions involved in the logic of the phenomenological method. Would you like to comment on this criticism of phenomenology: the question is whether the phenomenological method allows us to answer a 'why' question, a generic question.

Kuypers: The phenomenological method is a descriptive method. It stems from Brentano, who distinguished between the question of 'why' and the question of 'that'. But Husserl's method is devised precisely upon the basis of this distinction. The way he describes the phenomena shows what appears to us and explains simultaneously the way of appearing. The description of the way the phenomena are given to us is a means to answer their 'why'. The causality questions to the contrary are based upon the relation of causality; one can give a description of the way causality is given in experience, but one cannot answer the question of what it is that is given in experience, but only the question of rationality. The 'why questions' are generally empirical questions of science.

Questioner: Your answer does not justify your statement (or does not support your statement) that we can have an objective science of the cultural activities of the human spirit. How could we, if we are unable somehow to introduce causal explanations and predictive statements? The implication of the use of the terms *Science*, *Objective Science*, or *Cultural Science*, is ability to predict or at least to give a causal explanation.

Kuypers: That is my question at the end of my paper. What are the sciences of man? Should we restrict men to the phenomenologico-historical disciplines or extend to the social, economic views? I believe that we must restrict the science of humanities to phenomenologico-historical ones. Psychology and the social sciences do not belong to this sphere but they have been influenced by the idea of historicity which is elaborated by the phenomenologico-historical disciplines. This restriction is in no opposition to both Dilthey and Husserl.

Laskey: My question is directed not to what you say in the paper, but to what you didn't say. On page 195, section three, the statement appears: "ethical ideas, such as the idea of a true society and of true humanity, one characterized by the idea of a true and genuinely personal individual life (*Einzelleben*)". Are you thinking of ideals which could not be relative? In that case, from which level would these be derived?

Kuypers: The view of Husserl was a prophetic view, as Professor Gadamer said this morning. Husserl lived in the terrible times of National Socialism and in this period elaborated a view of the future of mankind, seeing it as an evolution towards a genuine new society and mankind. The ideal to be proposed for it and the responsibility for finding the way for its realization remains with philosophers. It this perspective of the future of humanity and at the same time responsibility for the philosophers (and especially for the phenomenologists) the proposed way for a genuine personal life and truly human community is that of the transcendental phenomenology.

Adamczewski: Perhaps I might take the liberty of supplementing what you just said by a personal recollection of my encounter last October with Martin Heidegger, who told me about a conversation with Husserl and his son, in which Husserl was arguing for an objective universally valid phenomenological examination of values and concluded that if this were possible, if this examination were explored sufficiently, it would make all men good. I imagine that the only moral Professor Heidegger or myself can draw from this is that Husserl was certainly an optimist in this respect.

Kuypers: At the end of Husserl's life, transcendental phenomenology took on for him profound significance for the future of mankind. It was like a religious conversion. When one is exercising the phenomenological attitude at the end it transforms his views of values, of humanity, of community, in much the way that religious conversion does.

Questioner: One of the problems that modern analytic philosophy faces is the distinction between two uses of the terms 'why' or 'causal explanation.' One can give a 'why' explanation by citing a covering law. That is, one can describe in mathematical terms what it is that makes the universe the sort of thing it is. But there is a second sense of 'why' explanation, namely, the giving of a motive, for instance: "Why did you come to this meeting?" "To learn something." This is to give a motive.

In Dilthey, the second 'why' sense that I gave is distinct from explanation. That's one of the reasons why he wants to distinguish between two kinds of science (Natural science and *Geisteswissenschaft*, sciences of man). Would Husserl accept this kind of distinction between causal explanation and the 'why' explanation? If he wouldn't accept it, there is surely some distinction here to be made. A universal science would be the same for the humanities and for the natural sciences; it would have to have a unified 'why' explanation. I would suggest, however, that there is a difference between 'why' explanations in the physical and in the social sciences.

Kuypers: Husserl also makes this distinction and he accepts as valid the question of the motivation of one's actions; it is necessary in the science of man to seek the motivation; also in the *life-world* we are always asking for the motivation of our attitudes and our actions. The foundation of the science of man is to find out the motivations of our actions. By excluding all personal forms of expression and personal motivations, one would limit himself to the mechanistic way of explaining the attitudes and actions of human individuals. The question of 'why' is originally excluded by Husserl, under the influence of Brentano; later Husserl recognized the meaning of the question of motivation especially in the sciences of man. *Motives* of one's action is puite a different thing from the laws of action as we find them in the causal explanations of natural science. Then you are excluding also the question 'why'. But when you give the law explanation of natural sciences in the mathematical fashion you also exclude the question 'why'.

Spiegelberg: I am delighted to see that we are finally beginning to get a dialogue between phenomenologists and not only the ordinary language analysts but the constructivist analysts like Quine and Goodman. My specific questions are: It seems to me that the attempt to bring in the distinction of the *noema* and the *referent* may be still a little too constructivist. And I would like to ask, grinding an axe, whether or not the sort of thing which Pfänder did in his logic where he distinguished between the formal and the material object and the formal and material state of affairs, couldn't do the same thing? The time might have come, I think, when this logic of Pfänder, which is completely developed and which Ingarden knew and used as far as I know, may be helpful in this attempt.

The second point I want to make is to agree with you that perhaps phenomenology has something to offer to the kind of Carnapian constructivism. It is already an historical fact that the *Logical Structure of the World* (I am referring to Carnap's early work) is full of references to Husserl. Lately I have found out that there is much more to this. Namely, Carnap's doctoral dissertation in Jena contains a considerable number of references to Husserl, even to the essential intuition. In fact, Carnap moved at this time to a little village near Freiburg, Bruchenbach, where he stayed and attended Husserl's seminars. This explains why there is so much reference to intersubjectivity and similar concepts in the *Logical Structure of the World.*

My real question is: While I'm happy about the fact that the term 'class' offers perhaps a bridge, I'm never quite sure what is meant by 'class'. This is comparable to what you said Ingarden calls 'essence' or '*Gestalt*'. But, after all 'eidos' comes into play and I have still somehow the impression that the class concept is a pretty nominalistic concept. This has to be cleared up. Can phenomenology do that? Can Mr. Küng do it?

Küng: Let's begin with the last point about classes. Of course, 'class' is an extremely technical concept in contemporary mathematics. For 'class' is a concept in logic and it is equivalent to the concept of 'set' in mathematics. But that is precisely the point. When people tried to make a system of set theory, just based on the guidance of their intuitions, then Kantor, who was certainly well trained in intuiting mathematical entities constructed a system which was logically contradictory. Only by finding this logical contradiction with the help of formalization could we modify our notion of a set. It's still a very unsettled question how symbols take on meaning; often logistic philosophers try to communicate with their symbols and don't explain what the symbols mean exactly. Goodman is frustrated by this failure to explain what the symbols in set theory really mean and therefore he says he cannot understand what they mean. Class would be an example of this difficulty; the distinction between a heap and a class is a distinction which couldn't be made before the development of formal set theory. To get back to Pfänder's distinction between formal and material object, I must confess that I haven't studied Pfänder's logic and so I can't answer the question. Carnap, in his dissertation on space, had three chapters. One chapter on physical space, one on phenomenological space, and one on mathematical space. You said he studied with Husserl *after* writing his dissertation, so I wonder why just after his dissertation he drops his interest in phenomenology and starts accusing Husserl of psychologism.

Pietersma: I would like to remark on an assortment of things. First, to the last point that Dr. Spiegelberg and Dr. Küng were talking about: I haven't made a thorough study of the sources but it seems to me that in these writings of Carnap, for example his treatise where he tries to prove that psychology could be written in physicalistic terms (I have forgotten the title of it), what bothered him and Schlick too, specifically, was the notion of intuition. I felt that Husserl himself was to blame for it because I don't feel that the notion of eidetic intuition is ever really presented adequately by Husserl anywhere until the *Phenomenological Psychology*. Of course this was preceded in terms of publication by *Experience and Judgment*. To Carnap it seemed as if it were a kind of mystical insight. It was something instantaneous and almost infallible. This is actually insignificant; that's just a reaction. Now, a specific question: On page 212 you say that the "*noema qua noema* has a certain 'heteronomous' mode of existence." What is meant by that? Does this mean that we can talk about the *Noema* as such without, in our talking, committing ourselves to the existence of anything else, so that we can talk about the *noema* as such? Is this what is meant by its having a *heteronomous* mode of existence?

Küng: A *heteronomous* mode of existence or maybe we should rather turn it around, a mode of heteronomous existence, according to Ingarden is a mode of existence of an object which depends in its content on mental acts, e.g. Hamlet. What is in the content of Hamlet, what colour of hair he has, what he does, and so on, is all in Hamlet not because of Hamlet himself but because Shakespeare ascribed it to him. Now once Shakespeare ascribed it to him then it becomes objective. And Shakespeare can't change what he has ascribed to Hamlet. If he writes it down, then other people can find the same thing. So it is not something in Shakespeare's mind only. It becomes objective, but nevertheless it was created by the mental acts of Shakespeare and if Shakespeare dies without writing it down then it disappears. If all the books about Shakespeare are lost and all the memories about the books are lost then Shakespeare's Hamlet is gone. In this sense Hamlet is entirely heteronomous in his mode of existence. And the *noema* would be the same.

Pietersma: I think that's quite correct. Another question that I'd like to explore: You ask here on page 213 "Does the *noema* have a referent?" That's a troublesome question: "Does the *noema* have a referent?" On one hand, I construe it as the question of a subject in a certain epistemic situation who asks himself "Does this thing which addresses me as a really existing thing, does it really exist, or alternatively does this thing really have the properties which it seems to have?" The subject asks that question and then proceeds toward a possibly more favorable epistemic situation in which he can decide about such questions. On the other hand, the question, "Does the *noema* have a referent?" can be the question of an external observer critic who stands outside the epistemic situation – he's not himself the epistemic subject – and who asks: "Can this person, who says there is something, really be speaking the truth?" Is he right about it? Does there really exist a thing which he says exists? It could even, if the observer critic comes to the conclusion that there isn't anything – that this epistemic subject about which he judges is completely mistaken – happen that we would have to say that, therefore, the epistemic subject is not referring to anything. We must remember that 'referent' is a term that can be used both in terms of intentionality but also in the sense of the actual existence of the object.

Küng: The expression, '*noema* having a referent' is not of the ordinary language. We would have to define what we mean by the phrase " a *noema* having a referent". The best thing is to teach it by examples. Namely, take Mr. X, who says that there are centaurs grazing outside this building. Certainly you would say that this Mr. X has an intended referent. He wants to talk about centaurs which are grazing outside the building. But in fact, I think there is no actual referent. Who can discover that there is no actual referent? Maybe nobody finds that out in certain cases. In other cases, Mr. X may find out. In a third case, you may tell him. And maybe he gets convinced, maybe he doesn't.

Van Breda: What is the language of painting? Of architecture? That is my first question. My second question: Foucault, in his book, *Les Mots et les Choses*, claims that each cultural period has a language of its own which has surged from the type of the lived experience, which incessantly changes. Consequently, living in one period we cannot understand the language of the preceding ones because the types of lived experience associated with those earlier periods are irreplaceable and irretrievable. This break in communication results, according to Foucault, from the revolution of fundamental categories. And yet there is some sort of communication across periods. How is this to be understood?

Gadamer: I made the same reflections about the plastic arts that led Fr. Von Breda to his questions. I begin in my book with the transitory forms of art and the architectural form comes to the focus of attention because there we have the greatest difficulty understanding in what sense architecture does not stay *era perennis.* However, I think that despite the flux of time, motion, and change, we always retain an experience of the indentity of the work. There is a margin of fluctuating difference, but there is also repetition of form. This repetition accounts for the fact that a building represents more than the cultural moment or period of its origin, the taste and ideas of the architect and the community who built and demanded this type of edifice and no other.

Concerning the second point about the structuralistic view of language: I think it is a good approach for expressing transformations within the human world. But real transformations occur within human beings. Changes still retain the message of the past by pouring it into a new language and giving it a new orientation. In Plato, the presocratic tradition is maintained and simultaneously transformed by means of a new language: the dialectic, the *logae*, the identity and the difference. In the same dialogue where these inner conditions of language of being one and the same and of resuming the difference from one and the other are laid bare, it appears now the *logos* statement cannot be formed without identity and difference. In the same dialogue Plato introduces also two other great categories: rest (stillness) and movement, and at the same level of concepts. Here Plato brings together what was and what is. It appears like a mirroring of an older language in a younger language when rest and movement, identity and difference are grasped together as the common conditions of the possibility of understanding. In this case, it is quite useful to realize that movement and rest come from an older world experience. But at the same time, in describing the inner change of language, we cannot overlook that there is a continuity – as the living experience of history – continuity between the stages of changing languages. The most instructive are instances of very sudden changing. Wittgenstein never reflects upon the forms of passing from one linguistic game to the next. He shows that a game is possible in which rules would change during the game. But he makes no use of, nor does he reflect upon the diatonical dimension of the linguistic experience. In this respect, I think that it is helpful to be an historian and a philologist. Then one cannot overlook this dimension.

Küng: I would like again to recall the contributions of Roman Ingarden. Actually, in 1931 he treated analytically, of course in this own style and outside the perspective of the *life-world,* all these questions concerning the work of art. Namely, he showed that the work of art has a life of its own and what its relation to the interpretations of it is. He distinguished the work of art from its different interpretations and distinguished the influences from the interpretations as well. He also showed how the work of art is connected with the material substratum to which it is tied, say the printed text, etc. He elaborated the question of the specificity of plastic art. What helps in this respect is to see that even in painting the work of art is not the material object hanging on the wall. That which hangs on the wall is actually very much like the book which stands in the library. The work of art is a purely intentional object which transcends the mere physical object hanging on the wall; the same holds true for plastic arts, e.g. works of architecture. There one can also distinguish between the work of art and the interpretations of it when one is judging it. Again you can judge it from various points of view... as an art historian, as an art dealer, or as a consumer who enjoys it. I would like to stress that Professor Ingarden has distinguished all these various ways of experiencing the work of art. I would like to ask Professor Gadamer how he feels about this academic or painstaking labor, putting the whole philosphy of the work of art into such rigorous ontological categories. Do you see this as a kind of distortion?

Gadamer: I think it is reasonable to learn wherever one can and I did learn very much from Ingarden, whom I cited on several points in my book. Of course, his work is a little restricted in its approach. Therefore Ingarden does not draw the conclusions from the results he obtains. My interests go further then his. I have other aims and I risk to draw conclusions. But I am very greatful for the great help he was to me in my younger years. For my first steps in the field of aesthetics his work was, I think, the only basis which I had.

Küng: Would you have some criticism with respect to his ontological categories?

Gadamer: Yes, in this respect I think we learn very much from the Heideggerian critique. Perhaps in answering your questions I will by the same stroke answer some points of Kisiel's paper. The best ontological category to account for the dialectical togetherness of the one and the many I think we have is the category of *participation* in the old Platonic sense. There we must turn away from the realm of ideas as well as from the realm of moving and apparent things and try to grasp the point which makes a bridge, the point of synthesis, that is in the Platonic view, the *Beautiful.* There we have the *eidos* and we have the appearance.

In the moment of the erotic enthusiasm we have a new approach to the realm of eternal or of spiritual being (entities), to the constant and permanent meaning; we have the possibility of regaining what we forgot in old pre-life times. The ontological categories of Aristotle seem to me insufficient for interpreting the experience not only of art but of meaning at large.

PART III

PHENOMENOLOGY AND NATURAL SCIENCE

ELISABETH STRÖKER

EDMUND HUSSERL'S PHENOMENOLOGY AS FOUNDATION OF NATURAL SCIENCE

If we have in mind to consider Husserl's phenomenology as foundation of natural science, we do not mean that Husserl's philosophy is confined to such a foundation. According to purpose and realisation Husserl's monumental opus is more than that. He aims at nothing less than phenomenological clarification of knowledge as such. His philosophy intends the critical elucidation and the phenomenological foundation of rational knowledge on the whole.

So, with Husserl, we have to understand scientific knowledge as one form among others, as a certain mode of knowledge only – as a specially distinguished one, however. For it was natural science where the idea of rigorous knowledge not only originated, but where it also was most purely preserved and most distinctly realised.

As we know, Husserl himself found his way to philosophy through positive science; he was a mathematician before he turned to philosophy. This motivates his great interest in natural science and his highest esteem for scientific achievements.

But Husserl always abstained from a one-sided overestimation of a single science or a special scientific method. He was far from regarding natural science as a prototype of knowledge at all, or as the only means to reveal truth. It will become obvious that the main feature of Husserl's conception of a philosophical foundation of natural science rests in the fact that he does not confine the notion of knowledge to the natural science alone.

This distinguishes his philosophy in a principal way from positivism, whose concept of knowledge is exclusively derived from empirical and experimental science, and which denies strictly any other possibility of knowledge. Therefore Husserl's phenomenology, as far as science is concerned, is something else than mere methodology, i.e. analytical clarification and description of scientific procedures.

Indeed, there cannot be any doubt about Husserl's high recognition of the philosophical importance of methodology. Not lastly it has been

Tymieniecka (ed.), Analecta Husserliana, Vol. II, 245–257.

his phenomenology which has brought the analytical thinking to high philosophical repute and which has prepared the mental tools, which positivism also benefited from. But in addition to that Husserl was also convinced that philosophy had a more comprising task with regard to science. His questioning is different from the outset, it is more profound than that of the positivistic epistemology. It is concerned with that notion which here is circumscribed by the term 'foundation'.

But in what sense do we talk about a foundation of science? Do sciences require a foundation by philosophy at all? Are it not the sciences which have provided their own fundaments being the basis for their works and having been able to support their progress up to now? It has been indeed so. The so-called basic crises (*Grundlagenkrisen*) which shake sciences off and on are not at variance with that. On the contrary, nothing shows more clearly than the term 'basic crisis' that scientists regard it as a part of their own task to privide the basis of their disciplines. Hence, in the view of scientists, it is not legitimate for philosophy to contend with science in its claim for self-foundation.

In fact it is a characteristic feature of science, and especially of natual science, not only to collect and pile up mere facts, as they are given by nature, but to impose a structure on nature determined by the scientific method. It is typical for this method to give a structure to the domains of science in such a manner that an infinite manifold of facts can be reduced to a few basic facts; or more precisely that principally an infinite number of statements about nature can all be reduced to a final number of fundamental laws or 'first principles', as physicists say.

This is an epistemological enterprise of some weight. On one hand here we find the essential criterion to differentiate between natural sciences and liberal arts (*Geisteswissenschaften*). On the other, however, it is the 'first principles' together with the hypothetical-deductive method which enables natural science to derive all its statements from such basic principles and thus to establish them.

So science stands for a system of knowledge connected by reasons and arguments in such a manner that each step is built on its predecessor in a necessary sequence. Such a connection of the steps requires extreme clarification of the first principles and a systematic order in deriving further statements from them.

So natural science is indeed able to provide its own foundations and insofar does not depend on any other support. The more urgent becomes the question, why and how philosophy is concerned with the basis of science at all.

Here we have to realise that the term 'foundation' has not been used in a unique sense. Not seldom in the run of history its equivocations have complicated the intercourse between philosophy and science, and has led to complications in the dispute on their competences. Nobody could be more qualified than Husserl to clarify this issue. He was enough of a scientist to confine philosophy to its proper realm. But he had the courage, too, to tackle problems which had been overlooked by science so far and to discuss them as such – primary problems even and lying deeper than scientific questioning. It had been Husserl who stated clearly and repeatedly that science – despite its own foundation – rests upon deeper but not yet recognised prepositions; and that science, though being on a fundament, did obscure its view to its true fundament – just by the special manner of its own providing of a fundament. Husserl drew attention to the fact that the concept of foundation in science always is of a relational character. Science provides foundation "with respect to" that what it tries to achieve with it. But it does not have any concept of foundation as such; it knows no ultimate and absolute foundation.

Husserl, in his phenomenology, however, ventures for such an ultimate basis of science. The different stages of his philosophical development are characterised by a search for last roots and origins. Starting from so-called 'given' (*Gegebenes*) – we have to explain this expression at once – he does not proceed to new insights but digs deeper and deeper for the ultimate sources of all knowledge. In fact, the term 'archeology' still available, Husserl would have liked to give this name to his philosophy.

Let us now, as far as the limited time allows, follow Husserl's way of thinking during its main stages and see (1) at which philosophical basis of science Husserl arrives, (2) by which method he strives to uncover it, and (3) which significance Husserl himself attributed to his fundamental work. These will be our main concerns.

(1) Visualising the scope of problems in the earliest phase of Husserl's phenomenology as it has found its programmatic expression in his *Logical Investigations*, we do not seem to find anything alluding to a phenomenological foundation especially of natural science.

The central insight of Husserl's phenomenology is the intentionality of consciousness. It is essentially characterised by events, called 'acts', which are intentional in the sense of referring to intentional objects. Consciousness as consciousness of... becomes the leading theme of Husserl's philosophy. According to its method it is a phenomenology in so far as it is interested in analysing and describing, in minute detail acts and their different types of intentionality, both with regard to the quality of these acts and also with regard to that which is given in the acts, i.e. the intentional objects.

Thus the turn to the object was at the same time supplemented by a turn to the subject by way of a new kind of reflection which discovers the phenomena as the primary 'given' in the sense of Husserl. Decisive for this was the insight that there is a parallelism between the structures of the subjective acts and of their objective references. This parallelism forms the basis for a correlative investigation under which both aspects of any phenomenon are to be studied and described in conjunction only. It would be an artificial abstraction to study one without the other, an isolation from the context of actual experience.

This actual experience is Husserl's starting-point. By analysing it phenomenologically in its specific correlations between the experiencing subject and the object which can be experienced, Husserl hits upon a world of entirely different structure than that of natural science.

How far is this world of concrete experiences in our everyday-life a different one, and how is it nevertheless related to the world of science?

One has to realise that the world, as represented in science, shows a certain peculiarity which, to be sure, is not very exciting for science, but which nevertheless entails one remarkable consequence. This world is negatively characterised by the fact that in it the human subject is reduced to an aggregate of objective data and hence is eliminated as a subject. Scientific thinking disregards the subject, excludes it methodically.

Of course there are branches of natural science which deal with the human being to a certain extent, for example anatomy and physiology. But they consider man not as a subject but as an object. The subject in its subjectivity is simply not of interest to the natural sciences.

Let us consider sense perception as an example. To be sure, physicists and physiologists are able to explain my act of seeing by means of scientific concepts and laws. But then I find seeing reduced to a physical-physiological process which is no longer my seeing the things in the world, because I as seeing subject am totally neglected. I being subject of seeing do never see light rays and inverted images on the retina, being the elements of scientific explanation; but I see things, people, events in the world which affect me and which I am engaged in. First of all, as a plainly seeing subject I see already without any physical and physiological theory concerning my seeing. And in my pre-scientific experiences of everyday-life it makes no difference at all whether I know something about theories of sense perception or not.

Husserl's phenomenology now investigates sense perception as such a pre-scientific process. He requires to seek for the phenomena, what they are in themselves. To put it differently, Husserl reminds us of a seeing which is a phenomenon of a consciousness of a subject, a seeing which originates from the subject; and the latter exists not only parallel to theories and besides them, but it provides already the fundament for the possibility of constructing a scientific theory of sense perception.

What here was demonstrated with the example of visual perception is according to Husserl principally valid for the relation between experiences which the subject owes to its direct consciousness and scientific theories. It is the phenomena themselves philosophy has to begin with; all study of theories, however significant they might be, must take second place. Husserl's phenomenology reaches back to the original phenomena of which science as well as our too superficial everyday-thought has lost sight.

Here we find the concept of everyday-reality (*alltägliche Wirklichkeit*) separated from the concept of reality as we find it in the natural sciences. This seperation, carried out by Husserl, was a philosophical procedure whose extraordinary difficulties can easily be underestimated, once it has been performed. But up to the beginning of our century everywhere in philosophical thinking the everyday-world was identified with the world as a scientific object. Their distinction became not apparent until Husserl, with his phenomenological method and with his analyses of intentional consciousness, discovered that the form of reality science talks about is not the world we live in as actual human beings.

With this Husserl made a rediscovery which may seem trivial to us, but which was of prime importance in the historical situation at the turn of the century. He rediscovered the subject in its subjectivity, i.e. the subject as it is given and present to itself.

This indicates the ensuing trend of Husserl's philosophy – the investigation of subjectivity. Husserl called it the wonder of all wonders. Its mystery finally led him with unequalled intensity to a new transcendental idealism.

Now there might arise doubts and scepticism. Why can a philosophy whose leading theme is something like subjectivity be concerned with foundation of natural science? Did we not state just before natural science to be exlusively engaged in objects – hence in that which is not something subjective at all, but which is opposed to the subject and separated from it?

But here Husserl's phenomenology of subjectivity shows us that the usual conception is wrong. Subject and object are not separated at all in a manner, the world of the objects being on one side for itself and on the other side the subject being detached from it and only occasionally interacting with the world.

Rather, the subject is nothing without the world; it is always "filled" with world, and this due to the intentionality of its consciousness. But consciousness is not a natural thing among things. It is essentially nothing else than intentional relation, directedness to objectivities. It is the subject which is not part of the world of objects and which cannot be derived from it. But nevertheless it comprises the entire world of objects. Hence anything which can become an object at all must be provable within the subjective consciousness. All possible objects possess the status of intentional immanence of consciousness; and this is the case for the trivial objects of our actual experience as well as for the scientific objects. These are also the intentional objects of a subjective consciousness: the scientific operations are in the phenomenological aspect nothing else than acts of consciousness – very complex in their structure and subtly differentiated, but nothing else than intentional acts which must be phenomenologically analysable.

Here we encounter the first conception of foundation of natural science in Husserl's phenomenology: The reality of science is constituted in specific acts of consciousness of the subjects, in a way however, that these are only given on the foundation of plain originary (*originär*) acts of consciousness.

It would be a matter of further investigations to show how the epistemological roots of essential basic physical concepts – for instance space, time, mass, force, energy etc. – can be traced back to the corresponding terms as they are used in the non-scientific world of everyday-life; and how, vice versa, the scientific concepts originate and are built up from these terms. In addition there are indispensable processes of modifications of meaning, abstractions, special types of generalisations and formalisations which in turn play the role of special characters of acts, constituting step by step scientific concepts until their precise scientific meaning is achieved.

For these processes Husserl himself gave no detailed analysis. But he knew how difficult and tedious this would be. It were these problems which made him aware of the significance of historical research. Here his idea of intentional history (*Intentionalanalyse*) found its motif – an idea whose far-reaching importance for epistemology has not yet been comprehended, not to speak of having been exhausted: to ask back for all implications of previous ways of thinking which still can be found in our present scientific concepts; to uncover the normally unconscious operations behind our consolidated world which represent the "sedimentation" of our past intentions, to use Husserl's telling phrase. Here the eminent importance of Husserl's phenomenology for the philosophical foundation of science is disclosed by his including non-scientific experience and history into his epistemology.

But with this, as stated, the very meaning of Husserl's foundation is not yet apparent. Anyway it has to be emphasized that Husserl did not consider the plain phenomena of our originarily-given everyday-world isolated from science, but that for him the phenomenological description of everyday-world had been the motivation to investigate the relation between this world and science. For it is obvious that the so-called world of science does not represent a second world besides or above the reality of everyday-life; but it is just this one same world which on the one hand is plainly experienced and on the other is objectivated by science. This led Husserl to the fundamental question: How can the objectivation of the world by science be comprehended at all? How can science be understood in general as achievement of the subject?

This represents basically the problems of the uniform structure of subjectivity. For it is the same subject living its life here and now maybe with effort, maybe just bearing it, but certainly fulfilling it one day, because it is a being finite in time – and at the same time being capable of constituting, examining, corroborating and abandoning relations in science, thus participating in truth and bringing forth accomplishments outlasting its life and everyday-world.

(2) This intricate problem of subjectivity led Husserl to a radicalisation of his suppositions and premises (*Ansatz*) and hence to the second phase of his philosophy, being the most characteristic one. Methodically it is characterised by a series of so-called 'reductions', among which the phenomenological reduction proper, leading to *epoché*, which is the most significant and the crucial point for Husserl's further development.

It means the suspension of all implied beliefs in our acts which are related to being and all modes of being of the intentional objects, the inhibition of belief in existence that accompanies all our acts in everyday thinking as well as in scientific thinking. Hence *epoché* is not – as often misinterpreted – a procedure to negate world and all existence. Rather its main function is to reflect on those beliefs and to concentrate in clarifying what we mean in saying: something is, exists, something exists in different modes.

Thus the problem of knowledge appears in a new form. Because in all insights somehow existence is confirmed, being is posed (*gesetzt*). Husserl's previous question as to the peculiar features of knowledge, concerning its types of acts, thus leads to the more profound problem how knowledge is possible at all. Where do we find the conditions for the validity of our propositions? How is objectivity possible at all? – It is this the classical transcendental way of questioning. With it Husserl's phenomenology becomes a transcendental phenomenology.

At this point Husserl radicalises his former analytical procedures to a method which he called intentional analysis (*Intentionalanalyse*). Penetrating far deeper into the facts than in any of his previous analysis Husserl forces open the intrinsic structures of consciousness, including all that is unexpressively given and implied in it.

Finally, Husserl adopts the view that all being, all entities and objectivities have their source in subjectivity.

Here we encounter a further conception of 'foundation'. Husserl goes so far as to ascribe to the intentions the function of 'constituting' the intentional object, thus becoming the achievement (*Leistung*) of acts, of subjective consciousness. Hence the intentional object is no longer to be conceived as pre-existent, as if the intending acts were simply referring to it as already given, but as something which originates in the acts; and therefore Husserl argues conclusively that being by its very meaning refers us back to consciousness and that it derives its intrinsic sense from subjectivity as transcendental source and origin.

But these are still entirely general statements and at first sight without specific relevance for the foundation of natural science. A further consideration of Husserl's transcendental phenomenology is mandatory to reveal the specific constitution of world in science.

As we say, natural science strives for knowledge of external nature. This leads Husserl to two transcendental-phenomenological questions: (a) How is it possible that something like external nature as such can be given to a consciousness at all? (b) By which way of thinking can the possibility of scientific objectivation of this nature be revealed?

Ad (a). One has to realise that the first question concerning the transcendental possibilities of an external nature is a very difficult problem especially in Husserl's later phenomenology. For it is the subject as a transcendental one which according to Husserl constitutes the being of the world. And this subject is an extramundane one, since it constitutes the world as an intentional object of its consciousness. In relation to the transcendental consciousness the status of the world is therefore that of an intentional immanence.

For a transcendental subject, however, the differentiation between "external" and "internal" does not yet have any meaning. But it must be possible to make this difference transcendentally comprehensible if Husserl intends to answer the question, how external nature can be comprehended at all. Because a solely transcendental subject could not even conceive something like external nature.

Since this difference is in general referred to a subject which is not only a pure, transcendental one but which at the same time is existing in a somatic way. The somatic existence is indispensable for a subject's possibility to experience a world as an external world. Thus the subject becomes able to pose the external world against its own 'interior'.

A peculiar ambiguity in the problem of subjectivity is exhibited by the fact that the subject can become conscious of this difference between external and internal; in other words, that it can pose as an object of reflection the external nature as well as its own self. That the subject in its somatic existence is on one hand part of the external nature and in so far being a thing among things; that it, however, on the other hand can transcendent all things of the world and by way of its own consciousness can oppose them to its own self –: This ambiguity of the subject in its somatic and conscious existence indicates that wonder of all wonders, that true mystery of subjectivity which fascinated Husserl.

These ideas might appear a little remote and not very significant in this context. Also, we cannot elaborate further on the problem of subjectivity in Husserl's phenomenology. But let us remember that Husserl – resting in a long tradition, at least of European philosophy – made remarkable efforts to fill up a considerable gap in philosophical questioning. He recognised clearly the demand for a transcendental conceiving of the subject also as a somatic being – not lastly because of the fact that the transcendental subjectivity is not a sufficient condition for the constitution of nature. Nature is relative to the factual subject in its somatic existence. Thus Husserl's phenomenology has to clarify that and how the bodily functions are essential for the constitution of nature.

But nature as 'achievement' of constituting acts of subjectivity does not mean any longer that nature had its transcendental origins in a pure, absolute subjectivity, but rather in a subjectivity as a human one. That means in addition that nature is nothing unless related to human consciousness and the somatic mode of human existing. It is both together which provide the conditions for nature as it is. For differently organised beings nature might well be an entirely different one. Thus nature in Husserl's phenomenology is a 'human' one in so far as the conditions for the possibility of its constitution are to be found in the human subject.

At the same time nature is the substrate of natural science. Because for the latter nature must firstly be given in order to be the material of scientific procedures. From this point of view natural science does nothing else than continuing the everlasting process of constitution which the pre-scientific subject already had been initiating.

This implies that according to Husserl's philosophy one has to regard the human subject as the origin not only of nature but also of natural science. Not only because and not only in so far as it is active in science, it is foundation of science; but it is foundation already by its specific existence as a human being and as such constitutive for the existence of nature which can be made a scientific object.

But this foundation is still imcomplete. Husserl did not only investigate how and in what sense nature can be given, but also how the science of nature is possible in its objectivity.

Ad (b) Objectivity in its proper sense does not only mean to intend certain objects, but it means a certain quality of proposition and statements about objects. These we call 'objective' if they are to be made and to be proved principally by everybody. Thus objectivity can only be constituted by a plurality of subjects. A solipsistic subject would not be able to understand the term objectivity; it could not even conceive it. Objectivity refers not only to any subjectivity but to intersubjectivity.

Here we cannot dwell on Husserl's attempt to comprehend intersubjectivity also in a transcendental manner, i.e. to elucidate how intersubjectivity is constituted. In doing this we would find that for this constitution too, the somatic existence of the subject is a necessary condition. For the plurality of subjects which is presupposed in intersubjectivity is a plurality on grounds of difference and distinctiveness of human bodies. In turn the possible commonness in which the difference and distinctiveness must be overcome without the plurality being eliminated, is commonness of their intentions and achievements of consciousness which refer to the same things and their connections. It is the proven constitution of intersubjectivity, the showing of its possibility in the transcendental consciousness which completes Husserl's transcendental phenomenology.

With this Husserl can claim to have discovered the phenomenological foundation of natural science. It is apparent that the claim of science itself to provide founded knowledge of reality is left untouched. But through Husserl we see that science has its ultimate foundations in subjectivity. Husserl tries to demonstrate that science is not simply constituted by the subjects by way of certain achievements of its consciousness, rather these achievements also find their roots already in the somatic existence of the subject.

(3) The problems which arise here we can only discuss shortly in connection with our third question on the importance Husserl himself ascribed to phenomenological foundation of science.

The answer is given in Husserl's last work, *The Crisis of European*

Science and the Transcendental Phenomenology. It is a work of surprising scope of phenomenological thinking in which Husserl reaches a new level for this former transcendental problems.

"Crisis" does not term a state of science here but rather the historical situation of mankind in the present era of science. According to Husserl this situation is a critical one in so far as science has lost its contact with the deeper concerns of man, his values and aspirations in life – in spite of the progress of science or even because of its progress.

Nature as the domain of scientific research is not any more a reality which man truly can comprehend as the domain of his actual living. Thus the progress of science is paralleled by an ever-increasing estrangement of man from the scientifically disclosed reality. In his work mentioned above Husserl demonstrates the genesis of this estrangement since Descartes and Galileo. This, however, is no assault against science. Rather Husserl tries to overcome this estrangement by analysing its historical development.

This is for Husserl one of the most imperative demands to philosophy today. It is the duty of philosophy and not of science to preserve and to renew in our present consciousness the knowing about where science derives from, i.e. about its factual and historical roots. For this retrogression is necessary into the originary *life-world*.

Transcendental phenomenology is needed to make explicit and thus save the concealed and endangered sense of European science; i.e. the transcendental method is needed to demonstrate the subjective conditions of scientific knowledge and the objectivity constituted therewith.

Here Husserl is struck by the insight that science is not only a theoretical structure but the highest and most extreme form of human activity (*Praxis*), having been historically developed as such. So finally the transcendental questioning comprises history also, because it comprises those forms of subjectivity too, in which man is living his actual life in time.

Thus Husserl demands for a philosophical foundation of science to investigate how science belongs to the *life-world* of man on the whole, how it has its roots in it historically and where science stands today.

In Husserl's work we do not find any detailed analyses concerning these problems. His last work also remained an introduction to phenomenological philosophy, as were all of his previous works. So Husserl handed down to those who follow him a wide range of open problems, relating not only to intrinsic difficulties of his own philosophy, but also to the infinite historical task of philosophy to clarify the foundations of science. To Husserl science as well as philosophical foundation of science were essentially enterprises whose goals lay in the indefinite future and whose task was consequently unfinishable.

One may argue with Husserl; one may even condemn his conception of philosphy. But the philosophical imperatives he set, the questions he put, they have to be embarked on, if we do not just want to do science but to grasp it as something to be permanently achieved by our subjectivity and thus being brought into existence.

HENRY ELKIN

TOWARDS A DEVELOPMENTAL PHENOMENOLOGY: TRANSCENDENTAL-*EGO* AND BODY-*EGO*

I

Husserl in his later years came to realize that his phenomenology, to become the properly foundational science that had always been his goal, would have to take on a genetico-historical focus. Freud, whose familial and cultural background strikingly parallels those of Husserl in some fundamental respects[1], formulated a theory of psycho-sexual development which greatly influenced Max Scheler's latest work, and upon which Merleau-Ponty and Ricouer have elaborated their own thought. More recent Freudian studies on infancy, however, if examined from a phenomenologico-existential viewpoint, may apply more directly to Husserl's goal by illuminating the very onset and development of consciousness. I wish here to present a theory, drawn mostly from Freudian investigations, which has served to clarify my own work as a psychotherapist, and which, I believe, may have an important bearing upon current phenomenological research and philosophy.

My starting point is Rene Spitz's findings on the infant's 'smiling response' to an adult's face, usually beginning in the *third month* of life.[2] He regards this as the initial onset of human consciousness, and variously characterizes this event, in Freudian terms, as the emergence of the 'rudimentary *ego*,' the separation of the *ego* from the *id*, the conscious from unconscious,[3] the psychic from the somatic,[4] the onset of the reality principle;[5] and in more inclusive terms, as the beginning of thought processes,[6] and as the point of transition from an "objectless state of undirected activity to that of *ego*-directed, structured and (intentional)[7] activity."[8] These determinative qualities are implied by such features as the infant's beginning to differentiate his own sound from that of others, as by imitating his own sound in babbling monologues, and by "volitionally and deliberately"[9] signalling his needs.

It is only at about the *sixth month* of life, however, that the infant's consciousness, along with the visually controlled motoric use of the

Tymieniecka (ed.), Analecta Husserliana, Vol. II, 258–266.

hands, enters upon the physical world of distinct objects in space. This explains why at this time the infant stops giving the smiling response indiscriminately to any stranger as readily as to the mother. For only now does the infant begin to recognize both himself and the mother, as well as other material objects, as distinct physical entities.

I have termed the period between roughly three to six months of age that of *primordial consciousness*.[10] Its onset, presumably in the smiling response, is a momentous ontogenetic breakthrough from the vital to the mental-spiritual life (or in terms which Teilhard de Chardin applied to phylogeny, from the biosphere to the nöosphere). Thereby man, lacking the animal's instinctive endowment of early postnatal motility, *acquires mental-spiritual autonomy before acquiring physical autonomy*. Through Spitz's findings on the infant's smiling response, psychoanalytically oriented investigation has thrown sharp light on this critical anthropological issue.

The smiling response entails the differentiation, or psychical disengagement, of visual from global sensory perception. Spitz implies this when he links the smiling response to the cognitive awareness that is founded on visual, distance perception in contrast to oral, contact perception. The smiling response itself, however, he regards as a biological 'innate release mechanism' (IRM), like those found in animal life, to the 'sign Gestalt' – this may be of cardboard or a Halloween mask – of the human forehead, eyes and nose presented straight on and in motion. Thus his naturalistic presuppositions deflect him away from truly understanding, or intuiting the meaning of, the smiling response. Taking into account the general meaning of the peculiarly human, non-animal phenomenon of the smile, a cognitive awareness of a seeming paradox, or ambiguity, which is accompanied by an orally relaxed, playful openness to diverse possibilities,[11] we readily perceive, as Spitz does not, that the breakthrough into human consciousness is here a sudden awareness of seeing a face, as the infant has always done while nursing, without simultaneously having the sensations provoked by having a nipple in the mouth.

Thus primordial consciousness is an initial upsurge of knowledge in the form of a recognition, having an inherent temporal structure. For the present is no longer a discontinuous moment, but contains within it a memory of past and anticipation of future experiences (of feeding). The adult's smiling face is the visual image of this temporal cognition, the initial sensory image of (mental-spiritual) Selfhood. However, in the non-material, non-spatial world of primordial consciousness, the human face cannot be experienced as an object in the sense of a separate entity, or discrete figure delineated against a background. The infant perceives the face in eye-to-eye contact (otherwise it is not experienced as a face) as an entity, to be sure, but in the particular sense of a mental-spiritual focal point of recognition within a shifting, often kaleidoscopic sensorial totality (encompassing what we, but *not* the infant, may distinguish as walls, ceiling, furnishings, light and shade, sounds and odors, etc.) LSD and other psychedelic drugs seem often to re-induce this sensorial as well as other non-sensorial aspects of primordial experience.

The primordial Self is both separated from and yet identified, in union with, the primordial Other. This polar relationship, doubtless best termed a co-union or communion of distinct Beings, forever remains, however attentuated, an earmark of consciousness. The diverse aspects of the primordial communion have been reflected, from time immemorial, in various mythical and religious symbolizations; and they inevitably reappear in all thinking – philosophical, scientific, or ideological - which aims to give meaning to personal life. Thus they are reflected in the writings of Freud and Jung. My treatment of primordial consciousness, in fact, serves to elucidate Freud's cryptic remark about the child's 'personal prehistory' when he makes his 'first and most important identification' with a 'Father,' a sexually undifferentiated parental figure – '*a direct and immediate identification... (which) takes place earlier than any object-cathexis.*" and which remains "the origin of the *ego*-ideal."[12] With regard to Jung, the concept of primordial consciousness provides an ontogenetic matrix for his life-work as a whole, and in this particular context, for his extensive findings on the mandala as a "symbol of the Self."[13] The mandala is a square or circular enclosure whose sacred or numinous quality radiates from its bright center. It seems clear that whereas Freud reflects the characteristic goal of Western religious thought, a 'mystical union' of Father and Son, Jung reflects more nearly that of Eastern religious-philosophical experience, that of Selfhood as identified with the Divine Absolute; e.g., Atman is Brahman. The experiential roots of both are located in diverse aspects of primordial consciousness. Permit me to hazard the remark that this very difference is also reflected in the thought of Husserl as compared to that of Heidegger.

II

At about six months of age, the infant's voluntary use of the hands in coordination with eyes and mouth brings a radical transformation of consciousness, to which Heidegger's concept of *Dasein* and Merleau-Ponty's of 'body-subject' now begin to apply. To the primordial temporal structure of consciousness is now added a spatial structure, or organizing principle, founded on the experience of bounded material concretion, as in the infant's own body-image and the images of distinct physical objects. The primordial *uni-verse* is thus transformed into an ever expanding *multi-verse*, within an infinite spatial horizon in which new objects constantly appear and old objects acquire new meanings. The infant's knowledge, moreover, is no longer absolute as in the primordial world but becomes highly problematical: that is, he is bewildered, if only because sensorially perceived objects often turn out to be phantasmal. At the same time, the infant is now problematical, bewildering to himself. For his sense of identity, despite its new association with a body-image, remains rooted in the primordial world, the matrix of mental-spiritual Selfhood. The infant thus enters upon *ex-sistence*: from the transcendence conferred by his primordial identity, now reflected in the infinite spatial horizon – in the physical world the primordial Other has become infinite *no-thing*, a horizon – he feels the radical strangeness of a world of manifold objects to which he also belongs as a corporeal being. Seeking to bring coherence into his experience, the infant searches for meanings to resolve the questions of *what* and *why* evoked by his participation in two orders of reality.

Although all genuine philosophy is ultimately rooted in this experiential situation, the infant himself does not produce philosophy. His fundamental questions of identity and meaning are immediately resolved, not by reflection, but by his images of the mother; i.e. in a context of *Eros* rather than of pure *Logos*. For with the mother, also a corporeal being like himself, he enjoys a continuity of experience from the stage of primordial consciousness. If he sees her as two distinct individuals, termed Good Mother and Bad Mother by Freudians, this is not the result of splitting an original image of mother, as Melanie Klein supposes.[14] These polar images, which give effective coherence to the child's now blissful, now terrifying experiences in the newly revealed physical world, arise from the matrix of primordial conceptions, or archetypes, that reflect a dual aspect of the primordial Other, best denoted by the terms *God* and *Devil*. Hence the child's present images can more fittingly be termed *Divine Mother* and *Diabolic Mother*, bearing in mind, however, that like Freud's primal 'Father' in the previous citation, these images are sexually undifferentiated. The Divine Mother revives feeling-memories of communion with the primordial Other as understanding and merciful Loving-Cognition, whereas the Diabolic Mother revives those of panic terror at the sight of the Other's satanic aspect, wrathful, mocking, inscrutable, or indifferent, from an intermediate, pre-culminating phase of primordial consciousness.[15] The period of roughly six to eight months, during which these numinous images provide the ultimate context of meaning for the child's experience, may be termed the *transitional* stage of infancy.

Here the child is experientially engaged in a struggle between omnipotent powers of cosmic Good and Evil, 'God' and 'Satan'. He offsets his 'holy terror' by "holy wrath"[16] (this may be repeated in later life in an acute outbreak of psychosis or in going berserk), thereby participating on the side of the Divine Mother, the concrete embodiment of the Good, or 'God', who is the bulwark of his existence. In his terrible opposition to the Diabolic Mother, however, as well as in his blissful communion with the Divine Mother, the Child remains an integral mind-body unity, in Winnicott's term, a *psyche-soma*.[17]

III

At about eight months of age, there opens a third stage in the development of consciousness. Melanie Klein rightly saw that the merging of the dual images into a unified, more realistic – I would add, mundane – image of the mother, ushers in a cataclysmal process which she named 'the depressive position'. She mistakenly placed its onset, however, at three to four months (confusing it with what I have elsewhere referred to as 'the primordial drama of spiritual regeneration'). At eight months, the depressive position corresponds not only temporally with Spitz's 'eight-month anxiety', but also functionally, when the latter is explained, not simply by the child's fear of losing the mother, but by the role of the unified, mundane image of mother in disrupting his integral Selfhood as a pure body subject.

Melanie Klein perceived the child's conflict of ambivalent love-hate towards the mother and his overpowering sense of guilt for having injured her, which then impels him to make reparation. This constellation reflects 'infantile omnipotence,' a concept which Melanie Klein's first teacher, Ferenczi, applied to the infant from birth. The depressive position also has a contrary aspect, however, more in keeping with Bibring's view of the depressive state,[18] which reflects the child's sense of his total impotence. For with the passing of the dual images patterned after the primordial archetypes (in Nietzsche's phrase, 'God is dead') the child is confronted with sheer, unmitigated physical power, by which he is puny, helpless and utterly dependent on the Giant Mother. This momentous shattering of his cosmic picture forbodes that the 'natural values' linked to physical power and weakness may nullify the primordial "super-natural" values of mercifully understanding and abidingly *faith-full* love – the ultimate foundation of Erikson's "basic trust"[19] – which have been the foundation of his very being.

It is not the question of impotence or omnipotence, however, that goes to the root of the depressive position, but rather *the infant's new capacity to distinguish physical omnipotence from mental omniscience*. The original polarity of dual mother images collapses, in fact, not simply because the child no longer sees them as omnipotent, as Melanie Klein supposes, but because they no longer give coherent meaning to his growing experience. For the child sooner or later becomes aware of the ineffectiveness of his 'holy wrath' against the Diabolic Mother (which in itself, moreover, would tend to confirm her omnipotence), and comes to realize that the Divine Mother is herself fallible: she shows frailty of mind in grasping his needs and feelings, if only by thinking he is hungry when he is otherwise distressed. Then, with the formation of a unified, more realistic image of mother, still physically omnipotent, follows the realization that he, the child, knows not only his own but also her feelings better than she does herself. For he perceives her most subtle expressions of fear, anger, impatience, or disregard when she herself is quite unaware of them. (This capacity, generally retained only by schizophrenics and 'childlike' creative personalities, is lost in what will be referred to later as erotic hypnotization by the mother).

This is the crucial point in the child's development which marks the shift from his being an embodied primordial Self, or pure body-subject, an essentially 'eye-hand-mouth-mind' linked primarily to cardio-respiratory affects centered in the upper torso or 'heart', into becoming an inwardly divided though more differentiated being. This primary division of integral embodied Selfhood results in, (1) a *body-ego* which, whatever its subsequent vicissitudes, remains grounded on the embodied primordial Self, but is now governed by the affects, libido and aggression, primarily rooted in the pit of the stomach; and (2) a *transcendental-ego* (T-*ego*), an observing 'eye-mind' centered in the lower forehead that, always aware of a general, all-encompassing mental-spiritual horizon, transcends in its cognitive detachment the body-*ego*'s affect-laden experience of concrete physical immediacy. The T-*ego*'s roots go back to the identification with the primordial Other at the very onset of mental-spiritual life. In broader anthropological perspective, the division of the embodied primordial Self into the T-*ego* and body-*ego* is the psychical outcome of the elemental fact that the human infant, while nursing, continiously looks at the mother's face while holding her nipple in the mouth.

IV

In its immediate event, the division of embodied primordial Selfhood into T-*ego* and body-*ego* structures entails a loss of Selfhood: the T-*ego*, identified with the primordial Other which now, following entry upon the physical world, is infinite *no-thing*, functions as purely anonymous, impersonal *Logos* or cognition, and the body-*ego* as purely anonymous, impersonal *Eros*, or desire. These qualities of the initially disjoined *ego* structures may manifest themselves in later human life by extreme forms of detached, objective cognition and of violent sexual or crowd behavior.

In normal child development, leaving aside childhood schizophrenia, the T-*ego* and body-*ego* structures re-integrate, though never fully, to constitute *secondary* forms of *body-subject*. Insofar as the depressive position is resolved – it is never perfectly resolved, least of all in childhood – by the mother and child forming a *co-union* or communion of distinct, mutually loving persons through both playfulness with, and concern for one another, the child's T-*ego* and body-*ego* structures directly re-integrate to constitute a form of body-subject that may be termed a *communal ego* structure. Insofar as this re-integration is not achieved, however, a corresponding portion of the T-*ego* changes its character from 'divine' to 'demonic', and becomes an *autistic ego* structure; and this normally re-integrates to a greater or lesser degree with the body-*ego* to constitute a second form of body-subject that can be termed a *psychopathic* (or *sociopathic*) *ego* structure. At this point the child's image of the father may have a significant bearing. Insofar as the child can identify with the father in his autistic *ego*, and this identification is lovingly accepted by the mother, the autistic *ego* dissolves to become an added segment of the communal *ego* structure. But whatever remains in the form of autistic *ego* brings on the problem of social adaptation. This is normally resolved by the gradual transformation of the autistic *ego*, through its identification with the frustrating aspect of mother in a process of erotic hypnotization, into the initial form of the *superego*. This collectivized mind-spirit then re-integrates with a portion of the body-*ego* to constitute a third form of body-subject which may be termed a *collectivized ego* structure. The communal and psychopathic *ego* structures of the body-subject are the dual modes of personal Selfhood; and it is in terms of the dialectical tension between them that the problems of freedom, creativity, and ethics can be properly studied.

Apart from the three secondary forms of body-subject, there inevitably remain unintegrated portions of the original T-*ego* and body-*ego* which were not able to be lived out as embodied Selfhood in the mundane world. On the one hand, stemming from the T-*ego* by way of the autistic *ego*, is a mental-spiritually detached, disembodied *schizoid ego*; and on the other, a cerebrally disengaged portion of the original body-*ego* that becomes a *libidinal 'ego'*, corresponding to Freud's *id*.[20] Psychotherapy aims to re-integrate these split-off entitites into body-subject structures. The most radical psychotherapy aims at the dissolution and re-integration of all other *ego* segments into the communal *ego* structure. This therapeutic goal can be pursued with clear understanding only by focusing on the ongoing dialectic between the T-*ego*, body-*ego*, and their derivatives.

What bearing does this theory, which reflects clinical practice aimed at radical psychotherapy, have on current phenomenology? For my own part, I wonder if it may not have helped Husserl to clarify his objections to Heidegger's work. To designate *human being* as *Dasein*, *being-in-the-*

world, is certainly true enough. But does this not too facilely slur over certain fundamental, primordial issues, which may then subsequently reappear within a confused or disbalanced ideational context, viz. Sartre? I refer here to issues that arise when, following Husserl's example, we question most radically the being of consciousness.

NOTES

[1] See my 'Freudian and Phenomenological Approaches to the Emergence of Individual Consciousness', in *Humanitas* **5**, No. 3, Winter 1970; and at greater length, 'Return to Foundations: An Existential-Phenomenological Re-Evaluation of Freud', in a forthcoming issue of *The Human Context*, London.

[2] Spitz, Rene A., *The First Year of Life. A Psychoanalytic Study of Normal and Deviant Development of object Relations*, International Universities Press, N.Y., 1965.

[3] *Ibid.*, p. 103.

[4] *Ibid.*, p. 224.

[5] *Ibid.*, p. 102.

[6] *Ibid.*, p. 103.

[7] *Ibid.*, p. 84.

[8] *Ibid.*, p. 109.

[9] *Ibid.*, p. 152.

[10] Elkin, Henry, 'On the Origin of the Self', *Psychoanalysis and the Psychoanalytic Review* **455**, No. 4, Winter 1958–59. – 'The Emergence of Human Being in Infancy', *Rev. of Exist. Psychol. and Psychiat.* **1**, No. 1. January, 1961. – 'The Unconscious and the Integration of Personality', *Rev. of Exist. Psychol. and Psychiat.* **5**., No. 2, 1965. – 'Love and Violence: A Psychoanalytic Viewpoint', *Humanitas*, Vol. II., No. 3, Duquesne Univ. Press, 1966.

[11] Plessner, H., 'On Human Expression', *Rev. of Exist. Psychol. and Psychiat.* **4**, No. 1, 1964.

[12] Freud, S., *The Ego and the Id.* Vol. XIX, Standard Edition, Hogarth, London, 1961, p. 31. Italics mine.

[13] Jung, C. G., *Psychology and Religion: West and Eeast*, Vol. XI. *Aion Rearches into the Phenomenology of the Self*, Vol. IX. *Collected works*, Pantheon Press, 1959.

[14] Segal, H., *Introduction to the Work of Melanie Klein*, Basic Books, New York, 1964.

[15] See note 10.

[16] These descriptively fitting terms are taken from R. Otto's phenomenological study of elemental religious experience, *The Idea of the Holy*, Oxford, 1929.

[17] Winnicott, D. W., *Collected Papers*, 'Mind and its Relation to the Psyche-some', Basic Books, New York, 1958, pp. 243ff.

[18] Bibring, E., 'The Mechanism of Depression' in *Affective Disorders* (ed. by Phyllis Greenacre), International Universities Press, New York, 1953.

[19] This concept, introduced by Erik Erikson, has been widely adopted by psychoanalytic writers who are satisfied by his genetic explanation in terms of 'good mother-child relations' in the oral stage. See *Identity and the Life Cycle*, Psychol. Mono. Vol. 1, International Universities Press, New York, 1959.

[20] For this inherently phenomenological reinterpretation of Freud's id, see Fairbairn, W.R.D., *Psychoanalytic Studies of the Personality*, Tavistock, London, 1952.

ERLING ENG

BODY, CONSCIOUSNESS, AND VIOLENCE

> Two elements of plot then, reversal (*peripeteia*) and recognition (*anagnorisis*), turn upon such incidents as these. A third is suffering or calamity (*pathos*)... A calamity is an action of a destructive or painful nature, such as death openly represented, excessive suffering, wounding and the like. (*Poetics*, ch. 11)[1]

To relate the theme of violence to transcendental phenomenology may seem an act of violence itself. For Husserl, though concerned with intentionality, is preoccupied with perception and awareness of perceiving. Violence lies closer to the theme of will or, at least, of action. Yet when Husserl has made the shift from *eidetic* to the transcendental reduction, and therewith to an emphasis on passive genesis, the way in which consciousness, through history, is already informed with implications of action becomes increasingly evident. Eventually, in the *Crisis*, modern science is revealed as a European form of action, even of violence within the horizon of the everyday world. Whereas Husserl had earlier turned his attention first to the *noeta* and then to the *aistheta*, he begins now to direct it toward the *poieta*.

The development of Husserl's thought appears to be marked by a series of reversals or peripeties, to use Aristotle's term. The Husserlian peripety is, in a sense, invited, it is not sudden, and it is implicitly considered a change of fortune for the better and not for the worse. In the *Crisis* the historical 'reversal' of the meaning of science from promise to fatality after World War I and before World War II, set the stage for considering intentionality in connection with action in the everyday world. In the *Crisis* Husserl reached the threshold of a possible treatment of violence within the framework of transcendental phenomenology. I would now like to try to venture a few steps across that threshold.

When I use the word 'violence' here I am first of all referring to violence within awareness itself, which I call "primary violence" to distinguish it from forms of violence I consider derivable from this

Tymieniecka (ed.), Analecta Husserliana, Vol. II, 267–277.

primary form, although I shall not go very far in attempting such derivations in this paper.

A passage from Aristotle and another from Husserl should bring us *in medias res*. The first is from the *Ethics*. After observing that "some men believe their own Opinions no less firmly than others do their positive Knowledge", and citing Heraclitus as "a case in point", Aristotle continues:

> there will be a difference between a man's acting wrongly, who though possessed of Knowledge does not call it into operation, and his doing so who has it and actually exercises it: the latter is a strange case, but the mere having, if not exercising, presents no anomaly.[2]

It is this 'strange case' that I shall try to elucidate as primary violence in terms of *homo violens*' self-justified intentional project, however 'wrong' he may discover himself to have been in a later, more inclusive context.

The next passage is from Husserl's *Ideas II*, from the supplements to that volume:

> Let us now speak of sensibility (*Sinnlichkeit*): by it we understand the 'psychic' underground of the mind, of the mind in all conceivable steps, or the underground of the mental acts in all conceivable steps: from the lowest which we represent to ourselves as the simple act of radiation of the mental glance onto something affecting the subject, terminating therein, grasping, contemplating it, or simply turning toward something with active satisfaction, etc., up to the highest acts of reason of theoretical thought or of artistic creation, of social-ethical action.
>
> Now what is this psychic underground? The primordial sensibility (*Ursinnlichkeit*) is the data of the senses, color data in their visual field of sensation, which is a unity prior to any 'apperception' and which can also be found to subsequently exist as moment of an apperception, namely as moment of the visual aspect. Likewise not only are the sensible feelings grounded in these sensible givens, but the sensible pulsional data are as well, the pulsions (*Triebe*) not taken as transcendant to consciousness, but as primordial experiences (*Urerlebnisse*), continually comprised in the psychic underground, which is a primordial store of sensibility... The original sensibility does not arise through association, primordial sensibility, sensation, etc., it does not grow out of *immanent grounds*, from psychic tendencies, it is simply there, enters. The primordially intellective (*Urintellective*) does not develop 'psychically' from associations, but rather as emanation from the I, which is not 'there' (as the *ego*-alien is) but which is just absolute. On the other hand the I presupposes sensibility as affection, as stimulus, sensibility first of all primordial and then secondary. The I also has a possession. The primordial sensibility is its primordial possession.[3]

In this passage Husserl first of all considers sensibility as the 'underground' of mind, of all mental acts. Then, within the sensibility he distinguishes primordial sensibility, present within perceptual sensibility as a unity or moment in the field of conscious perception. Although Husserl does not say so, such a unity or moment is probably passively given in the field of consciousness. He then takes an important step in characterizing primordial sensibility in observing that the 'pulsions' or 'drives' (*Triebe*) are not transcendant to consciousness, but have eidetic character as primordial experiences.

To primordial sensibility he then contrasts primordial intellection, which does not derive from 'associations', that is to say 'psychically', but from the I as absolute. Nevertheless this I does not exist outside nature, since it presupposes primordial and secondary sensibility. The I is absolute as position. Thus, despite the way in which the I presupposes primordial and secondary sensibility, it is also evident that the I possesses primordial sensibility as its original possession.

Yet realization of the I as absolute position is not accomplished from birth. Only within time does the I realize its original possession within the primordial sensibility. Or, consider the 'reversals of fortune' in which the power of the I is weakened, as in conditions of shock, intoxication, or psychosis. In these instances the I suffers dethronement by the primordial sensibility which attempts to become absolute, as it were. However, it is only to the degree that the dispossession of the I secures the cooperation of the I that we may speak of primary violence.

Such cooperation can only occur, passively and actively, if the position of the I is considered a calamity. The I can consider its own position to be a calamity in a situation which suddenly and powerfully reminds it of its conditionality which it had partially forgotten. Now this is the situation of the person (or group of persons), who has suffered an abrupt reversal of fortune, with recognition of his real situation and all the attendant changes in his awareness of the world and self.

This of course is the situation Aristotle has described. And I have used his description because it helps us to grasp the problem of the phenomenological reduction in terms of everyday life.

The order of events in the phenomenological reduction tends to differ from the order of tragic drama as described by Aristotle. In the phenomenological reduction there is a kind of recognition that precedes the 'reversal', in the sense that the epoché or suspension of belief as voluntary precedes the reversal through which transcendental subjectivity is realized. This differs from what commonly occurs in everyday life where, as in tragedy, reversal alone precipitates recognition and suffering. Now an important consequence of the reversal of fortune coming first to consciousness is that the concomitant suffering appears to have emerged as the merely natural consequence of the reversal, without any responsibility on my part, even for my forgetfulness. In everyday life, in contrast to phenomenological reduction, efforts to re-cognize the situation tend to *follow* the reversal, however closely; thus the effort to grasp what has happened attempts to explicate the tie of the calamity with the reversal as a purely natural one not involving the I. An attempt may even be made to consider the recognition itself responsible in some sense for the *experiencing* of the reversal of fortune. This involves a refusal to see that reversal following *partial* recognition – similar to that in the transcendental reduction – is of a different character from the sudden and unanticipated reversals of fortune occurring in the natural attitude. Although Aristotle is speaking of the situation of tragedy, what he says is also of relevance for the problem of achieving phenomenological reductions, belatedly so to speak, under the conditions of everyday life. Is it possible to achieve a phenomenological reduction in the wake of a sudden reversal of fortune, even if without the intellectual distinctness and clarity sought for of Husserl?

To see what is involved in this problem we must first consider Aristotle's meaning more closely, especially the meaning of the peripety. It is not just any reversal which results in recognition and suffering. The reversal involves a special kind of situation:

> Bur when the sufferings involve those who are near and dear to one another, then we have a situation of the kind to be aimed at. (*Poetics*, ch. 14)[4]

In this situation the person is endangered. He may neglect himself, may give himself over to intoxication and addiction, become careless of his life, or even attempt to end it, injuring the own body. With loss of awareness of the I as absolute position, consciousness becomes confused with the sheer bodily. When this happens that "radical cleavage" by which Husserl in *Ideas I* said that "consciousness is traversed"[5] becomes a cleavage in the sphere of the body. The contraposition of I and world within the encompassing horizon of experience and meaning tends to shrink to an opposition of the lived body and the physical body as phenomenon.[6] The lived body is simply our own body as we are aware of it in action, while the physical body as phenomenon is simply the body as it tends to appear with something of the quality of a physical object, in particular, heaviness, within consciousness itself.[7] Now the physical body as phenomenon tends to monopolize the meaning of world in a fashion contrary to consciousness as norm, incompatible with the ruling position of I in consciousness. The lived body tends to usurp consciousness at the same time, incompatible with the I as original possessor of the primordial sensibility. Now it is just a situation which has emerged as a contradiction within consciousness itself that provokes, for the compromised I, what is construed as a legitimate attack by the lived body on the physical body as phenomenon, both to gratify the lived body as well as to remove the physical body as phenomenon. The lived body and the physical inanimate body tend to become rival vicars for a deposed I. In this connection Husserl in *Ideas*, Vol. II observes:

> the human body ... is an *organ of will*, the *only object* which is *immediately*, *spontaneously mobile* for the will of my pure ego, as well as means to generate a mediate, spontaneous movement of other things which my immediate, spontaneously moved hand, for example, pushes, seizes, lifts, etc. Merely *material things* are merely *mechanically mobile*, and *spontaneously mobile* in *merely mediate fashion*; only lived bodies are spontaneously mobile immediately, and indeed through the free I that belongs to them and its will.[8]

To the extent of the lapsing of the I, the physical body as phenomenon and the live body now stand in opposition to one another, evidence of failure to realize the norm of consciousness. This very possibility of conflict of the lived body with the phenomenal physical body is potential in the I-world relationship, insofar as the physical body is also a part of the world as *koinos kosmos*. With the shift toward the *idios kosmos*, reversing the entire situation, the situation of primary violence emerges. This is a situation in which the legitimacy of violent action is derived by the weakened I from the unanticipated reversal of fortune in the order of world events, perceived as radically incompatible with the rule of the I in the sphere of the *idios kosmos*. Therefore such a reversal is not only illegitimate but also unnatural in the ancient sense of *violens*, not only toward the 'pure I', but toward us all since it is the pure I that is the position from which the logos is realized, and therewith intersubjectivity. This is the point from which we could begin to derive violence as a social phenomenon, but which I will forego here.

Some confirmation of the foregoing may be seen in the peculiarly convulsive character of violence as it manifests itself in the everyday world. In its less pronounced forms or as it may appear to a spectator it assumes a fascinating, dreamlike, even iridescent quality, not unrelated to the fascination exercised on us by fire and leaping flames. For its curious spontaneity has *both* mediate and intermediate characters, as Husserl discriminated them in the passage above. This situation of convulsed intentionality with its strange confusion of freedom and necessity exercises an understandable spell over us in the effort and responsibilities of consciousness.

But we do not need to invoke extreme situations to illustrate primary violence. It is already implicit in the simple observation of Husserl, when, after observing the limits set to our ability to look at our own body, he concludes:

> The selfsame human body which serves me as the means of all perception, stands in the way of my own perception of myself and is a remarkable, imperfectly constituted thing.[9]

Here the lived body and the physical body as phenomenon can already be found in a kind of *sotto voce* contraposition to one another.

But it is the more acute convulsions of violence that draw our attention and at this point lead us to raise the question: is it possible to emerge from the situation of secondary, convulsed violence, even though the naive suffering of the reversal hinders effective recognition of the tacit affirmations and assents that preceded it?

To consider this question more carefully I wish to introduce a clinical phenomenon that has often appealed to the curiosity of philosophers. I refer to the 'phantom limb' phenomenon. The 'phantom limb' appears after the sudden loss of a bodily member, by accidental or deliberate amputation.[10]

The missing portion is subsequently experienced as if it were actually present, although of shrunken dimensions. In a small number of cases the presence of the phantom limb is accompanied with an intractable pain whose nature remains a mystery.[11]

In Husserlian perspective the loss of a portion of the body requires a phenomenological reduction from the side of the amputee if he is to effectively live within his altered body. In Aristotelian perspective the sudden loss of organ is a reversal of fortune. Combining perspectives we may say that the moment of discovery or recognition follows, recognition of the body as constituted within consciousness in the form of the phantom, though not clearly and fully acknowledged as phantom. So that this eidetic 'body schema' remains an incomplete reduction; one might rather term it an 'exposure.' Now the phantom limb itself stands in the way of full 're-cognition' which, as transcendental reduction, can occur only within consciousness at once transdescendant to the I and transascendant to every activity of my own intentional, historically constituted presence. It is also possible that the intractable pain is rooted in this convulsive structure of consciousness, as tends to be suggested by the way in which exacerbation of psychic conflicts may increase phantom pain.[12] Similarly, when in the everyday world reversal and loss precede recognition, attribution of the suffering to the sudden reversal as a kind of natural outcome weakens the stance of the I, tending to exclude awareness of the loss from consciousness, treating it as if non-existent, irreal, or, at the very most, unimportant. The person who has become psychotic, slowly or suddenly, often does not recognize what has happened, and except for murmured complaints of 'nerves', actively and even violently avoids any efforts to confront him with an account of what has 'happened' to him.

The preceding analysis is supported by recent observations on treatment of the phantom limb and phantom limb pain by a method that has been reported as effective in some cases. The method has relevance here for the way in which it involves alteration of consciousness reminiscent of the following passage from *Ideas*, Vol. II:

> What is educative in the phenomenological reduction lies also in the fact that it makes us above all receptive to the grasp of changes in attitude, changes on the same level as those of the natural (*natürliche*) attitude or, to put it now even more clearly, the attitude by nature (*naturale*). Such changes in attitude, like the attitude by nature, constitute correlates of being and sensibility which are merely relative and limited.

A group of Japanese psychiatrists has reported on the effect of LSD-25 on the phantom limb and phantom limb pain with seven different patients.[13] In a more complete discussion of this paper we would of course need to concern ourselves with some of the variations in treatment outcomes among this group of patients. Here I shall merely take up the principal findings and those observations relevant to our present topic.

The first three main conclusions are stated as follows:

> (1) The process of change in the phantom limb due to the effect of LSD follows a definite sequence; (2) A movement factor is involved in the body scheme; and (3) LSD treatment is effective for phantom pain.

The generic pattern of changes in the phantom limb reported by these patients was as follows:

> first there arises an awareness of those portions of the phantom limb which had not been felt until then, following which the phantom limb vanishes, and this seems to be a pre-condition for the disappearance of the phantom limb as such.

With this *re-constitution* of the full phantom in consciousness, reaching back beyond the peripety, the I is now able to intuit the possibility of a reduction of the phantom limb in the discovery of presence succeeding to the tacit suffering of absence. In short, the possibility of a passage from 'natürlich' to 'naturale' may be grasped. Only it is one thing to perceive the possibility, another to realize it.

In order to accomplish the 'undoing' of the phantom with its sometimes attendant suffering, it was necessary, in the above reported treatment, for the patient to recover the fullness of I-world consciousness superior to and inclusive of the body in its lived and physically phenomenal modes of appearing. But the drug alone was not sufficient for this, for in actual fact the effect of the drug is to 'wound' a second time, with fresh *pathos*. The drug enables the peripety to recur, this time with less surprise. In addition, a unity of context favorable for the achievement of a *subsequent* reduction must be maintained.[14] Unless this condition exists chances for re-cognition to prevail over pathos and naive views of what has happened will be no better than they were originally, resulting in merely an increase of the already existing disorder.

As I have indicated, the 'reversal of the peripety' requires the movement of consciousness beyond the peripety, behind it, so that the 'attitude by nature' can now be realized belatedly from the convulsion of the 'natural attitude' following the sudden reversal of fortune. That this in fact is what happens is shown by the reappearance of the full phantom limb in consciousness before it completely disappears, sometimes along with the pain. But the limitations and uncertain effectiveness of such therapy, for example in the treatment of alcoholism using LSD, indicates that the possibility of the reduction is insufficiently realized in actuality.

I believe Husserl throws some light on this. In his description of phenomenological procedure in *Experience and Judgement* he discusses the importance of the span of comprehension for the phenomenological reduction. In describing "free variation as the basis for envisioning the essences" he first takes up the importance of 'inclination' (*Beliebigkeit*) in the forming of the variations, and then he speaks of "the comprehending of the entire manifold of variations as basic for envisioning the essences." It is only in the course of the sequence of variations that the generic emerges as eidos, in a manner 'passively preconstituted'; the envisionment of the eidos lies in the grasp by the active gaze of what is thus preconstituted. Then he continues:

> Naturally what this presupposes is that the manifold as such, i.e. as multiplicity, is conscious and never lost from grasp. Or else we do not secure the eidos as the ideally identical, which exists solely as *hen epi pollōn*. If we actively fancy something or some figure in everchanging forms, we always find something new, and always one alone, namely the last fancied. Only when we maintain a hold on the earlier *ficta*, as a manifold in open process, and only when we focus on the congruency and the purely identical do we secure the eidos.[15]

It seems to me that it is this particular com-prehension necessary for realizing the reduction that is missing in the drugged condition. The drug state itself may furnish materials for subsequent possible synthesis, but it is itself deficient in synthesis, which is just what accounts for its phantasmagoric abundance.

This is the same span as that in which the I lifts itself out of the internecine division of lived body from body as phenomenally physical, to realize the more accomplished consciousness of the I-world relationship in which the body as such is less rather than more apparent, either as actor or as acted upon.

A question which lingers from our digression into Aristotle is this: Is the moment of peripety missing from the phenomenological reduction? Is the realization of the transcendental subjectivity itself without pathos? Perhaps the only consideration which can save us from the thought of the epoché as a kind of self-inflicted wound is that, after all, consciousness itself is a wound that in its way takes precedence over all others, thus curiously serving to reduce all others. Perhaps it is in this way that occurrence of pathos enables that reversal which is a change of fortune for the better.

SUMMARY

The attempt has been made to elucidate a difficulty arising within consciousness as temporal and as constrained by the character of the human body. This is the difficulty of maintaining consciousness as norm despite the incessant lapsing of experience into the merely factual. When consciousness suffers contraction within experience as a consequence of reversal of fortune or intimate loss, the situation of primary violence tends to be sharpened. In this situation the distinction between the body as lived and the body as physical phenomenon becomes one of mutual opposition. I have suggested that other forms of violence are grounded in such failures in consciousness and position of the I stemming from the impact of reversals of fortune on men in the forgetfulness of the natural attitude.

NOTES

[1] Dorsch, T. S., *Classical Literary Criticism*, Penguin, 1965, p. 46–7. – Else, Gerald F., *Aristotle's Poetics: the Argument*, Harvard Univ. Press, 1957, pp. 342–3.
[2] Aristotle *Nicomachean Ethics*, Everyman Edition, Dent, London, 1947, pp. 157–8.
[3] *Husserliana*, Vol. IV, Martinus Nijhoff, The Hague, 1952, pp. 334–5.
[4] Dorsch, T. S., *Classical Literary Criticism*, Penguin, 1965, p. 50.
[5] Husserl, E., *Ideas*, Vol. I (Transl. by W. R. Boyce Gibson), Allen and Unwin, London, 1958. Par. 114, p. 319.
[6] For a phenomenological clarification of the I-world relation see Straus, Erwin, 'Psychiatrie und Philosophie', in *Psychiatrie der Gegenwart*, Vol. I/2, Springer, Heidelberg and New York, 1963, pp. 926–96 1963. (Forthcoming in English translation.)
[7] This usage is to be distinguished from the common reference to 'physical body' (*Körper*) in phenomenology to refer to body as anatomical or physiological structure in biological science. Cf. Plügge, H. 'Körper als Phänomen', in *Der Mensch und sein Leib*, Niemeyer, Tübingen, 1967, pp. 34–42 (to appear in translation in a book of readings *The Philosophy of the Body* – ed. Stuart Spicker).
[8] *Husserliana*, Vol. IV, pp. 150–1.
[9] *Ideen* Vol. II, p. 159. *Husserliana*, Vol. IV.
[10] It may also follow the removal of other organs, although that is not crucial here. See for example: Weinstein, S. *et al.*, 'Phantoms Following Orchiectomy', *Neuropsychologia* **6** 1968, 1.
[11] Kolb, Lawrence C., *The Painful Phantom*, Chas. Thomas, Springfield, Ill., 1954, pp. 5, 17–8.
[12] Cf. Kolb, Lawrence C., above.
[13] Kuromaru, S. *et al.*, 'Effect of LSD-25 on the Phenomenon of Phantom Limbs; the

Problem of the Body Scheme', *Psychiatrica et Neurologica Japonica* **64** (1962), 604–13. For aid in securing a translation of the original version in Japanese I am gratefully indebted to Roland Fischer of the Department of Psychiatry, Ohio State University.

[14] Masters, R. E. L. and Houston, Jean, 'The Varieties of Psychedelic Experience', Dell, New York, 1966, Ch. 5.

[15] Husserl, E., *Erfahrung und Urteil*, Claassen Verlag, Hamburg, 1964, pp. 410–4.

H. PIETERSMA

THE CONCEPT OF HORIZON

The term 'horizon' became prominent in Husserl's later writings and it has been conspicuous in phenomenological works ever since. Though some aspects of the idea can be seen adumbrated in the *Logical Investigations* and the first volume of the *Ideas*, the clearest statements are found in the following texts: *Husserliana*, Vol. VIII, pp. 146–52; *Husserliana*, Vol. XI, pp. 3–15; *Experience and Judgement*, Par. 8. My remarks are principally based on these passages.

I. ANALYSIS OF THE CONCEPT IN ITS PRIMARY APPLICATION TO THE PERCEPTION OF MATERIAL OBJECTS

Perception of a material object is phenomenologically characterized in two ways: (a) The percipient takes his situation to be one in which he has achieved the highest degree of cognitive proximity to his object; the mode of givenness is self-givenness. (b) The percipient accepts the object as certainly existing, *because* of what he takes its mode of givenness to be.

Now in such a perceptual situation that which is 'actually perceived' (*das zur eigentlichen Wahrnehmung Gekommene*) is not isolated and given status as an object in its own right. It is accepted as 'part' of a whole which is not as such actually perceived. What is not so perceived, of course, are such items as the inside and the other side of the perceptual object and other objects in its environment. Speaking strictly in phenomenological terms, however, that which is not actually perceived cannot be delimited in purely objective expressions (e.g. as that which does not lie in the direct line of vision); to delimit it reference has to be made to what the subject himself holds to be an item not actually perceived.

Because what is taken to be actually perceived is not treated in isolation from what is not, the certainty characterizing the doxastic attitude of the percipient embraces both (*Husserliana*, Vol. XI, pp. 39/40). But, although the *Seinsgeltung* of the whole – its status as a reality for the subject – is given with a perceptual situation, it is not a *Geltung* which derives from

Tymieniecka (ed.), Analecta Husserliana, Vol. II, 278–282. *All Rights Reserved.*

that situation as such. Perceiving is an experience in which the object is treated as certainly real. And giving it this status is rationally required by its mode of givenness; it would be unreasonable to refuse to give it this status. But giving the same status to what lies precisely beyond the achievement-scope of the perceptual situation obviously cannot be justified in the same manner, namely by reference to self-givenness. The kind of justification needed must come from elsewhere and be in some sense prior. The perceiver who takes this justification for granted is thus not a subject who starts *ab ovo* but one who enters *in medias res*.

It follows that perception cannot be defined exclusively by reference to its achievement-sense. As an intentional experience in which a material object is apprehended 'in one stroke' (*in Einem Schlage*) and taken as not the only object in the universe perception is intrinsically characterized by both achievement and limitation.

Proceeding now to a description of limitation as a phenomenological feature of perception we find it characterized as a set of implicit, general references to what is held to exist beyond the scope of the subject's present achievement. The references to what lies beyond are said to be implicit, inasmuch as the subject may be attending only to the perceptual object in its actually perceived aspects; in such a case the subject refers to the beyond only in the sense that he treats what he primarily attends to *as* actually perceived aspects of the object. The references are general, inasmuch as the character of the beyond is indicated only in outline. The above-mentioned, all-embracing certainty therefore extends only to the *general* character of what is not actually perceived.

Since these references are constituents of the percipient's doxastic attitude they embody certainty in regard to the existence of what is referred to. And by this certainty the perceiving subject can be 'affected' (*kann das Ich intentional "affiziert" werden*, *Husserliana*, Vol. VIII, p. 147) so as to direct his attention to what is referred to, to wonder about its unknown detail, to speak about it, to visualize it, and finally to proceed along lines already indicated towards its actual perception. In concatenations of perceptions the general references function as rules: only those forthcoming data can be accepted by the subject as, for instance, giving detailed information about the initially perceived object, which fit the outline and specify its generality by individual details. Because perceptions can be experienced as being related in such a way that a later perception fills out, so to speak, what was in an earlier one recognized as indeterminate a percipient can be directly aware of the numerical identity of an object of different perceptions.

What Husserl called *belief in a world* is this certainty regarding the existence of a wider context than that achieved at any given time – a context which is in some ways known and in other ways unknown, and therefore constantly inviting the subject to continue exploration. It is called belief, because the claim involved has not been justified. And it is further qualified as *passive* and *universal*, because its justification is taken for granted in *all* perceptions without *any* being charged with the task of giving this justification. In fact it is not a belief which *can* and therefore *ought* to be justified, as is the case with beliefs pertaining to particular things in the world. And this is perfectly intelligible, because it is not a belief about a particular thing but about the existence of a context or matrix of things that is wider than whatever it is about which we have attained justified beliefs.

Thus in any perceptual experience, which is as such primarily directed to an individual object or a small complex of objects, the world is always already experienced as the wider context or matrix within which its object(s) is (are) taken to have a place. What this wider context comprises is to some extent, of course, determined by personal, past experiences. But the domain of the familiar in this sense does not exhaust what a percipient believes to exist. At its boundary the subject still takes himself to be confronted with an invitation to further exploration.

The world is in principle that totality whose parts are so related that without incompatibility they can all be treated as really existing things spread out in a spatial and a temporal way (*all das, was als einstimmig Zusammenstehendes im Geltungscharakter "daseiende Wirklichkeit" auftritt, Husserliana*, Vol. IX, p. 60). Treating particular objects as really existent can therefore lead a cognitive subject into a situation of conflict with new data that demand similar treatment. But such a conflict is always experienced by him as in principle decidable because of his conviction that he can enlarge the scope of his experience and resolve the conflict in a wider context. That is to say, further investigation is meaningful. And in virtue of the presence of general references such investigation is always methodical.

The concept of horizon as presented in the above analysis involves the idea of a beyond that can be reached. This implies the idea of a subject who is aware of himself as endowed with capacities to change his relation to his object. Perceptual experience, in particular, presupposes a subject aware of himself as capable of bodily comportment; correlative to the horizon in its noematic aspect there is a noetic "practical, kinesthetic horizon" (*Husserliana*, Vol. XI, p. 15).

II. THE CONCEPT OF HORIZON IN ITS APPLICATION TO EPISTEMIC SITUATIONS IN GENERAL

As *experienced* limitation horizon can only be *strictly* defined by reference to achievement as such, namely as that which is taken by the subject to lie beyond the scope of his present achievement. I already touched on this in connection with the distinction between what is, and what is not, actually perceived. Now achievement must be understood contextually, namely in the context of non-achieving intentions, e.g. those which Husserl called empty intentions. Already in the *Logical Investigations* elaborate analyses were devoted to the teleological order in which various intentions directed to a given object or state of affairs can be placed. Various intentions are here considered with a view to what I call the subject's cognitive proximity to *x*, placing them, so to speak, on the scale marking degrees of proximity. That intention in which the subject takes *x itself* to be given, to be given in the mode of self-givenness, stands out as that to which there is nothing superior on this scale. It is clearly an achievement and must accordingly be seen strictly in relation to this scale and not in relation to some other ideal. Here too, we might note in passing, the cognitive subject is understood as aware of his capacity to better his epistemic situation in relation to his object.

One of the terms Husserl uses to signalize such situations of cognitive achievement is *Evidenz*. Epistemologically defined, it is that kind of situation in which one cannot reasonably disbelieve in the existence or being of the object. Husserl himself often defines it as the experience of truth. His attempts to exclude from the characterization of *Evidenz* all doubt-creating ideas (e.g. the idea that it is only a feeling) are well-known. Yet he also affirms that in some sense the truth is *an sich*; it is *was ein für allemal gilt* (*Husserliana*, Vol. I, p. 96; *Husserliana*, Vol. VIII, p. 363). Therefore one must somehow distinguish between *Evidenz* and truth;

alternatively formulated, a distinction is called for between that-which-is *an sich* and that-which-is in its mode of self-givenness. It seems to me that we do indeed want to make this distinction. The epistemically most favourable situation referred to by the term *Evidenz* has been described in terms of intentionality. And this means that, although the situation does not include reasons for doubt, it does not logically *entail* the being or so-being of its object. However, the distinction should not be made in such a manner as to destroy the cognitive worth of *Evidenz* and to lead us to scepticism.

Now I think that it is precisely the notion of horizon that makes it possible to draw the distinction in *phenomenological* terms without opening the door to scepticism. On the scale marking degrees of cognitive proximity *Evidenz* is an achievement and its cognitive worth may be defined epistemologically in the above indicated manner. Yet as such, i.e. as an epistemic situation, it has its limitations, since its objective performance or achievement (*objektivierende Leistung*) is incomplete (*Formale und Transzendentale Logik*, Par. 59; *Husserliana*, Vol. I. p. 96). But it should be clear that making this point does not at all imply a qualification or devaluation of its achievement-character on the scale marking degrees of cognitive proximity to the object. It does not disqualify the epistemic worth of *Evidenz*, leaving the subject without further resources and scepticism as the only attitude to, adopt because as a *phenomenological* description the assertion that *Evidenz* has horizons implies that the subject possesses a method for further investigation. What is evident to him in a given situation and therefore accepted by him as true may in fact be false. But if so he will find out and, what is more, he already knows which direction to take, because his present situation already contains references to future situations that will either strengthen or weaken the present claim. Limitation being a phenomenological fact the subject of an epistemic situation in a sense knows what is lacking and how to proceed. The distinction between *Evidenz* and truth is an intrinsic, descriptive element of the experience. The sceptical way of making the distinction, however, reduces the subject to helplessness, not knowing what is lacking and therefore having no method for further investigation. Because Husserl's manner of making the distinction involves reference to horizons, it steers a navigable course between infallibilism and scepticism towards an acceptable doctrine of empirical knowledge.

ULRICH CLAESGES

INTENTIONALITY AND TRANSCENDENCE*

On the Constitution of Material Nature

The idea of intentionality becomes fruitful for the first time when seen in connection with the methodological first principle of Husserl's phenomenology, namely the phenomenological reduction. Husserl developed this principle in 1907 in the context of the epistemological problem of transcendence. In the following we shall examine what the idea of intentionality can accomplish with respect to the problem of transcendence. We want to limit ourselves, however, to a special problem sphere, namely, the constitution of material nature[1].

The epistemological problem of transcendence was certainly not posed by phenomenology[1]. Rather, phenomenology found the problem in the self-assurance of natural cognition (*Selbstverständnis der natürlichen Erkenntnis*) and in the manifold attempts to solve this problem. The problem of transcendence originates through reflection on the distinction between genuine and simply presumed cognition. Taken as experiences of the perceiving subject, both acts are in the same way real and accordingly as such indistinguishable. Rather, the distinction exists in the fact that genuine cognition of an object corresponds to that which is cognited. The object is not immanent in the act of cognition, rather it is transcendent to this act. The simply presumed cognition, on the other hand, corresponds to no such transcendent object. The epistemological problem of transcendence could then be formulated thus: How can an experience immanent to a subject correlate to an object not contained in this experience?

This question is usually not raised by the natural consciousness (*natürliches Bewußtsein*) which takes the transcendent objectifying function of knowledge as a simple matter of course. When, however, natural consciousness does raise this question it becomes involved, as Husserl demonstrated, in contradictions. These contradictions arise through the fact that what constitutes the problem, namely the transcendent function of cognition, is simultaneously taken for granted. Phenomenology avoids precisely this problem, and does so by excluding the implicit self-assurance of natural cognition. This is achieved through the phenomenological reduc-

Tymieniecka (ed.), Analecta Husserliana, Vol. II, 283–291.

tion. Transcendence is not solved in this way, rather it remains as a claim on natural cognition which must then be examined. The examination of this claim begins with an analysis of the intentional structure of consciousness where it becomes obvious that natural cognition did not adequately pose the problem of transcendence. Namely, we must distinguish in a twofold way between immanence and transcendence.

The first difference between immanence and transcendence is present to every consciousness; it is that between the intentional act and that which is intended as such. The intended is not contained as an immanent aspect of the intentional act, rather it is transcendent to this act. This difference between *cogitatio* and *cogitatum*, between *noesis* and *noema*, is known as the difinition of the nature of intentionality, and doesn't need to be explained here.

The second difference between immanence and transcendence involves the realm of the noematic itself. With reference to a definite *noema*, a complete identity can exist between the given and the intended object. The intended object is intended precisely as it is given. This identity of given and intended is referred to by Husserl as self-givenness, a givenness which excludes any doubt. Husserl calls this perception.

Usually, however, a *noema* involves more than what is self-given in the implied object. This excess of meaning beyond the individual self-given is referred to by Husserl as apperception. The implied in this way through apperception vis-à-vis the self-given in perception is now seen as being transcendent in a new sense. The difference between immanence and transcendence is therefore none other than the difference between perception and apperception.

With this distinction in the sphere of the noematic itself a definition of intentionality has been formulated that permits an adequate posing of the problem of transcendence. Inasmuch as the distinction between *noesis* and *noema*, as well as the transcendence of the noematic as such, is present in every consciousness, the epistemological problem of transcendence must be seen as the problem of the relationships between perception and apperception.

In general we can say about the relationship of these two at least that they stand in a necessarily regulated interconnection and in such a way that apperception is only possible when grounded in perception. What this means in particular will become clear when we explain the problem of apperception with reference to the constitution of material nature. But a preliminary remark is necessary.

The constitution of a region of objects (*Gegenständlichkeiten*) is explained in a so-called intentional analysis, which in turn seeks an intentional guideline. This guideline is established in a regional ontology which returns to the manifold actual or possible modes of consciousness (*Bewußtseinsweisen*) in which the object functioning as guideline is given. The guideline for the constituted analysis of material nature is provided by the field ontology of the object. As we know, Husserl distinguishes three strata in the nature of a thing, which stand in a definite fundamental relationship to each other. These three strata are the *res temporalis, res extensa*, and the *res materialis*, of whose essential characteristics the following is to be remembered. A thing is a *res temporalis* to the extent that it is necessarily defined as a unity of duration in a temporal flux. A thing is a *res extensa* to the extent that it is necessarily defined as a physical and qualified *Gestalt* (form) in a certain place (position), and can preserve its identity through changes in *Gestalt* and position. A thing is a *res materialis*, that is, a substantial-causal thing, to the extent that it can act on other things and can be acted upon by other things, in other words, to the extent that it possesses active and passive qualities.

These three strata stand in a one-way fundamental relationship (*Fundierungsverhältnis*). Every *res extensa* is necessarily a *res temporalis*, but the converse is not true. And every *res materialis* is necessarily a *res extensa*, but the converse is not true. The intentionality analysis with regard to these three can't be dealt with in detail here. The stratum *res extensa* I have tried to explain elsewhere[2]. Here we want rather to examine the relationship between perception and apperception in these three strata in order to approach a solution to the problem of transcendence with regard to material nature.

Even in the constitution of the res temporalis (take, for example, a melody) we have a certain relationship of perception and apperception. In listening we concentrate on the melody as a whole. But what is actually given in each phase of the perception is the originary impression that I have now. The melody, however, necessarily encompasses the preceding phases retained by consciousness. The mode of givenness of the phases retained in this way can be called apperception. They are necessarily implied, although not given in the impressional presence. It is more or less the same with the future phases of the melody. They are implies by protention in the entire melody, but they are not themselves given.

The relationship of perception and apperception is necessarily bound by a certain principle. This principle is in the case of the *res temporalis* present in the structure of the inner time consciousness, and consists in the fact that retention and protention are necessary components of every originary-impression present (moment). However, this structure also governs the transformation of apperception into perception. In the distinct, i.e., the active recollection the past phase, simply as past phase, is rendered as self-given. It is different, however, with the future phase of the melody apperceived through protention. This is raised (*überführt*) to perception through the passive flow of the series of present moments. The implicit future phase of the melody moves of itself into the originary-impression present.

This outline may suffice to clarify the structural principle of the relationship of perception and apperception: the same structure (and here we mean the structure of immanent time) which governs the relationship between perception and apperception, or which governs their necessary unity, also governs the transformation of apperception into perception. In a *res temporalis* we can therefore only call in a strict sense immanent that which is given in the individual originary-impression present. Past and future temporal extension are in comparison transcendent. Because of the above-mentioned fundamental relationship what we have said here is true *mutati mutandis* for the other strata of the object. But for simplicity we shall leave that aside.

The above-mentioned principle becomes clearer in connection with the *res extensa* where the governing structure which controls the relationship of perception and apperception, and concomitantly guarantees the transfer of apperception into perception, or the transfer of the simply implied to a self-given. For an example we can start with the perception of a sphere. One 'side' of the sphere is perceived on the basis of a definite kinaesthetic situation, or naively formulated, on the basis of a certain position in relation to the observer. The sphere is, however, intended as a sphere, i.e., as a uniform and closed surface. But that means that the non-given sides of the sphere are necessarily intended through apperception. It is clear how this apperception is transformed into perception: I only need, naively formulated, to look at the sphere from all sides.

But what is the governing structure that determines the relationship of perception and apperception? It is a kinaesthetic total system, a system of subjective potentiality. The actual kinaesthetic situation is indeed only the actualization of one of these potentialities. But the kinaesthetic total structure must be regarded as an ideal structure: it has the character of a-temporality (*Allzeitlichkeit*) because individual situations can be realized in the structure at any time. Inasmuch as what we call space is necessary as a corollary to the kinaesthetic total system, one can also say: The structure of the space itself governs the relationship of perception and apperception, and then in the above-mentioned two-fold way. According to this principle the consciousness of space is structurally a necessarily indissoluble unity of perception and apperception. There is, of course, the ideal possibility to transform each apperception in an a priori set way into perception. But that does not mean that the connection between perception and apperception could be dissolved. Every transformation of apperception into perception is at the same time a transformation of perception into apperception. With this law we establish the transcendence of the res extensa.

Let us now turn to the stratum of the *res materialis* with which the constitution of material nature is first realized. And we depart once again from our example of the sphere. Such a perceived object is not simply as a *res extensa* intuited, or to use Husserl's term for the qualified space form, not simply as a phantom intended. Rather it is a substance-causal thing as we have said. In order to define the relationship of perception and apperception on the stratum of materiality we must take cognizance of the relationship of *res extensa* and *res materialis*. (I am following here paragraphs 15–17 of the *Ideas*, Vol. II, eliminating the references as I have been doing all along to save time.) We have said that every *res materialis* is necessarily a *res extensa*, but not vice versa. That means that phantom, when given, then principally as concrete, are not intuited as material things or can be intuited as such. An abstraction in this sense was the basis of our previous consideration. However, we can look at this abstraction from still another side.

We take the thing precisely as it is first apprehended, namely as a material thing. If we exclude all attributes which are literally or expressly material then we see that with regard to what is self-given of the thing nothing changes. The thing remains before our eyes as the same qualified space form seen from a definite side. We can look at it from various angles, and transform the previously apperceived into a perceived, and so forth. With that, however a distinct insight has been gained: if nothing changes with regard to the given, if the strata of materiality is eliminated, then it would seem that the strata of materiality has no mode of perception peculiar to it. That which belongs to this strata cannot be self-given in perception, as e.g. the aspect of a phantom. In other words, in the transformation from *res extensa* to *res materialis* no new perceptive stratum is gained as is the case in the transformation from *res temporalis* to *res extensa*. The stratum of materiality is pure apperception. What, however, does that mean?

We have said that no apperception is possible without perception. And that applies here. In the case of the *res materialis* the entire stratum of the *res extensa* assumes the function of perception. This thing is intuited as a *res materialis*, although it is only given as a *res extensa*, i.e. as a phantom. If the stratum of the *res materialis* is of a purely apperceptive nature then the nature of transcendence in such apperception is maintained and the *res materialis*, which, of course, is essential to material nature, achieves transcendence in an especial way. In examining the *res temporalis* and the *res extensa* it was shown that there is a governing structure that the relationship of perception and apperception controls. It is the same that permits the transformation from apperception into perception. If, however, the stratum of the *res materialis* is of a purely apperceptive nature then apperception is in no way to be transformed into perception. We can therefore not have a governing structure in the previous sense.

Our question has now become more precise. What governs the costituitive relationship of perception and apperception for the stratum of the *res materialis*? And we know Husserl's answer to this question. The materiality of the thing identifies itself through its dependence on circumstances. If we consider an isolated thing we have no possibility to determine if that thing is really a material thing. This possibility exists only when we see that thing in connection with other things. But whatever is perceived as actual or possible in such a context of changing things belongs only to the phantom realm. What belongs to possible or real perception in this realm can be described as follows: a certain movement, e.g. change of place or form, or a certain qualitative change in phantom A is concomitant with a corresponding change in phantom B. I can also determine whenever A is altered in a certain way, B undergoes a corresponding change. On the basis of this givenness, Husserl maintained, A and B are grasped as material things, i.e., as substance-causal unities. That means, however, in short, that all events in the phantom realm have their ultimate ground in the interplay of basically non-visible substantial unities. That which can achieve self-givenness within the phantom realm is grasped in the sense of a substantial or causal property, which in turn is only a radiation, as Husserl said, of a unified being.

That such a materializing apperception, which in fact is more complicated than we could present in the context of this paper, takes place and has always taken place may be beyond doubt. Beyond doubt may also be the fact that this conception is only relevant where a governed dependence exists within the phantom realm. But it is not a question of whether this act of apperception takes place, rather why it takes place, or, more precisely, how it is possible. Husserl did not answer this question in *Ideas II*. He describes apperception without explaining it with regard to its possibility and necessity. This is, however, necessary in order to avoid the sceptical attack that causality and substantiality are deceptions of habit.

Let us finally ask why Husserl did not give us such an explanation of materializing apperception, or why he couldn't give us one. The reason, it seems to me, lies in his concept of aesthetic or sensuous synthesis. This is, as Husserl expressly said (*Ideas*, Vol. II, p. 20 Note), also responsible for the constitution of the *res materialis*. What is the nature of the aesthetic synthesis? It is first of all defined in a negative way through distinction from the so-called categorical synthesis which is by nature a purely spontaneous act. Such a synthesis is present, e.g., in every judgement. The aesthetic synthesis is, on the other hand, receptive by nature, joining elements that either originate in an aesthetic synthesis or are ultimate sensuous properties or givenness. What is important is that the aesthetic synthesis be regarded as a synthesis of things of the same kind (*Synthesis des Gleichartigen*). In the synthesis of the *res extensa* (i.e. in the synthesis of the completely qualified space form) the aspects and facets which are already qualified extentional givens are unified. And the whole as result of the synthesis is likewise a qualified extended given. In this synthesis perception and apperception are blended together so that the apperceived, to the extent that it is united with the perceived, is of the same nature as the perceived, i.e., it is determined *a priori* as a qualified extended matter. And for this reason apperception is to be transformed into perception.

But now the synthesis that constitutes the *res materialis* must also be an aesthetic synthesis. The substance-causal thing as unity is, Husserl maintained, a result of precisely this synthesis. But that is impossible. What is pre-given as material to the synthesis that constitutes the substance-causal unity belongs to the phantom realm, and is therefore of a qualified-extended nature. That which is the result of this synthesis, namely the real thing as substance-causal unity, does not belong to the phantom realm, rather it belongs to a completely new realm established through this synthesis. The synthesis is in this case therefore not a synthesis of similar things in the sense that the unity has the same structure as that which is unified through the synthesis. For that reason there is no transformation from apperception into perception here. The synthesis of the *res materialis* must therefore have another structure, one which Husserl did not provide us with.

We do find a suggestion in Kant. He distinguished between mathematical synthesis and dynamic synthesis. The mathematical synthesis is a synthesis of similar properties, and constitutes quality and quantity of the space time manifold. Or in Husserlian terms, the *res extensa* as qualified form in a certain position. The constitution of the substance-causal thing is realized by Kant, however, in the dynamic synthesis. But one should not overlook the fact that in Kant quantity and quality are also categories and as such belong to the pure understanding.

Without getting involved in a discussion between Husserl and Kant, the question seems to me to be legitimate whether Husserl doesn't overlook a distinction with regard to the constitution of material nature, a distinction which Kant with his distinction between mathematical and dynamic synthesis took into consideration. In any case the relationship between perception and apperception in the constitution of the *res materialis* remains obscure; obscure as well is the sense of its transcendence. Whether Husserl's phenomenology provides another way out of this difficulty must remain here an open question.

NOTES

* The German version of this paper has appeared in volume I of the *Analecta*.

[1] *Note from the Editor:* A special study by Roman Ingarden of four meanings of 'Transcendence' in Husserl has appeared in *Analecta Husserliana*, Vol. I.

[2] Ulrich Claesges, *Edmund Husserl's Theorie der Raumkonstitution*, The Hague, 1964.

DISCUSSION

Strauss: To set an order to the discussion, I propose first to ask if there are some disagreements with the speaker's presentation of Husserl's doctrine, and then we may enter into discussion of those doctrines as presented or supplement them.

Triantafillou: I find the paper quite illuminating but something occurs to my mind. A scientist for example, a biochemist like Watson, would say: I accept your phenomenological analyses of subjectivity but do not think that the phenomenological method will help me to make a discovery. In other words, it is the purpose of discovery that characterizes a scientist's work and which the phenomenological method does not serve. A scientist may accept the phenomenological description of what goes on in a subject but this description does not yield clues towards discovery. The scientific method has proven to be more helpful to that effect. How could the phenomenological method contribute something to the progress of scientific discovery?

Ströker: We must understand the difference between the aims of philosophy and the aims of the positive sciences. The problem of discovery is an intrinsic problem of scientific methods and it is not the task of philosophy and not the profession of philosophy to help scientists in discovering their positive facts. It is not the task of the philosopher to explain nature, but to understand what scientists do when they are explaining nature. Consequently, concerning discovery, we cannot put science and philosophy on the same level. Philosophy is not a field of discovery because it is another way of questioning.

Triantafillou: I accept your explanation but I would insist further upon this point, putting it differently, namely, that the phenomenological method proves somehow to be unable to explain creativity in human nature. It seems that the transcendental reduction does not allow us to describe how creativity takes place in the human *ego*. I include discovery within the meaning of the word creativity. Scientific discovery is a part of creativity. In that sense it seems to me that there is a conflict.

Tymieniecka (ed.), Analecta Husserliana, Vol. II, 292–313. *All Rights Reserved.*

Ströker: I do not agree in all points with Husserl's philosophy. It is indeed a principal difficulty, especially in Husserl's transcendental phenomenology, how to understand the creative functions of subjectivity.

Strauss: In Husserl there seems to be one blind spot: How is the constituting consciousness able (and this is in line with what the gentleman says) to ask questions? This is an old philosophical problem – enough to remember the *Meno*. How is questioning possible for the transcendental subjectivity?

Ströker: For the possibility to ask questions we must presuppose that we undertake the phenomenological reduction. It is extremely difficult with Husserl and I don't believe that Husserl was successful in reaching what he claimed, because in all his transcendental steps he really presupposed that which he was trying to comprehend by means of the constitution of the transcendental subject, namely, the body and the factual subject. Consequently, Husserl should have said that the transcendental reduction (this grand idea of reducing something in order to be able to reflect upon it) showed its own absurdity.

Kuypers: It is, I think, illuminating to reflect upon the history of the notion of foundation itself. The concept of foundation originated with the Greeks in mathematics. It was Plato, who was searching for an absolute hypothesis, an absolute foundation of science. Husserl goes back to this origin and he believes that we have neglected the transition made by the Greeks from *doxa* to *episteme* insofar as all science rejects subjectivity and will not reach back to the *doxa*. And yet science is grounded in the *doxa* of everyday life in the human world, in the subjectivity. It is difficult to show the scientist the necessity to go back to the *doxa*, as is obvious from the remark of Mr. Triantafillou, who assumes that one cannot find anything new at the level of the subjective, naive, pre-theoretical human existance. The reason for this difficulty lies in the fact that the notion of foundation in science itself is different from the notion of foundation in philosophy. The scientist is persuaded that he himself can reach back to the theoretical foundations of science, while the philosopher sees the foundation of science in the subjective conditions of cognition, specifically in the transition from *doxa* to *episteme*.

Tucker: I would like to hear some more about your notion of the foundation of a science within the science.

Ströker: What does "foundation" mean? For the physicist or for the scientist in general, 'foundation' means: We have some basic ideas, some basic laws, e.g. The axiomatic basis in mathematics stated in such a manner that one can derive from them further statements in a necessary sequence, so that all the statements a science contains may be derived from its first principles. Scientists use the terms "foundation" if they are able to reduce their statements and propositions to these *basic* statements. The very remarkable feature of the structure of science is that a few basic ideas are sufficient for ordering an infinite number of statements. It is true not only in mathematics but also in physics, although there we do not speak of an axiomatical foundation but of a hypothetical foundation. In physics it is more difficult because there are two different sources of legitimation for the statements in natural science. All statements in natural science must, firstly, be derived from theoretical statements of basic principles; secondly, they must be proved in experience. Otherwise they would not yield a scientific law.

Eng: We have so far in our debates omitted one thing. Despite the fact that we have been concerned with the late Husserl, we have virtually neglected passive intentionality and passive genesis, which is one of the signal points in the development of the later Husserl and which will become ever clearer as more of the *inedita* are brought out. Of course, with these notions we enter the realm of the 'impure ego'. It was interesting in your paper that at one point you speak of human efforts and you even speak of the task we have of bearing things. With this we are reminded that 'subject' literally means *unterworfen* not *ingeworfen*. It means 'subjected to' and the 'subject' so far has largely been used in the sense of *subjectus*, of the active subject. But to be a subject is to be 'subjected to' as well. And I submit that this comes out in the overtones and undertones of the Husserlian *opus*, particularly clearly in *Philosophy as a Rigorous Science*. Now it seems to me that science is one great effort whereby man alleviates his predicament in a situation where he is subjected. To come a little bit closer we would have to consider the *experimentum*, and in the *experimentum* we would have to consider the nature of the *instrumentum*. We know that some of the most remarkable advances in science (and this comes to your point) have emerged through the development of new instruments, the microscope in particular. What I submit here is that science has alleviated man's condition, e.g. we do not have to write down everything that is being said at this conference because this instrument 'suffers' for us. It adsorbs some of the pathos. I think that these three terms: the nature of the *experimentum.* the nature of the *instrumentum,* and the scientist as subject, that is the experimental subject as *subjectum,* need, in a concrete analysis of science to be elucidated phenomenologically. But there remains a basic question. In what sense would you say that the scientist subjects himself to nature, lays himself open to it? Ordinarily we say he poses a question and the experiment answers it. That's not the whole story. The question is primary but the scientist always finds things that go beyond it. And of course the classic story there (and I think it is a significant myth) is the myth of the fall of the apple. Scientists have established there was such an apple tree, it did happen and so forth. Alexander Koyré has investigated all the material on that. There was that fall. So here's this contingency and the exposure to contingency and here's where suffering and pathos must somehow enter, however quietly, in the Husserlian perspective.

Strauss: May I first perhaps give a little comment on ancient terminology. The term *subjectum* is the latinization of the Greek *hypokeimenon.* It means the ground of things. *Hypokeimenon* never appears completely but only in its aspects. I think this idea is pretty close even to Husserl. In Descartes' thought, the *hypokeimenon,* the ground of things, was sought in what we call today 'subject'. And with this historical turn, what was first called *hypokeimenon* became now the *objectum,* which is opposed to the subject.

Dagenais: If I understand the paper correctly and if I understand Husserl correctly, the transcendental subject constitutes the being of the world as object, but not the being of the world in its reality.

Ströker: Yes, when you are 'constituting' something that means for Husserl always constituting something as an object and nothing else.

Mayer: I was going to comment on your account of the attempt to found sciences as not being radically new with Husserl, but really very much part of the philosophical world in which he lived. In his attempt to elaborate a foundation for science, Husserl draws upon a double tradition. It seems that he combines the insights of the logicists like Bolzano and Lotze, who attempt to found sciences universally and aprioristically, and the insights of the psychologists, like Wundt, who recognize that the foundation of the sciences has to refer to being human, to the modes of human thinking. In this sense it seems to me that Husserl combines two schools of his contemporaries or immediate predecessors into something that preserves the insights, the positive contribution, of both of these schools.

Ströker: To some extent there is in Husserl's thought such a combination, but his idea is so radically fundamental that it surpasses in its originality all possible historical influences. The essential point in his proposal is new; it is his transcendental manner of questioning, seeking the ultimate conditions for the human enterprise in subjectivity.

Strauss: Is there not a danger of considering with Husserl consciousness as constituting time and space? The world is finally absorbed in that stream of consciousness; the whole of reality, even the body, is submerged in it. I wonder if in the later works consciousness is not taken as *Ding an sich*, or *ens per se*, so that it always seems for me (and for Husserl) difficult to come 'out of consciousness' into the *life-world.*

Ströker: What does it mean that something is 'in' my consciousness? Something is 'in' my consciousness as an intentional object, not as a real content of my consciousness. 'I have things in consciousness' means I am referred to things in the world.

Strauss: Is not the attempts of Marcel, Sartre, and Merleau-Ponty to account for the embodiment the consequence of this situation? Would it not be more appropriate to reverse the attitude to: How is a body capable of consciousness? Not: How is consciousness embodied?

Ströker: Yes, there are two questions, depending respectively upon two directions. One can in fact, approach this question by asking about the ways in which my consciousness is embodied (that is a former question of Husserl in the *Ideas*); and one can also ask the question, How do I become conscious of my body?

Strauss: However, it is not just a body but a *living* body. Is there not an additional problem now to account for the *living* body? It is only as a *living* body that a body is mine. Has Husserl any way of accounting for the living body, for the foundations of biology, for DNA (chemistry)?

Ströker: No, he hasn't. In the second volume of his *Ideas* he tries to show how the different dimensions of the world are constituted on different levels; among these levels, we find the biological sphere too. But the constitutive analysis of living things is not 'transcendentally clean', not clear (or clean) with respect to Husserl's methodological principles.

De Boer: If permitted, I would like to come back to the concept of foundation. You have deliberated on the two significations of that word in science and in philosophy and I think that is very important, especially in our discussions with scientists. It's always a delicate point. The scientist feels hurt when you say that phenomenology will found science. He can found it himself. I think to avoid misunderstanding it is very important to clarify this concept of 'foundation'. I find it always very difficult in my discussions with more positivistically minded colleagues to explain what is meant by the concept of the *life-world.* And since you come from a technical university I suppose you must have had the same experience. One time, in a debate, I said that you can compare the foundation in the *life-world* with protocol sentences. That's more correct than to say that the foundation in the *life-world* is a question of founding science in opinions.

Kuypers: What the *Crisis* is really doing is going back to the most basic propositions that you can verify.

De Boer: Yes, but the most basic propositions are the protocol sentences.

Kuypers: But Husserl goes back to the everyday world, which he identified with the *doxa* of the Greeks. Then what can be clarified can be clarified by the elemental form of all propositions, because the protocol propositions are the last intuitive elements of science.

Strauss: If you say this is brown you mean that it is not black or green. There is no 'simple' statement which is wholly simple. There is no pointing and nothing else. Pointing and nothing else is empty. If we point at something and convey a meaning – as we do with the sentence 'This is brown' – we already have the potentialities on the basis of which the meaning is determined.

Ströker: It is a very interesting thing to compare the neo-positivistic and Husserl's own use of the term 'foundation'. It is so important that it would need a new paper. Concerning the other question, I must confess that I have no difficulties with my scientific colleagues at Braunschweig because we always confine our questions to their proper realm and they understand that we in our attempt to give foundations do not touch their aims. They see a difference.

Kisiel: I believe we really needed this development in the direction of a hermeneutical phenomenology, since Husserl really developed from the conception of truth as correspondence, towards truth as discovery By. moving in that direction, he offers help in this particular problem. Perhaps then something in *The Origin of Geometry* might be helpful here. But there is something very peculiar about *The Origin of Geometry*. Namely, it is this process of looking at a situation which can in some sense be traced back to the Egyptians, when they discovered geometry out of the practical arts. Certainly that dimension of discovery is very much subdued as far as the present situation is concerned, which is precisely where discovery is occurring. In what Husserl called his historical present (*historische Gegenwart*) which in some ways is an absolute to be compared with the living present (*lebendige Gegenwart*), and a more intersubjective situation is involved, in terms of right now, the present situation in which the process of discovery is occurring. Now that process of discovery in the present situation is much more complex than that very simple archaic *life-world* in the ancient Egyptian period. Among other things the scientific idealizations are in some way incarnate in our present *life-world*. There it is all around the scientist in the form of computer equipment, highly idealized equipment embedded in the *life-world*. It is that very complex *life-world* that has to be examined in terms of the process of scientific discovery which I would suggest is a hermeneutical process, and to which I think some of the innuendoes of Gadamer's hermeneutic can contribute. Husserl is not very positive about the fruitful possibilities of a mathematical language. One senses that an impoverishment is involved in moving to a mathematical language. But, for instance, Bachelard points precisely to the fruitfulness of mathematical language itself. It has a metaphorical quality just like the lived language, the natural language. And it is precisely in this situation of a thinker involved in his mathematical terms in which all kinds of possibilities suggest themselves to him, in which the process of scientific discovery has to be examined. Perhaps you have some thoughts on this: Why is it that Husserl has so very little to say about this kind of problem? Why is it that transcendental phenomenology doesn't help out in terms of the real issues in the philosophy of science?

Ströker: I don't know.

Strauss: Well, I think Aristotle gave the answer in the *Metaphysics*, when he said that all men wonder. They first wonder about everyday life, banal things; but then they begin to wonder that things are as they are, and that means to have discovered other possibilities, at least in hypothesis. And then comes the next step, to define methods to solve these problems.

But I think that the first task of discovery is to discover precisely what the problem is. And this is not to take the world as it is, but also to ask how it could be otherwise.

Sokolowski: In regard to protocol sentences: the only reason they serve as evidence is that you have a theory about protocol sentences. If you simply went back to the ordinary person who had no theory about protocol sentences they wouldn't serve any function at all in verification because they wouldn't be understood as such. In the same way, the *life-world* is a *lived-world* only for Husserl, or for somebody like him who has moved to another level of analysis where the *lived-world* appears as a *lived-world.* And that is why *doxa* is not yet *doxa* to anyone who doesn't have *episteme*. Only Plato knows that *doxa* is doxa: only Parmenides knows that the way of opinion is a combination of the good and the bad ways. The people practising it naively do not know it.

Kuypers: Why has Husserl used the words *doxa* and *episteme*?

Sokolowski: He used *doxa* for the state of mind of a person who naively believes in the world without having criticized it. Then *episteme* arises when Husserl himself questions this *doxa* about the world.

Strauss: And why does he question?

Sokolowski: For the same reason you question, "Why is there something rather than nothing?" Or why Aristotle asked: 'What is Being'?

Strauss: In the *Meno*, it is said that you cannot question because either you know, in which case you need not question, or you don't know, in which case it won't occur to you to question.

Sokolowski: That's why philosophy doesn't give new facts. It only explains what one already knows, except that it changes it from *doxa* to *episteme*.

Kuypers: What is at stake is the elucidation of the transition from *doxa* to *episteme*. *Doxa* is an opinion only; Husserl is saying that in everyday life, we are always in the situation of doxa and this is the foundation also of science.

Strauss: It should be the situation of 'wonder' (*taumadzein*), not of *doxa* nor even of *episteme*.

Ströker: To make the transition from *doxa* to *episteme* is to thematize *doxa* as *episteme*. This belongs to the phenomenological reduction.

George: I think that this meeting needs a positivist, and for the purposes of discussion I shall assume that role. However, I would want to ask first of all what Professor Stroeker means by a positivist here? Mach was a positivist. Carnap was not a positivist, in one sense of the word, but he was a physicalist. In another sense of the word, the positivists are all those who use horse shoes and existential quantifiers and do other nasty things. In this sense, of course, Carnap *is* a positivist. He speaks about verification. But one gains the impression from the first page of this paper that positivists want to restrict the range of intellectual and philosophical inquiry to science. Now this is very true, but it is because they have a very generous conception of what science is, rather than a very poor conception of what there is outside of science. You see, when the child says, "The cat is on the mat," he is functioning as a scientist, although one in a very non-elevated position. But the statement is after all one that can be verified, it is meaningful. And so in this sense if what Husserl says is meaningful and true then he also is doing science, according to the positivist conception of what science is. On the other hand, if he doesn't then he isn't.

Now, I have a few other things to say on this subject. First, it seems that in this discussion there is a confusion between the structure of a science considered as a system of hypotheses and sentences of various sorts, on the one hand, and on the other, the structure of a science as a system of concepts. These are not the same thing. They are related, but two different ranges of problems open up depending on whether one concerns oneself with the conception of the things and entities with which the scientist is concerned or the question of how he knows and to what extent he knows whether something that he said was true or well confirmed. These are, I think, things that one should at least in a preliminary way keep separate in order to explore their relation. And in this connection, just as an aside, it is very important to realize that the primitive terms of science are often and quite frequently not the interpreted terms; that is, those places at which the scientist connects with reality, as it were, are not always the places where the primitive terms by means of which he defines all his other terms occur. This puts a particularly sharp edge on the problem of foundation insofar as it involves concept formation, because the foundation of the concepts is not where the scientist makes his connection with reality but elsewhere.

Then further, as a last point, I would like to have a sentence on page eleven clarified: "A solipsistic subject would not be able to understand the term objectivity." Now, it seems clear, and I suppose anybody would have to admit, that to have a concept is to master something. If to have a concept is to have a mastery of something then to have a concept is to have a mastery of its converse as well I can't master a concept without at the same time mastering its converse. So this man, the solipsistic subject, who is not able to understand the term objectivity is also unable to understand the term, 'solipsism' or the concept of solipsism. He would be a man, poor man, of whom one could say, "Look over there at Jones, he's a solipsist." But *Jones* couldn't say, "I'm a solipsist," because if he had no concept of objectivity then he would not be able to recognize himself as a solipsist. If he has a concept of solipsism and of objectivity then, of course, he can say that he is a solipsist. He has the concept but he denies that it applies to him. But if he has no concept then he is not only unaware of other people but he is also unaware that he is a solipsist.

Ströker: Yes, that is true, but it is not relevant to what I said. If I were talking about the possibility of solipsism, I would need to show how Husserl constitutes objectivity. And therefore, I brought out that a purely solipsistic subject could not conceive the term objectivity because the conditions for the possibility of coining this term are in the subjective realm. That is not merely a description of mine; you can find it in Husserl's papers. In the *Cartesian Meditations*, he showed in detail how objectivity is constituted and that it can only be constituted by a plurality of man.

George: But I'm concerned with *this* man. Could he say, "I wish there were other people?" You see, if he can say that then he has a concept of objectivity. But if he cannot even say it then he has no such concept. In other words, it is not a question of what has to be the case but of what concepts he masters. If he could even express as much as a wish for other subjects, he has a concept of objectivity, although perhaps he does not believe that it exists or that it is exemplified.

Ströker: We have to understand the problem of constitution of objectivity in relation to the constitution of subjectivity in a transcendental manner. This question is in the transcendental field, not in the field of reality.

Tucker: I would like, by an example, to put in doubt the distinction which the debate assumes throughout, between the objective foundation of science and subjective experience. In the first paragraph on page five, you talk about the visual field and you draw a contrast between what we learn scientifically about the visual field and what we ourselves experience. This is not a viable contrast. What we know scientifically about our visual field includes, for example, what ophthalmologists or oculists tell us, and this involves an immense accumulation of subjective data because people going to their oculists give them reliable subjective data; they are worried about their eyes and about going blind. If one examines the way in which this datum is in fact collected, I think that one can show that this side of visual experience can be in fact incorporated into the canons of objectivity so that the division between naive subjective experience and the objective approach of scientific inquiry can't be kept up. It does seem that upon a careful scrutiny, which cannot be made here, we aren't left with a contrast between, let's say, my vision *as I live it* and *as I experience it*, and the scientific account of it, as supposedly totally different things. The *scientific* account of vision includes, to the contrary, every aspect of my *lived* vision.

Ströker: That's exactly what I meant, I did not intend to introduce a separation, I only wish to differentiate and to explain this differentiation between my plain seeing, on one hand, and theories about seeing, on the other hand. I just try to explain that I as a seeing subject am already a foundation for the possibility of a theory about my seeing.

Sokolowski: I find Dr. Elkin's paper extremely interesting because it seems to me it's a type of cross-analysis which we always talk about but which hardly anybody ever does. I'd like to raise the following problem. First it is the question concerning the terms you use which we all recognize from early mythology and in many ways early Greek mythology. Then I would like to ask about the transfer between the mythological expression and what one could call a phenomenological expression. I find for example in the two syntheses you talk about as the first two stages (identification of the self and then three or four months later the differentation of the mother and the father into *good* and *bad*) a curious, if you will, mythical statement of the anticipation of a unity of meaning and the presence-absence intentionalities of Husserl. The first one would be the anticipation of a unity of meaning and the second would be what Dr. Brand brought out very nicely in his paper, that every intention has a balance of givenness and absence. This would be a phenomenological way of saying the good and bad of the mythical. But now the question is: when you say that, when you call it a good and bad factor, it's you who are talking, it's not the infant, at the time, so that the comparison is between your speech about the infant and then the phenomenological speech about intentionality. Is the activity that the phenomenologist describes just a repetition of the child's experience so that in talking about meaning is just a refined modern way of talking about this brute thing that keeps repeating itself in a constant attempt of the *ego* to keep itself identified? Or is the phenomenologist the one who has the higher view point, and is that early stage the first access to what the phenomenologist talks about? I wouldn't even demand that you choose between those two options. But I think that the problem of the alternatives is a very serious philosophical problem.

Elkin: In my mind, the whole of Husserl's reduction, Husserl's whole scheme of reduction, appears within the broader framework of the dialectic between *Logos* (mind) and *eros*. Reduction for me is prefigured in what the infant undergoes; he practices a certain reduction, practices a certain *epoche* to save himself from being lost in an erotic fusion. All of human *logos* is in a sense an attempt to establish our selfhood and to avoid being swallowed up in erotic fusion the way *das Mann* is swallowed up in erotic fusion.

Sokolowski: I would say then that you are trying to move phenomenology into myth. Can it be done?

Questioner: This is precisely the point because he is moving away from what Husserl says, that we *go to the things themselves through our own experiences*. The question is what experience do we have, in Elkin's approach?

Sokolowski: He's naming the experiences in his statement.

Questioner: And by what right can he relate experience to myth?

Elkin: About my own use of myth I would like to say that this is the Freudian myth clarified.

Sokolowski: It's like the *Timeaus* myth.

Eng: I'm in somewhat the same position as Dr. Elkin. I'm both a psychotherapist and a philosopher, and I try to work these two roles together. The first thing that one has to ask is: How does Dr. Elkin know all these things he presented to us? And I'm willing to consider his statements as stemming from naive, pretheoretic observation, at the level of *doxa*. I'm not questioning the concrete validity of his statements. But we would want to ask: What are the modes of consciousness through which he knows these things? Can he refer to, can he perform, some of these detailed analysis of constition and subsequently even intentional analysis whereby he knows these things? These are the basic questions we would ask from the phenomenological point of view. Instead of accepting arbitrarily some myth or other as a point of reference to organize the data of the patient's behavior, one needs intentional analysis, first of all, of the experience of the person who comes to a psychotherapist. People who come for help are in distress; they are in pain. What are the constitutive analysis of their pain, of what occurs to them? The therapist must continuously ask himself what he perceives of the patient's experience and he must try to reconstitute the process within himself. Unless he does this, he is in danger of either suffering from the myth which he arbitrarily accepts and under which he operates or/and inflicting this myth violently on his patients as a principle of explanation. With this postulate in terms of psychoanalysis, the whole question of Freud's self-analysis is reopened. We know something of the path Freud took while trying to discover the patient's illness from within himself. He began to have dreams about Anna O. He was also horrified when one of his patients threw her arms around his neck. Then he saw the limit: this isn't medicine anymore. When he began to dream about his patients he realized that the basic question is: How am *I* suffering? And in trying to discover the ways of suffering, he goes back and back into the constitutive unfolding of his life. However, instead of following it up, he seeks the ultimate reason for suffering in the *Urtrauma*. Freud asks himself, in 1938, very unphenomenologically: What is the original *trauma*? Husserl, by contrast, attempts to follow up the constitutive stream of consciousness to its origins. The whole question of therapy brings us to Freud's self-analysis, which has assumed mythical proportions. Nobody does self-analysis after Freud; you always go to someone else. Your paper really points to these analyses by the therapist, and I think you've quite directly showed us a little bit of how you have dared to approach the results of such analysis and to interpret them.

Dagenais: I would like to make some remarks on the relation of Dr. Elkin's paper to current phenomenology. I agree with Professor Wubnig that the biological questions ought to be considered. You've spoken of the body-subject of Merleau-Ponty. That very concept would have been impossible without the work of Jacob von Üxkull, Bertalanffy, Goldstein, and perhaps Driesch. But the questions of the genetic processes of organisms posed by Driesch have not yet been settled. They've only been swept under the rug called 'vitalism.' For a phenomenology or a genetic psychology, I would rather refer to Piaget, or if I may be heretical, to Carl Rogers, who forms his theory of the self on the perceptual principles which are germane to the context of phenomenology where the model is after all one of experience. Since one cannot observe anything from within the experience of a child, neither can one get a direct insight into the experience of the reconstruction of a shattered personality which the child in the course of time has developed. One can reinterpret, given the process of the reconstitution of personality, what way the genesis has taken; but to this purpose I think we have a far better and more solid basis in the phenomenological genesis of the *self*. I would prefer to speak here of a self rather than an *ego*.

Triantafillou: I have the following question: It seems to me that Dr. Elkin tries to make a synthesis of two methods: behaviorism and phenomenology (or transcendental phenomenology). It seems to me that the synthesis of these two methods could very well be possible and fruitful when we are dealing with human beings who are conscious of their physical state. But, when it is applied to explain the behavior of a child, this synthesis of the two methods breaks down. In that case, the person who attempts an explanation of the behavior is going to construct a hypothesis to explain the phenomenon. Whether the hypothesis of the scientist is valid or not would not be confirmed by phenomenological method. It would refer for confirmation to some experimental method, which would investigate the occurrence of a regularity between the stimulus and the response.

Elkin: I am struck by the fact that in the remarks and statements that have been made there is a failure, understandable enough, to appreciate the wealth of knowledge that psychoanalysis has acquired from clinical experience. However distorted this knowledge may be there is still within it a tremendous wealth of information about infancy and about early childhood through the vast experience of transference relations. Winnicott, whom I quoted here at one point, a psychoanalytic thinker in London, was a distinguished pediatrician. In his mid-life he had psychoanalytic training and has now become one of the most important, significant, and fertile minds in psychoanalytic theory. He maintains that he has learned much more about the infant's experience in his psychoanalytic practice than he learned during his many years of working as a pediatrician. Also, Merleau-Ponty recognized that within the psychoanalytic experience, however distorted it may be, with whatever negative criticisms it should meet – I will agree with them all, especially concerning the theoretical formulations – there is still a wealth of human experience dealing with primal elemental operations of the human *psyche*. We should learn from that. What I attempted here was to make psychoanalytic findings accessible to the world outside of psychoanalysis, in order to throw some light on the problems of phenomenology. One should accept that there is a wealth of knowledge gained in the psychoanalytic realm that cannot simply be disregarded.

Sokolowski: Husserl talks about scepticism as the thing that shocked man into doing transcendental phenomenology. It is very similar to Kant's little history of philosophy at the end of the *Critique of Pure Reason*, where one has the dogmatic consciousness first, then the sceptical, finally the critical. Husserl also says in the *Ideas* and in *First Philosophy* where he gives sketches of the history of philosophy, if one stays in the natural attitude problems arise.

Tymieniecka: We are missing here the voice of phenomenological psychology, which, in the most recent development in Dr. Ey's and Dr. Lanteri-Laura's thought, may offer a basic challenge to Husserl's claims regarding, on one hand, the absoluteness of consciousness, and, on the other, the total explicitness of the constitution. Indeed, Professor Pietersma's subject is the target of this criticism, which I have presented elsewhere, of the doubt in the absolutism of consciousness. The concept of horizon, as is well known, is supposed to account ultimately for the explanation of the human cognition. With reference to the horizon of the past, of the anticipated experiences, of the formal possibilities of structures as such, etc., as I have pointed out in my lecture, Husserl claims that whatever can be constituted, the complete universe of cognition, can be explicated precisely with reference to the horizon. Yet it is precisely with respect to the horizon that the basic doubt arises. As I pointed out in Vienna, at the colloquim on Brentano and phenomenological psychology and again in my lecture, interpreting the research of Dr. Ey and Dr. Lanteri-Laura, under pathological conditions there appear within the field of consciousness phenomenologically inspectable data for which the phenomenological horizon cannot possibly account. This provides a radical challenge to the self-explication of the constitutive genesis and to the "absoluteness" of the transcendental consciousness.

Pietersma: Perfect explication is then, in your view, the transcendental account that we give of it. We give an account of experience from the transcendental point of view, we reflect on it transcendentally, and we find there that experience is characterized by this, as Dr. Brand pointed out so well: The present refers to an absence, the here refers to the beyond, and there is always a further ongoing process by which we find out more and more about the world.

Tymieniecka: Transcendental phenomenology raises the claim that any question concerning human cognition – its origin, its meaning, its final explication – that could arise can be answered within the system of the constitutive genesis of experience. The concept of horizon means for Husserl as much the horizon of the lived world in progress as the system of reference for the structurizing possibilities and virtuosities of the genesis of the transcendental constitution. We have seen today, e.g., in Dr. Ströker's paper and the ensuing discussion, that whatever question is raised concerning the foundation of biology it is supposed to be adequately taken care of by the theory of the lived world and the constitution of it. The challenge which I present to the claim of complete self-explicitness of the constitutive genesis comes from within the constitutive realm, the realm of actually lived experience, e.g., pathological cases of the type of phobia constitutively analysed show within the noematic content qualitative elements, which do not occur within the *noemata* of the corresponding normal experience and consequently cannot be followed up through the constitutive genesis since it bears no reference to the constitutive horizon.

Strauss: I would like to ask a question about Dr. Claesges paper. One of the many questions concerning the *res extensa* is its perception. What is in Husserl the status of causality regarding perception? In practical life when I see the drummer come down with his drumstick and I hear the sound I do not doubt that the beat was 'caused' by the swinging and the stroke. However, in a Humean approach this would not be interpreted as a caual relation, but as one impression following another.

Furthermore, there seems to exist a distinct difference between such a succession of impressions within the same field of experience, e.g., the visual field alone, and two different fields, e.g., the acoustic field and the visual field; or the seen and the heard; or what is optically and acoustically given. In the latter cases, the causal relation seems to be quite different. While I am convinced that the beat finally causes the sound, upon reflection I see that the optical stimuli do not 'cause' in me the acoustical stimuli. Is there an explanation of the 'causal' relation for both, for the experience of causal efficacy and for the contrast between my experience of the connection among things as well as of the connection of my being affected in two different ways? I see, I experience, a connection within my very experience, and the causal nexus over there, in the object, appears as the polarization of the connection within my experience.

Claesges: It is correct that in the mundane, natural attitude one relates with things of one's surroundings in a rather special way. One grasps things and by "stating" that they are heavy or light grasps their materiality, but once the transcendental reduction is performed human beings belong to a different realm from things. The fact that this thing here is a material causal thing has to be stated (*ausweisen*) within the realm of the experiencing subject. In the natural attitude, what is given is solely temporal co-presence. When I abstract from the problem of the difference between the visual and the acoustic what there is is only the temporal co-presence (*beisammen*) of the encounter of the hammer with the anvil and the hearing of the sound. In the mundane, natural attitude I 'know' the relation to be of the cause to the effect, but in the transcendental attitude, how would I know that the movement of the hammer upon the anvil produces the sound?

Strauss: The stroke did not produce the movement.

Claesges: How do I discover that the stroke produces the movement?

Strauss: Do you believe that you have answered my question? What you say sounds very Jungian: What we deal with is only a succession of impressions, among which no causal relation can be established. However, this position stays in the most striking contradiction to the whole of our experience with the world in the natural attitude. Can Husserl solve this problem?

Claesges: Husserl cannot show the nature of this relation. But would you tell me where you obtained the concept of causality, that you obviously assume as natural? What is the origin of the concept of causality with which you operate? It is correct that when I hit this thing on the table I assume that it was hit. But where did I obtain this concept of causality?

Strauss: First of all, I would say, and this is not only my own view, that it is not the *concept* of causality but a much more primordial experience of happening (*Geschehen*). To reduce it to a concept might already make the answer impossible. It would be astonishing and terrifying if the natural attitude were not a natural foundation for the human world, since it is upon its basis that all is constituted. If this is correct, then my own interpretation of the causal nexus in experience is as follows: All our sensory experience is *polar*. In the act of seeing, I grasp the object seen together with my own seeing. But there are many instances of seeing the same object. In the theatre a thousand people see the same representation. The representation is one and common to all. There is a polarity between the multiplicity of the instances of seeing and the one and the same object seen. I do not know whether this has been already remarked by Husserl, but it appears that this polarity is differentiated in several variations according to different senses. We all hear the same word; should I now call on someone, we would hear it all together. But when I now touch my cigarette the situation is different: It is a private occasion. It is obvious that my bodily being is involved in a different way when I hear than when I see, am touched or taste. The consideration of the polarity and diversity of ways in which sensory experience occurs seems to me indispensable in the consideration of the body.

Claesges: I would like to point out that in my paper I have tried to present carefully all the levels of the constitution through which, according to Husserl, the apperception of the world and materiality comes about.

Strauss: However, this constitutive approach assumes that within the realm of my constitutive *ego* I do not have first myself and only then the world, that instead of the two being intertwined, they stay rather in the relation of succession, the primary factor being the experiencing subject while what is experienced remains secondary. But, in fact, I experience myself *within the world*.

Claesges: We seem to carry on the discussion on the level of the natural attitude without sharing the level of the phenomenological reduction. Back in the natural attitude, I must agree with everything you say, but upon the level of the reductive attitude the steps of the constitution which I have attempted to indicate are precisely the conditions for the possibility that we might in the natural attitude come to an understanding about these things. When we stay in the natural attitude, this community between us lies in the background, we have it behind us, we have accomplished it. The task of the constitutive analysis is then to clarify, to disentangle and bring into evidence in a progressive unfolding of the intentional analysis all this that in the natural attitude appears as quite understandable and obvious.

Ströker: This is not enough for Husserl.

Strauss: Is it not a *hysteron proteron*? First we live in the natural world and only from there we rise to reflection. In Husserl it seems, inversely, as if we had to start from an immaterial world of apperception and perception and then descend to the question of *res materiales*, asking how this immaterial subject of constitution with his world is 'embodied', 'incarnated.'

Claesges: The constitutive analysis is based upon a certain abstraction. It is the task of the intentional analysis to clarify what we otherwise accept as obvious. But in order to perform such a clarification, we have to distinguish abstractively some elements from the others in order to establish what is logically prior and what is logically posterior. We live in this world, we want to see how this world and our life in it is. The life within the world is in a progressive unfolding; it is transcendental history. Husserl attempts to grasp this unfolding.

Ströker: However, Husserl did not solve the transcendental question of materiality. He simply does not manage to constitute materiality by means of the intentional analysis and his various phantom things. The reason of this failure is as follows: (1) While Husserl performs the reduction, he is not aware of his own hidden presuppositions. (2) His intentional analysis does not cover all the steps of the reduction; in particular, it does not cover the constitution of the material thing. The differentiation of the levels of the constitution is made in the way in which it was possible for Husserl to do it, starting from the transcendental *ego*. But this starting point proves limited and insufficient to treat the question of the material thing. Consequently, the problem of causality remained also unsolved.

Tymieniecka: Husserl's unsuccessful attempt to account, by means of the constitutive analysis, for the material thing, which came to light in this discussion, seems to re-introduce in a novel perspective our initial debate about *idealism – realism*. It seems, indeed, to indicate additional, and perhaps insurmountable, difficulties inherent in the transcendental position of phenomenology.

De Boer: I cannot wholly understand Dr. Claesges' last remark. In your paper, if I understood you, you intended to correct Husserl by Kant. I understood that, according to you, Husserl takes the same position as you and that he cannot explain the causal connection between phantoms; he cannot explain the constitution of material things. It is difficult to correct this by Kant's views because precisely for Kant the thing is a construction, it is not phenomenologically given. But in your last remark you said that the constitution of the material thing has proved that the whole endeavour of Husserl's transcendental philosophy, as such, was a failure. Did you mean it?

Claesges: I do not believe that Husserl can be corrected by Kant. I only said that Kant in his concept of the dynamic synthesis saw something which Husserl didn't see; namely, that there is something specific which must enter into the constitution of the material nature; Kant called it "dynamic synthesis" and Husserl overlooked its necessity. Neither is it my opinion that the impossibility to account for the constitution of material nature in the intentional analysis shows that Husserl's phenomenology is wrong, as a whole. I have only emphasized that I believe the relationship between perception and apperception is a very important problem, and we must think it over, in Husserl. We may find there a new way to explain the constitution of material nature. This important problem of the relationship between perception and apperception comes to the fore also in some other points of Husserl's phenomenology, e.g., in the constitution of intersubjectivity. In the constitution of intersubjectivity, perception is only a body which moves in a special way.

Eng: This introduction of Kant into the debate is very interesting because in the *opus posthumus* Kant refers to the light and he talks about it in a way that suggests it is the basis of the synthetic apperception, thus solving some of the unresolved problems in the *Critique of Pure Reason*. This is in the *opus posthumus* where Kant speaks about the light and he talks about it in a way that suggests that it may possibly lead the way toward a solution of the problem of the *res materialis*.

Sokolowski: Would you say that Husserl fails by going too quickly from the material temporal thing to a mechanical thing, jumping over a realm of what we could call 'density'? Would that be the synthesis that is lacking?

Ströker: I don't think so. The problem in Husserl, is not that he goes too fast from the things of the everyday world to mechanical things, but there is still a lack in constituting the things of the everyday world.

Sokolowski: Yes, I agree with that. I didn't mean mechanical in the scientific sense. It seems to me that Husserl first has the temporal level and then the material level and then moves immediately from the spatial level to something that simply interacts with other objects; mechanical in *that* sense, not mechanical in the scientific sense. But in between there should be this realm of something which in Kantian dynamics would be the thing filling space and resisting an intrusion into its own space. Would that be the level that you think Husserl misses?

Claesges: I would like to say that within the framework of the Husserlian theory, we cannot take the body into consideration because the body, insofar as it is a material thing which within the world stays in contact with other material things, should be constituted in the same manner as the material thing; otherwise they could not, the material thing and the body, stay in the causal relation. But, in fact, although Husserl believes it, the body cannot constitute itself as a material thing. Consequently, it is no use to say in this connection that I may hit this thing with my hand and then it moves.

Sokolowski: I come back to my question: What, do you believe, Husserl overlooked?

Claesges: Husserl has not grasped the difference between the occurrence of two things together within the phantom sphere, within the world, and their moving according to certain rules, and what we mean when we say that a thing moves because it is hit by another thing. Husserl has overlooked this difference. He believes that since I experience that a thing stays in connection with other things, I already know and may affirm rightly that the thing moves because another thing hit it.

Sokolowski: Kant accounts for this situation by his category of causality. What then would it be that is missing in Husserl compared with Kant?

Strauss: The heart of the matter is that one cannot assume singular things as originary. The singular is never originary. It might be a different situation in the realm of sound, but within the optic realm it is certain that what we do first of all is to find, to differentiate, and to isolate the singular. The horizon of experience is not something that surrounds singular things, but inversely, the singular thing is located in a horizon.

Ströker: That is precisely what Husserl affirms. He emphasizes that a thing is what it is with respect to the circumstances. However, where the thing comes from, that Husserl, within the sphere of the transcendental judgments, cannot explain.

Elkin: Whenever the problem of materiality comes up, I suggest, for the clarification of the issue, we turn to Merleau-Ponty's recognition of the importance of the *Eros*. All materiality and certainly intersubjectivity (the problem of the constitution of the material realm enters into Husserl's problematic of intersubjectivity) is related to the question of the *Eros*. That's particularly marked in Eng's paper. In a way I was waiting for you, Eng, to bring up the relation of the problematic of Claesges' paper to your own; the problem of the body as dynamic, violent materiality over against the body as pure materiality seems to show, let's say in the Freudian lingo, a sort of 'delibidinalization.' Where the body is only materiality it appears as a delibidinalization; that body you can kill off, you can commit suicide, etc.

Strauss: Lessing once said: "If God would present to me in his right hand all knowledge and in his left hand all striving for knowledge I would choose the left hand." And I think we happily are in the condition of Lessing also. Even if we wouldn't take his choice, we too are always striving and always a little bit short. So our glorious meeting has come to an end. It was really a great experience for all of us, so much so that the day when I arrived in Canadian weather last Wednesday seems far remote, the distance we have traveled since then is considerable, and I am sure we are all enriched by what we have heard. On behalf of the conferees, I wish to express our appreciation for the warm welcome and hospitality that our hosts, the members of the Department of Philosophy, have shown us, and especially to Professor Tymieniecka, as the program chairman, to Professor Haworth, who as Chairman of the Department assisted her in the organization of the conference, and to Professor Richard Holmes, whose attention to day-by-day detail smoothed the way for us so that we could carry on our discussions in comfort and ease.

COMPLEMENTARY ESSAYS

J. N. MOHANTY

A NOTE ON THE DOCTRINE OF NOETIC-NOEMATIC CORRELATION

In this note I propose to draw attention to some aspects of the well known doctrine of noetic-noematic correlation as it is to be found in the *Ideas*, Vol. I of Husserl, and also to one serious difficulty connected with its statement.[1]

I

The doctrine of noetic-noematic correlation has been summed up by Husserl in one sentence: "No noetic phase without a noematic phase that belongs specifically to it".[2] But this is not as simple a statement as it looks to be. In fact, Husserl is aware that there are several parallelisms, and that they all are beset with great difficulties.[3] I will now try to make the situation as clear as I can, making use of Husserl's various analyses and taking the more convenient case of perceptual acts and perceptual noemata.

Let me consider my perception of the tree over there. In every perception, as I move around or orient myself differently or take a different perspective, I have both a different perceptual act and a different perceptual noema: however in all these cases I am still perceiving the same tree, the intended object is the same. Let these acts be A_1, A_2, A_3, $A_4, \ldots$, and let their respective noemata be N_1, N_2, N_3, $N_4, \ldots$. Thus there is *both* a noetic multiplicity and a noematic multiplicity. But as Gurwitsch well insists, the two multiplicities are on different levels.[4] For to each noema there corresponds a multiplicity of actual and possible acts; in other words, many different acts may have identically the same noema. Thus let A_1 be the act of perceiving the tree from the perspective (to simplify matters) p_1. Now from the same perspective many different perceptions of the same object are possible, all these acts will have the same noema N_1. If these acts are A_1^1, A_1^2, A_1^3, $A_1^4, \ldots$ then all these acts have the same noema N_1. These acts may differ amongst themselves in their temporal positions. I may have a whole series of successive perceptions of one and the same object, the *noema* remaining the same.

Tymieniecka (ed.), Analecta Husserliana, Vol. II, 317–321.

Thus though each act has its *noema*, and different acts may have different *noemata*, yet it is also possible that different acts have identically the same *noema*. Further different *noemata* may contain reference to identically the same object.

The situation may be represented in the following manner:

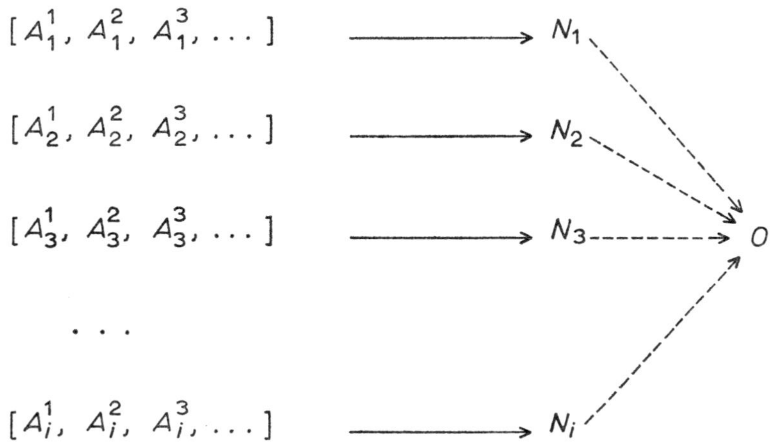

As should be obvious, there are various relationships involved in this situation. These are:

(a) Each act has its own intended object.

(b) Each act has its own *noema*.

(c) Different acts may have the same *noema*.

(d) Different *noemata* may refer to the same object. To these we add a fifth one:

(e) Each act-*phase* has its own noematic *phase*.[5]

Let us consider briefly these five relationships. These relationships show that it would be an error to identify the noetic-noematic parallelism with the parallelism between the unity of the intended object and the constituting formations of consciousness.[6] Further, the relationship between *noesis* and *noema* is not a simple one-to-one relationship. It is at one level one-one, at another many-one. Though each act has its *noema*, to each *noema* there corresponds a multiplicity of acts – so that at the noematic level there is achieved a unification which raises the *noema* above the temporality and particularity of acts: one sense in which the *noema* may be said to be irreal. ('Irreal' is here to be distinguished from ideal being. The ideal being, for example, of the essences is still being. The *noema* is neither being nor non-being, as Gaston Berger has emphasised.) We begin to see the affinity of this category with that of logical meaning, and a partial justification of calling the *noema* the meaning (*Sinn*) of the act.[7]

While the unitary *noema* in its irreality stands over against a multiplicity of real acts, a multiplicity of *noemata* – curiously enough – refers to, through passive synthesis of identification, to an identical real or ideal object. Thus the irreal *noema* mediates between real intentional acts and the real or ideal referrent of those acts. In relation to the correlative multiplicity of acts, the *noema* is an identity; in relation to the correlative noematic multiplicity, the intentional object is an identity. The constitution of the intentional object by its correlative noetic acts may then be said to be mediated by two constitutions: constitution of the identical *noema* by the correlative noetic acts, and constitution of the identical intentional object through synthesis of the correlative noemata.

II

Let us now turn to the last of the five relationships listed above. The source of the difficulty here is the conception of 'noetic phase' and also the correlative conception of 'noematic phase'. If Husserl's statement to the effect that there is no noetic phase without a noematic phase that belongs specifically to it means simply that to the structure of the *noema* with its content, quality and the central core there corresponds a noetic structure, and that to every noetic modification – attentional, temporalising etc. and their interpenetrations – there corresponds a noematic, we understand it. But it would seem that Husserl means something more. This is what he has further to say:

> ... to every component which 'objectively' directed description picks out in the object, there corresponds a real (*reelle*) component of the perception ... Also we can indicate all these noetic components only by falling back on the noematic object and its phases, and referring shall we say, to the consciousness ...[8]

The tree that I am now perceiving has its colour, the colour of its trunk let us say. The *noema* of the act of perceiving has, amongst its component predicates, the noematic colour. Corresponding to the noematic colour there is a whole multiplicity of sensory colours. These sensory colours or *hyle* are 'animated' by noetic meaning-bestowing acts which transform them and unify them into objective, noematic predicates. So in the case of each of the noematic predicates.

Let A be the act of perceiving, N its *noema*, O the object perceived. Let $S_1, S_2, S_3, \ldots$ be objective properties of O. (S_1 may designate the colour of O, S_2 its texture, S_3 its shape, and so on.) Further, let $s_1', s_1'', s_1''', \ldots$ be the hyletic data correlative to the noematic predicate N_{s_1}.. Likewise let $s_2', s_2'', s_2''', \ldots$ be the hyletic data correlative to the noematic predicate N_{s_2}. Similarly with the other noematic predicates. Now $s_1', s_1'', s_1''', \ldots$ need a noetic meaning-bestowing act or act-phase in order to constitute the noematic predicate N_{s_1}. Let us call this noetic phase a_1. The act-phase which constitutes the hyletic plurality $s_2', s_2'', s_2''', \ldots$ into the noematic predicate N_{s_2} is a_2. We have thus noetic phases $a_1, a_2, a_3, a_4, \ldots$. In general, the noetic phase a_i animates the hyletic data $s_i', s_i'', s_i''', \ldots$ and constitutes the noematic predicate N_{s_i} which corresponds to the objective property S_i apprehended in the original act of perceiving O. In fine, every generically homogenous group of hyletic data (sensory colour, sensory shapes, etc.) requires a specific noetic phase to animate it and to constitute the correlative noematic phase.

Now if this be Husserl's theory, rather succinctly stated in §97 of the *Ideas I*, several points need to be taken care of. In the first place, assuming that every noematic phase N_{s_i} has a correlative noetic phase a_i, one may wonder if an over-arching noetic act A is needed which would unify the phases $a_1, a_2, \ldots$ and which would be required to account for the unity of the noematic predicates in one *noema* N. Even leaving this apart, one is bothered by the thought, to what extent such an analysis may claim to be strictly phenomenological. May it not be reasonably suspected that this analysis is an unwitting victim of the much maligned constancy hypothesis and objective prejudice which phenomenologists like Gurritsch and Merleau-Ponty have so effectively criticised. For, is not Husserl assuming that there is a noetic component corresponding to every component of the object? Of course, he adds: "but only in so far, be it observed, as the description holds faithfully to the object *just as* it 'stands before us' in this perception itself". This suggests that what he has in mind is not a component of the real tree I am perceiving, but a component of the *noema*: but I think his doctrine makes use of both. Even if we take the noematic colour, the noematic shape etc., to suppose that corresponding to each of these there is an appropriate noetic act-phase is, in a way, to bring in the constancy hypothesis and the objective prejudice. It is to miss the unity of the perceptual act and the unity of the *noema*, and to dissolve them into the supposed multiplicity of noetic and noematic phases. There is no doubt that the notion of the *hyle*, as stated by Husserl in §97, requires this. The notion of the *hyle* I believe has a certain relevance and need, but it cannot be retained in the form in which Husserl formulates it here.

To what extent Husserl in the paragraph quoted above is under the influence of a sensationalistic psychology is borne out by the fact that the act-phases a_1, a_2, a_3,... are not all as such given. I am aware of perceiving the tree, and not of having all these acts of apprehending the sense data. The noetic phases can be identified, talked about, even defined by no other way save in relation to the noematic phases. I suspect here a remnant of the theory of *unbemerkte Empfindungen* in the form of act-phases lived through but not noticed!

There is a difference between saying that an act is *founded on* other acts, presupposes other acts, and saying that it has some other acts as its component phases. Is an act-phase itself an act? Is it itself intentional? How many act-phases are there in a given act? How to identify and enumerate them? I think there is neither phenomenological nor other theoretical reasons for assuming them, unless of course one is initially committed to an atomistic theory of *hyle* like the one suggested by Husserl's statement in the *Ideas*, Vol. I.

NOTES

[1] References to the *Ideas*, Vol. I will be to the English translation by Boyce Gibson in the Collier Books edition of 1962.

[2] *Ideas*, Vol 1, p. 250.

[3] *Ibid*, p. 266.

[4] Gurwitsch, A., *Studies in Phenomenology and Psychology*, North Western University Press, 1966, p. 138.

[5] *Ideas*, Vol. I, pp. 250, 262.

[6] *Ibid*, p. 267.

[7] Cp. Mohanty, J. N., *Edmund Husserl's Theory of Meaning*, Martinus, Nijhoff, The Hague, 1964. This book is primarily concerned with meaning in the logical sense.

[8] *Ideas*, Vol. I, p. 262.

TH. DE BOER

THE MEANING OF HUSSERL'S IDEALISM IN THE LIGHT OF HIS DEVELOPMENT*

It is well known that Husserl's transition to transcendental idealism around 1908 came as a great surprise for his students. It was alleged to be a repudiation of the earlier realism and its concentration on the object (*Wendung zum Objekt*). And many, if not all, students of the Goettingen-period were unable to follow the master at this point. It is remarkable, however, that precisely the later students from the Freiburg-period, e.g. Heidegger, Fink, Landgrebe and the French existential phenomenologists, likewise rejecting idealism, adhered precisely to the late Husserl and reproached the earlier students for not having understood the real intention of the master. In the first part of this paper we shall try to understand Husserl's transition to idealism as a perfectly natural outcome of his development since the time he published the *Philosophy of Arithmetic* and the *Logical Investigations*. The idealism characteristic of the first volume of the *Ideas* does not signify a break with the past, but, on the contrary, a solution to problems that remained unsolved in the *Logical Investigations*.[1] In a second part I hope to make clear that the existential phenomenologists, in spite of their rejection of idealism, could nevertheless link up their ideas with Husserl.

I. THE MEANING OF HUSSERL'S TRANSCENDENTAL IDEALISM

In the early writings the following two problems remain unsolved: the epistemological problem of transcendence and the problem concerning the relationship between explanatory and descriptive psychology. In the first two paragraphs of this section I shall show that this is the case, while in a concluding paragraph transcendental idealism will be discussed as a position offering a radical solution of these problems.

§1. When the *Logical Investigations* are taken as constituting a turn towards the object (*Wendung zum Objekt*) this interpretation is usually

Tymieniecka (ed.), Analecta Husserliana, Vol. II, 322–332.

justified by reference to two themes: (A) the intentionality of consciousness and (B) the intuition of essences (*Wesensschau*).

A. 1. A careful analysis of the *Philosophy of Arithmetic* shows that Husserl still adheres to Brentano's doctrine of intentionality. Intentionality here means that in opposition to classical British empiricism a distinction is made between 'idea' as act and 'idea' as content. One distinguishes, for example, between the hearing of a tone (mental phenomenon) and the tone itself (physical phenomenon). This tone is an immanent object. 'Intentional' as an adjective of 'content' is synonymous with 'mental'. Brentano's 'physical phenomena' are not contents to be directly identified with the extra-mental objects of physics. They are rather subjective representations of an external world of objects. In physics these phenomena function as points of departure for a construction of a physical world 'behind' them. It is Brentano's view that only this physical world really exists, i.e. independently of consciousness. Modern physics and physiology have shown that phenomenal contents such as colour and sound have only a subjective existence. "That which really is does not appear, while that which appears is not real".[2] For this reason *Wahrnehmung* is said to be *Falsch-nehmung*.

'Intentional' also occurs in Brentano as an adjective qualifying the mind's directedness to the object. From the above it appears that this is not a directedness towards an extra-mental object. There is no question here of turning to the object in the sense of realism. As we will see in a moment Brentano is a realist, *but not in virtue of his doctrine of intentionality*.

The difference between Brentano and Scholasticism can be formulated as follows. In Scholasticism the intentional object is an instrument, a medium through which one comes to know the real object; an *id quo cognoscitur*. In the spirit of modern philosophy and its emphasis on the mind as primary entity (*Satz der Immanenz*) there is in Brentano's view no ontological tie between the intentional and the real object. The former does not point to the latter but has become in the literal sense a representative; an *id quod cognoscitur*; it is not an image but a sign. Instead of realism it would be better to call this view representationalism.

With Brentano the doctrine of intentionality functions in a psychological, rather than an epistemological context. It serves to delimit the domain of psychology. A psychology intended to be only descriptive does not need to be concerned with the question of the external object as cause of the immanent content. This is a question regarding genesis and receives an answer through a causal inference. What we find practised already here is a first example of what Husserl later calls the psychological reduction. That is to say, the investigator limits himself to the immanent object and puts aside the question regarding the "real" object in the external world.

2. To what extent do Husserl's *Logical Investigations* go beyond the view stated above? Husserl's discovery of the meaning-conferring role of consciousness is documented for the first time in an article of 1894. The intentional relation to an object is now no longer taken as a 'mere having' (*blosses Haben*) of contents. It is now taken as an active interpretation of what is there as content. In an analysis of reading and of the comprehension of spoken words Husserl states that it is due to the activity of consciousness that a certain design (or arabesque) or a certain sound appears to us as a meaningful word. This is the first appearance of the concept of constitution: consciousness is shown to play a role in the constitution of that which appears. Since the *Logical Investigations* it is clear that by this time Husserl has also recognized the constitutive activity in perception. He says that the transcendent phenomenon appears to us only when immanent sensations are interpreted. (The words used are *Deutung*, *Interpretation*, or *Apperception*.) When we understand words we pass through the words to that which is meant.

Similarly, in perceiving the subject is directed to the perceived object *via* the sensations (*Empfindungen*). We perceive, not sensation of sound, but, so we read, the singing of the singer. Husserl places a great deal of emphasis on the difference between immanent colour-sensations (*empfundene Farbe*) and objectively perceived colour. The former is '*erlebt*' and immanent, while the latter is transcendent. However, this does not amount to a turn to the object in the sense of philosophical realism. For the perceived colour is nothing but a sensation interpreted in a transcendent or objective sense.

Nothing is said about the origin of the sensation. That is a 'metaphysical' question which does not fall within the domain of what is phenomenologically given (see above 1.). From the first edition of the *Investigations* it appears that Husserl, like Brentano, considers the real, independently existing thing to be the object of physics (in later terminology, *das physikalische Ding*). It lies 'behind' the perceived phenomena and is the cause of the sensations. Consequently, Husserl is realist at this point, *but not in virtue of the doctrine of intentionality*. The intentional object is not identical with the real object in the sense of the independently existing object.

(The occasional assertion to the effect that there is no difference between the real and the intentional object is to be understood as a criticism directed against the Scholastic view (in particular, the view of Twardowski). In this view there are two objects: the intentional object (*Inhalt*) and the object toward which the intention is directed (*Gegenstand*); the intentional object is viewed as a means, while the end-term of the intention is the other object. The import of Husserl's criticism is that the intentional object is the real object of the intention. In Scholastic terminology, the intentional object is *id quod cognoscitur* and not *id quo cognoscitur*. In other words, he rejects the idea of mediation in cognition as non-phenomenological. In perception the thing is 'self-given', not presented through an image or a sign.)

As with Brentano intentionality is a phenomenon of descriptive psychology. The Introduction and other places in the text clearly show that Husserl wants to remain faithful to the Cartesian starting-point by proceeding only from that which is phenomenologically given, namely the sphere of consciousness, while the questions about extra-mental things are relegated to metaphysics. Only in this manner can one satisfy the conditions laid down by the principle of *Vorurteilslosigkeit*. This means that, though the technical terms are still absent, there is in the *Logical Investigations* an epoche and a disconnection of the existence of the extra-mental object.

And here 'reduction' has the meaning which it is so often wrongly said to have in the first volume of the Ideas: putting within brackets the real existence of the object. Here one does indeed lose something (see §3.1).

Only in *one* respect does Husserl free himself from the *Cartesian* starting-point. There is the implicit recognition that also the intentional object, though not lived through (*erlebt*), is nevertheless given. But this does not become explicit until the lectures of 1908, entitled *The Idea of Phenomenology*, where it is admitted that something can be given without being really immanent. The object as outcome of interpreting sensations in an objective sense is then referred to as *cogitatum qua cogitatum*. Of this object it can be said that its existence is not affected by the burning of the real object.

This intentional object, as has already been said, may not under any circumstance be confused with the real, i.e. the independently existing, object. When Husserl in the *Investigations* speaks of self-givenness of the object he is speaking of what is distinctive of the intentional object and not of the extra-mentally existing thing. And similarly, when Husserl speaks of the cases where, motivated by self-givenness, the subject takes position with regard to the being of the object in what he calls 'belief' or *Setzung*, the true being (*wirklich sein*) of the object as correlate of belief is nothing more than a distinctive mark of the intentional object. (That Husserl does not follow Brentano in calling perception *Falschnehmung*, rather than *Wahrnehmung*, is due to the fact that he – rightly – does not evaluate perception in the light of non-phenomenal entities.)

3. In a certain sense this signifies also the epistemological impasse of the *Investigations*. The theory of knowledge is what Husserl will later call a psychology of reason (*Psychologie der Vernunft*). The *Ding an sich* is still concealed behind the phenomenal object. The problem of transcendence, formulated in the question 'How can I gain access to an external world?", is evaded rather than solved. The epoche does not solve the problem but eliminates it. It is an emergency measure which limits itself to the given (consciousness plus cogitata), because the real question cannot be answered. *Later Husserl will come to see that this impasse springs from a fundamental presupposition that is not at all critically discussed in the* Investigations, *namely the supposition (hypo-thesis) of a world existing in itself* (see below, §3). To be sure, for methodological reasons this world is put within brackets, but the supposition that it really exists remains.

From the preceding we may conclude that there is in the *Logical Investigations* an underlying realism. But it is not a realism based on the doctrine of intentionality. It is rather based on what Husserl will later call the prejudice of an independent, absolutely existing world; a prejudice which will then be attacked and rejected as absurd[3] because it cannot be phenomenologically justified. *Thus it is true that the first volume of the Ideas constitutes a repudiation of realism. But in making this move Husserl remained faithful to original intentions.* In fact, only thus could these intentions be worked out consistently. The realism of the *Logical Investigations* is a presupposition which must be surrendered if the theme of intentionality is to be fully carried through. This means that he discards the non-phenomenological distinction between a given, "subjective" appearance (*Erscheinung*), on the one hand, and an unknowable, 'really existing' *Ding an sich*, on the other. Only now can phenomenology change from a descriptive psychology into a *universal* science.

B. As I remarked at the beginning of this paragraph, the interpretation of the *Logical Investigations* as a *Wende zum Objekt* is influenced, not only by Husserl's emphasis on intentionality, but also by his teaching concerning the intuition of essences (*Wesensschau*). Let us briefly consider this second point also. The interpretation in question seems to gain plausibility here. After all, essences have being independent of our thinking. It should be noted, however, that in the *Investigations* Husserl defends the being of essences with psychological arguments; and this brings about a certain tension between his 'Platonism' and his subjective-psychological starting-point. The most important argument in favour of the being of essences is to the effect that in consciousness there are kinds of acts that are peculiarly directed upon essences. In contrast to a perception which is directed upon an individual object, an act of this kind constitutes a distinct form of consciousness (*neuartige Bewusstseinsweise*) that classical British empiricists failed to recognize for what it is.

Yet, of more importance is that in the controversy between idealism and realism the more crucial issue centers around the relation between consciousness and nature. And here the decision falls in the analysis of sense perception and not in the analysis of the intuition of essences, as becomes clear from the so-called Fundamental Meditation of Phenomenology in the first volume of the *Ideas*. In his *Formal and Transcendental Logic* Husserl speaks in this connection of a *"Seinsvorzug der realen vor den irrealen Gegenständen"* (p. 150).

§2. Another unsolved problem in the *Investigations* is that which has to do with the relationship between explanatory and descriptive psychology. The former is a natural science and is concerned also with physiological factors. Descriptive psychology, on the other hand, is "pure" in the sense that it abstracts from the relation of the mental to the physical.

As with Brentano this descriptive psychology is seen by Husserl as also serving as a basis for the normative sciences of man (*Idealwissenschaften*).[4] It serves to fulfil a task which cannot be fulfilled by the method of the natural sciences, namely to secure the foundations of the ideal norms of ethics and logic. Descriptive psychology is a psychology dealing with essences; on this point Husserl goes beyond Brentano. The laws of logic are founded on ideal meanings; the latter are here understood as essences of particular acts of meaning.[5]

This battle against what Husserl's Logos-article designates as a 'naturalization of ideas' (one exponent of which is psychologism in logic) signifies, to be sure, that one restricts the claims of positivistic naturalism, but it does not amount to a definitive overcoming of it. In the *Logical Investigations* can still be found what Husserl, in his *Logos*-article, calls the 'naturalizing of consciousness'.

The consciousness analyzed by descriptive psychology is still said to have its basis in material nature. Man is a psycho-physical unity. To be sure, in descriptive psychology one abstracts from the nature-side of man, but the psycho-physical unity is not as such subjected to criticism. By means of the body consciousness is still part of the whole, called 'real nature' (*reale Natur*) and as such subjected to its causal laws.

Husserl is still left with the following unanswered question: How can consciousness at the same time stand under ideal norms (logic, ethics, aesthetics) and yet be a part of nature, namely as psychological phenomenon? Like the neo-kantians of the Baden-school he speaks of supra-temporal values, but he does not answer the question as to how consciousness as a part of a causally determined nature can realize such values. He is still faithful to what he later calls the naturalistic prejudice *par excellence*: the independent existence of the material world. Consequently, consciousness is still apperceived, psychologistically and anthropologistically, as part of the world, as a part of a psycho-physical unity (psychological apperception).

§3. To what extent can the transcendental idealism of the first volume of the *Ideas* be considered as an answer to the problems discussed in the first two paragraphs of this section?

1. As to the epistemological difficulty, idealism amounts to a radical

solution of the problem of transcendence, because it undermines the foundations of the presuppositions that defined the problem in the *Investigations*. In the Fundamental Meditation of Phenomenology Husserl shows that the material thing and material nature as a whole do not enjoy an independent existence. They exist only as a dependent correlate of consciousness. This means that the presupposition of a naturalistic ontology – i.e. the (hypo)thesis of the natural attitude – has no phenomenological foundation. It is an illusion. The true presupposition (= ground) of the world is consciousness. This is the Copernican revolution in ontology which Husserl desires to bring about by the transcendental reduction.

The effect of this transcendental reduction or epoche is not that being is put within brackets, as was the case in the *Logical Investigations*. What is put within brackets is a particular, absurd interpretation of being! According to this interpretation nature is taken as absolute. And it was this way of taking nature which in the epistemological parts of the *Investigations* made the psychological reduction necessary as an emergency measure. Now at last is Husserl able to say that the reduction does not imply the loss of anything at all. What is lost is a mere naturalistic prejudice and its insoluble problem of transcendence.

In a special paragraph (52) he shows that the thing of physics (*das Ding der Physik*) is not an independent reality behind the phenomenal world, but a particular formal interpretation of the latter; it is a conceptual cloak (*Ideenkleid*) which should not be taken ontologistically as 'true nature'. Thus he overcomes the traditional view of primary and secondary qualities, according to which the former are regarded as true reality, while the latter exist only subjectively or psychologically.

The distinction, made in the *Ideas*, between the noema and the object (*Gegenstand*) is not a return to the traditional, non-phenomenological (Scholastic) distinction between intentional and real object. The object is the central point of unity in a series of noemata; it is not something behind the noema but an intrinsic feature of it.

2. The new turn in Husserl's thinking also implies a radical overcoming of the naturalization of consciousness. He shows that consciousness and reality cannot form a unity in the sense of a psycho-physical unity, so that consciousness would be founded upon material nature as the ultimate

foundation (*Fundamentalschicht*) of everything else. The recognition of the true ontological status of nature as a dependent correlate of consciousness implies that we free ourselves from the 'psychological apperception', according to which consciousness is taken as a part of nature. This apperception is now seen to be a misleading self-interpretation on the basis of the thesis of the natural attitude. The consciousness which founds the world would take itself to be a region of that world. The psychological reduction operative in the *Investigations*, in terms of which the soul is considered in abstraction from nature, now shows itself to have been a reduction *within* that natural attitude.

Husserl thus concludes the Fundamental Meditation by saying: "Thus the meaning of our talk about being is exactly the opposite of what it ordinarily is".[6]

Matter is not the foundation of consciousness; on the contrary, the relation is the reverse of what it is ordinarily thought to be. "The existence of a nature *can not* condition the existence of consciousness".[7] This thesis indicates a radical victory over naturalistic ontology. In the new, phenomenological ontology nature is viewed as the result of the teleologically related achievements (*Leistungen*) of consciousness. The comprehensive view of naturalism (consciousness being a part of nature) is replaced by the comprehensive view of idealism based on phenomenology. The explanatory sciences of nature and consciousness are not rejected but freed from their implicit, naturalistic philosophy.

II. TRANSCENDENTAL AND EXISTENTIAL PHENOMENOLOGY

1. The point of agreement between Husserl and his existentialist followers lies in the comprehensive, phenomenological view just mentioned. To this extent they are the true heir of transcendental phenomenology. Existentialism also opposes the naturalistic interpretations of man. Man is not a (mere) entity (*Vorhandenes*), thing (*chose*), but open-ness, existence, negativity.

2. Yet they reject idealism, as became first apparent with Heidegger. One might say that when they adopted the Kierkegaardian concept of existence in their phenomenology the recourse to idealism became unnecessary. Already man as situated and incarnate being possesses in

himself the capacity to transcend nature. This development is not a relapse into the anthropologism and realism of the *Logical Investigations*, as Husserl observed with bitterness.[8] For it is clear that nature is not viewed as a given independent from man. Also the existential phenomenologists speak of constitution. As in the Investigations, constitution is not taken to be creative, though there is a difference from this latter work. One does not project beyond phenomenal givenness a reality *an sich*, to be described by physics. *Cognitive consciousness is not measured by that which lies beyond the phenomena.* Hence there is no reason to devaluate perception (*Wahrnehmung*) to the status of a *Falschnehmung* (Brentano); nor is there a reason to think, as Husserl did in the *Investigations*, that existence is inaccessible and therefore to be placed within brackets, while we restrict ourselves to the Cartesian domain of *cognitationes* (and *cogitata qua cogitata*). The existence of a nature antecedent to human subjectivity is, as for Husserl, the product of the natural attitude. Merleau-Ponty calls this the bias for the objective world (*le préjugé du monde objectif*). The *life-world* is not a merely 'subjective' product of our constitution, concealing from us a non-phenomenal reality.

3. This does not mean, as it does with Husserl, that nature is a dependent correlate of consciousness. The opposition between reality and appearance is artificial, so that we are not compelled to choose one or the other horn of this dilemma: *either* the phenomenon is only a subjective interpretation of being (Brentano, Husserl in the *Logical Investigations*) *or* being exists only as phenomenon Husserl in the *Ideas*).

Reality unfolds itself only *in* the phenomena. Things come into their own true being and destiny only when the dimension of human consciousness opens up with the miracle of Man's emergence.

Amsterdam 1969.

NOTES

* My sincere thanks are due to Mr. H. Pietersma for translating this paper into English.

1 Further elaboration of this thesis can be found in my doctoral dissertation *De Ontwikkelingsgang in het Denken van Husserl*, Van Gorcum, Assen, 1966. A twenty-six page summary in German is provided at the end of this work. An English translation is in preparation.

2 *Psychologie vom empirischen Standpunkt*, Vol. I, p. 28.

[3] *Ideen*, Vol. I (1st edition), pp. 95, 97–8; Boyce-Gibson transl., pp. 155–6, 158–9.
[4] See my paper 'Die deskriptive Methode Franz Brentanos: ihre zwei Funktionen und ihre Bedeutung für die Phaenomenologie', *Akten des XIV. Internationalen Kongresses für Philosophie*, Vienna, September 2–9, 1968, pp. 191–9.
[5] See my article 'Das Verhältnis zwischen dem ersten und dem zweiten Teil der *Logischen Untersuchungen* Edmund Husserls', *Filosofia* **18** (1967), 837–859, published separately as *Saggi Filosofici* 27, Edizioni di 'Filosofia', Turin, Italy, 1967.
[6] *Ideen* Vol. I (1st edition), p. 93; transl., p. 154. The translation has been modified by the author.
[7] *Ibid.*, p. 96; transl., p. 157 (slightly modified).
[8] *Ideen*, Vol. III, *Husserliana*, Vol. V, p. 140.

LOTHAR ELEY

'LIFE-WORLD' CONSTITUTION OF PROPOSITIONAL LOGIC AND ELEMENTARY PREDICATE LOGIC

According to Bochenski and Menne, 'Symbolic Logic' can be characterized as the theory of logical calculus, its premises and its application. In their opinion it is the task of 'Metalogic' to investigate the formal premises of logical calculus. The fields of syntax, semiotics and semantics belong primarily to this discipline. They further distinguish between Metalogic and, what they term, *Metalogy*, which is the discipline which investigates the philosophical fundaments of the logical calculus.[1]

What constitutes the relationship between Metalogic and Metalogy? Bochenski and Menne write: "Da zum Aufbau des Logikkalküls keinerlei Ergebnisse der Philosophie als Voraussetzungen benötigt werden, überläßt der Logistiker die Metalogie im allgemeinen dem Philosophen."[2] The division into philosophy and logic is a division of labor. It is not the job of logic to legitimize itself by means of philosophical reflection in the form of a universally binding language and method. Rather Philosophy is a specialized branch discipline, just like logic. Philosophy is even subordinate to logic in that the philosophical inquiry into the fundaments of logical calculus presupposes an already existing system of logical calculus. Logic is no longer an integral component of man's act of living (*Lebensvollzug*) and is consequently no longer preserved in a universal and binding language. For logic and philosophy have separated and gone their own ways. And it is precisely this separation which Husserl considered to be the crisis of contemporary mankind.

Philosophers must face up to the fact that logic has developed into a positive science. Our present age is indebted to this positive trend for the progress and emancipation which has thus far been achieved. It is the task of contemporary philosophy and philosophers, therefore, to critically examine our consciousness for which logical positivity constitutes the measure of truth. I term such a point of inquiry *Metacritique*. It is *Meta*critique in that it comprises a particular type of consciousness for which positive logic constitutes the measure of truth. It is Meta*critique* in that it critically examines this consciousness with regard to its own

Tymieniecka (ed.), Analecta Husserliana, Vol. II, 333–353.

measure of truth. Metacritique can be considered to be Transcendental Phenomenology in the sense defined by Husserl, for Transcendental Phenomenology investigates the constitutive conditions in which the world, especially the positive world, the technological world, exists.

Transcendental Phenomenology is consistent with the Metalogy of Bochenski and Menne in the sense that both disciplines make a difference between logic and philosophical reflection on logic. But Metalogy fails to recognize that the difference between the two fields is a difference of identity, because it converts that which is differentiated into two independent, isolated levels of inquiry. Metalogy does not recognize that the difference between logic and philosophical reflection is one of *genesis*.

Husserl attempted to define this *genesis*. According to him logic is the logic of reflection [3] by the very reason of its self-oblivion in the act of reflecting. He writes:

> Also das Rätsel der Voraussetzung bekümmert den Wissenschaftler nicht, weil er wie jeder Könner sich im Können weiß; in seiner Arbeitseinstellung davon Gebrauch machend, ist er auf die Tat, auf Ziel und Weg gerichtet und in dieser Richtung liegt nicht die Vernunft als Thema, und zwar im Sinne jener beständigen Voraussetzung.[4]

And more explicitly:

> Positive Wissenschaft ist Wissenschaft in der Weltverlorenheit. Man muß erst die Welt durch *epoche* verlieren, um sie in universaler Selbstbesinnung wiederzugewinnen.[5]

It must be demonstrated how and to what extent logic is 'Science lost in the anonymity of the world' (*Weltverlorenheit*). For the purposes of demonstration I have chosen the examples of propositional logic and elementary predicate logic.

Lorenzen has elaborated propositional logic and elementary predicate logic as a Dialogue-logic [6]; although this cannot be taken up in detail at this point, none the less I will draw on the example of the junctors in order to demonstrate the genesis of logic in the world. At the same time it will be shown that not only does logic need the world but that the world also needs logic.

Every objective science develops within the framework of its own terminology. As much as a science is bound up in its own terminology, it nevertheless remains a language, and more exactly, science is immersed in our everyday language. The very introduction of a set of terminology satisfies a certain linguistic need.

Logic is itself a language-game. In the *Logical Investigations* Husserl emphasizes that logic must begin with a consideration of language[7]. And in the *Krisis der europäischen Wissenschaften*, he writes the following about the method which he utilises in it:

Es gehört selbst zu den großen Schwierigkeiten einer Denkweise, die überall die 'ursprüngliche Anschauung' zur Geltung zu bringen sucht, also die vor- und außerwissenschaftliche Lebenswelt... die naive Sprechweise des Lebens wählen zu müssen, sie aber auch angemessen zu handhaben, wie es für die Evidenz der Nachweisungen erforderlich ist.[8]

The logical language-game refers back to the everyday world. Our first problem is to account for the language of the *life-world.*

I

Language is an expression of our 'life-in-the-world'. In his *Philosophical Investigations* Wittgenstein remarks: *"Und eine Sprache vorstellen heißt, sich eine Lebensform vorstellen."*[9] And in another place he writes that language is *"ein Teil ... einer Tätigkeit oder einer Lebensform"*.[10] For the moment we can say that the *life-world* is our everyday language-world, our everyday language-game. The concept *life-world* must now be defined more precisely in the sense intended by Husserl. We must now explain the statement: language is an expression of our life-in-the-world (*Weltleben*).

We live in the everyday world. Husserl writes:

In der Welt wach lebend sind wir ständig, ob wir darauf achten oder nicht, der Welt bewußt, ihrer bewußt als Horizont unseres Lebens, als Horizont von 'Dingen' (realen Objekten) unserer wirklichen und möglichen Interessen und Betätigungen. Immer ausgezeichnet im Welthorizont ist der Horizont unserer Mitmenschen, ob irgendwelche anwesend sind oder nicht... Eben zu diesem Menschheitshorizont gehört die allgemeine Sprache. Menschheit ist vorweg als unmittelbare und mittelbare Sprachgemeinschaft bewußt.[11]

According to Husserl language and the world are without doubt bound together (*in untrennbarer Beziehungseinheit gewiß, obschon gewöhnlich implizite, horizonthaft*[12]).

What constitutes this inseparable relationship between language and the world? Obviously language is the guiding principle of our activities and therefore it is the world-horizon. The relationship between language and the world refers back to the relationship between the horizon and the world.

Now how should the relationship between the horizon and the world, between the guiding principle and the world be defined? The world-horizon is "Horizont für seiende Objekte ... und kann ohne sonderbewußte Objekte nicht aktuell sein."[13] Language is: "Die Differenz der Seinweise eines Objektes in der Welt und der Welt selbst."[14]

The world-horizon is only real in the consciousness of particular objects. With respect to the consciousness of particular objects (*sonderbewußte Objekte*) this means: the horizon as empty horizon (*Leerhorizont*) precedes the consciousness of particular objects. Husserl calls this emptiness 'empty intention' (*Leerintention*). This emptiness must be filled with a consciousness of particular objects. By reason of the fulfillment of the intention an object constitutes itself as that which is the invariant of two perspectives.

An object is 'something in the world'[15], it is *in* the world-horizon. That is, the horizon is that which is the invariant of two perspectives in the sense that it *precedes* them. The emptiness, which is the world-horizon, is not only antecedent to the consciousness of particular objects, but more importantly, the world-horizon is as an indication that the consciousness of particular objects must fill the emptiness. the horizon is identity in its indicative function.

How should the fulfillment of the empty-intention be defined? The empty-intention is indication (*Anzeige*). In the deictical act, i.e. in the fulfillment of the empty intention, the horizon reflects on itself.

So, as was said above, the index is an empty-intention. The intention is realised insofar as the intentum directs the intention. In the realisation of the deictical act, the intention is expressed as the intentum by means of language. The intention, as index and invariant, remains anonymous. And the intention, as emptiness, precedes in turn the intentum. The reflection is iterated in ascending stages. The horizon is the limit of the possible arbitrary choices of iterations. Identity is the determination of choice. The possibility of choice itself constitutes subjectivity.

The special nature of iterative reflection will be treated later on. First of all, the three constitutive factors of the *life-world* must be discussed. Husserl writes:

Immer ausgezeichnet im Welthorizont ist der Horizont unserer Mitmenschen, ob irgendwelche anwesend sind oder nicht ... Wir sind vor allem Daraufachten des offenen Horizontes unserer Mitmenschheit bewußt ...[16]

The factor which makes it possible that subjectivity becomes intersubjectivity (*Mitmenschen*, as Husserl says in the above quotation) is the difference between the real life of the world, and the world-horizon. The difference between the mode of being of an object in the world and the world itself imposes upon both of these two essentially different correlative types of consciousness.[17] These two different modes of consciousness will be developed further below as proponent and opponent.

The world-horizon is empty with respect to the consciousness of particular objects. As emptiness, it is the empty intention which remains to be fulfilled. Now, the empty intention is meaning-intention in the sense intended in the *Logical Investigations*[18]. In the same work Husserl also calls pure meaning-intention a 'signitive act'. The signitive act as such is *language*. Language becomes knowledge if the meaning-intention is fulfilled. And the signitive act acquires a correlative *intuition*. Language evolves itself as the difference between meaning-intention and fulfillment.[19] The meaning-intention is the intention which is to be fulfilled. It is therefore of its very nature *determination*. Knowledge, as the unification of intuition and intention, has therefore the nature of a task, of a problem to be solved.[20] The difference between the mode of being of an object in the world and the world itself is a temporal one. The consciousness of particular objects stands in a referential relationship to the preceding horizon. The above mentioned difference is called 'protention'. The world-horizon is itself the future. Only because the world-horizon as the future is real in the consciousness of particular objects, does it present itself. The future presents itself and then fades away, thus becoming part of the past. The two aspects of protention are the act of presentation and the act of receding into the past.

Now we can define exactly what the *life-world* means.

With regard to the consciousness of particular objects, the world-horizon is the empty intention which must be fulfilled. In the everyday life in the world the act of presentation obviously prevails. Intention means that consciousness is directed to the *intentum*. Consciousness is thus absorbed by the *intentum*. The intention itself is not of thematical interest. Intentions are the thematical manifestations of the very same *intentum*. Subjective actions are performed, posited, but not questioned. They remain mute and express themselves only by means of their products.

Husserl calls the absorption of the intention by the *intentum*, i.e. the fact that the act of presentation generally prevails, '*Geradehin*', and defines this attitude more explicitly as the 'natural standpoint'. What does the expression 'natural' mean?

'Weltleben' is first and foremost the act of presentation. The difference between the horizon and the consciousness of particular objects is *intuitional*, not conceptual. We will now discuss the three main characteristics of this intuition.

(a) The meaning-intention is fulfilled in the act of presentation and the act of fading away. Fulfillment is therefore *demonstration* and *reference*. Ostension is a special type of demonstration, as Husserl has shown.[21]

Ostension points at the This. It cannot be separated from the This. The This is only the This in reference to another This. For ostension always refers to an empty-intention. The numerous and varied instances of This are always united by something intuitional. (For the empty-intention is index). I term this intuitional something *synopsis* with reference to the concept as proposed by Kant in his *Critique of Pure Reason*.

(b) The ostension of the Here and Now is determined by the situation. It is situationally determined, occasional. The problems to be solved are numerous and varied to the extent that they are conditioned by situations. Ostension means denotation and designation. For instance, I can point to the Here, that is, to the tree and I can point out the Now, that is, this day. Ostension does not only imply the identity of the object, that is, of the This and of the act of ostension, but also implies the identity of identity and non-identity. Here and Now are not abridged denotations or indicators as Kamlah and Lorenzen claim in *Logische Propädeutik*.[22] The Here presupposes the Now. The Now is an expression of a distinction-system. It is an expression of the horizon which presents itself in the act of denomination and presentation.

(c) Because identity includes a distinction-system, the world-horizon is already posited in and with the act of ostension. Husserl calls this 'being posited' the "general thesis".[23] The world is not expressly posited by one single act. Rather every act is accompanied by the positing of the world. This is the meaning of the 'general thesis'. This thesis was first elaborated more exactly in Husserl's later works.[24] The world consists only of contexts which are continuously developing; language exists only in languages. It is meaningless to talk about the *one* language as such. But doesn't the 'general thesis' mean *one* world, i.e. *one* language? Is this not a hypostasation of language? Although the one language exists only in the many languages, there is nevertheless a unifying element in the many language-games. Every plural presupposes a singular. This singular exists *in* every plural but not for itself, not in itself. This singular which exists in every plural is the 'general thesis' of the world.

The *life-world* can therefore be defined as the "natural standpoint". The language of the *life-world* is an ostensive referring. Its fundament is the task as defined above in the sense of a problem to be solved.

Not only does our knowledge acquire its empirical content by means of the natural viewpoint, but more importantly it is this viewpoint which constitutes the basis of all our knowledge. Only through the 'natural viewpoint' do we become familiar with the world and its various manifestations and does it become for us a matter of fact. Life in the *life-world* is therefore something occasional. For the *intentum* predominates. We are conscious of the intention, but it always remains in the background. The tasks acquire a determined identity to the extent that the intention becomes thematic.

To the degree that the intention remains intuitional and affirmative, identity, *qua* identity, remains unexpressed. And the constitution of the unity, that is, the constitution of the task remains indeterminate.

Ostensive referring is a language-game. In the chapter entitled 'Sense-Certainty' in the *Phenomenology of Mind*, Hegel mentions this type of language-game.

> *Das Hier* ist z.B. der Baum. Ich wende mich um, so ist diese Wahrheit verschwunden und hat sich in die entgegengesetzte verkehrt: *Das Hier ist nicht ein Baum*, sondern vielmehr *ein Haus*.[25]

But why does Hegel write at the end '*ein* Haus' and not 'das Haus'? Obviously, the motion which is implied in the change of perspective is here essential to the intention and becomes a constitutive moment of the language-game. The Unity, that is, the synthesis, which is lacking in the natural attitude, now acquires more importance. But then we have to ask: Is the natural attitude an uncompleted language-game? Is the natural attitude only the beginning of a genesis? And how can we define this genesis? All these questions will be answered later on.

We must note that Husserl's definition of the *life-world* is self-contradictory. I cannot afford to dwell on this contradiction too long. Nevertheless, I would like to give a brief account of the problem.[26] According to Husserl, the anticipation of the horizon is an ideal and universal standard of validity. The one world is in itself, always existing, as Husserl emphasizes in his *Ideas*.[27] Protention is determined by the present. The present is divided into the occasional Now and the present as non-temporality, that is, eternity. According to Husserl, this difference makes idealisation possible. Mathematical and logical truths become eternal validities; the present becomes ideal supra-temporality.[28]

But eternal validity cannot be expressed by *language*. Language is the embodiment and the instrument of this eternal validity. Time succession is not essential to eternal validity; it is considered to be merely psychological in nature. In the *Logical Investigations*, Husserl eliminated *deixis* from logic. The communicative function of language is not essential to logic. It only forms a part of the solitary mental life.[29] But in his later works Husserl states that language is the difference between the world-horizon and the world, as I said above. Therefore logic can not have its fundament in the idealisation of the structures of the *life-world*. But just *this* is the thesis which Husserl propounds in his later works: Mathematical truths are ideal entities.[30a]

II

We must now define the junctors and quantifiers in their roles as elements of an elementary, logical language-game. In the logic of proposition junctors are very often introduced as truth functions. Each junctor is defined by a truth table. This type of logic presupposes that the meaning of a proposition is independent of the thinking subject. This fact, namely that the meaning of a proposition is "an sich", i.e. independent of the thinking subject, is expressed in the principle of the excluded third: that is, a proposition is either true or false, *tertium non datur*. Propositions are truth-definite (*wahrheitsdefinit*).

According to Husserl, the meaning of a proposition is its intention. The determination of meaning is not founded on a kind of ontology which nowadays is called 'Semantics'. On the contrary, meaning is performance, act. It is meaning intention. Meaning is an act, and more specifically, it is iterative reflection. The 'performance' of meaning constitutes subjectivity. Husserl states that this subjectivity is a factual one.[30] All acts are factual, that is, effective as Lorenzen says. Therefore the principle of the excluded third cannot presume general validity. Lorenzen has evolved an effective logic, the transcendental-phenomenological horizon of which will now be dealt with.

I said above that propositions about the life-world have the character of acts. The propositions in our logical language-games are acts as well. However, we must not overlook the fact that a logical language-game is an artificial language. Acts are performed according to exact rules, according to an exact schema. Acts are schematic operations. These schematical statements are not occasional but are objective, as Husserl writes.[31] Particular acts of synthesis are produced according to a schema.

On the other hand, conjunction as well as adjunction are rules which are admissible in all language-games.[32] But this is only possible if synthetic schemas define the task which is posed by the *life-world*. The tasks must themselves be schematic operations. By means of certain basic operations, other schematic operations, for instance conjunction and adjunction, can be introduced into every logical language-game. The task becomes a definite one, when we not only perform acts of synthesis but also reflect on these acts.

But in this case then, is the language-game of the natural world, which is characterized by acts of ostension and reference, not itself an artificial language? We are none the less familiar with schematic operations from our everyday experiences. (For example, the construction of a wall according to a plan is a schematic operation.) And it is only because we are familiar with such operations that we are able to set up particular rules and that we know how to operate according to such rules once they have been set up. Schematic operations thus presuppose the "general-thesis" of the world. Evidently a special kind of dialectic holds between the initial ostension and the positing of a rule, according to which an act advances, progresses into the future. This dialectic will be taken up once more later on. First we must set up the rules of conjunction and adjunction. For the time being we will assume the definition of a task as a schematic operation.

To begin with, 'et' and 'vel' are 'senses requiring completion' (*ergänzungsbedürftige Sinne*) i.e. meaning-intentions which are to be completed by the fulfillment of the tasks. The requirement of completing a sense can be indicated by parentheses or variables, for example:

$$(\ ) \wedge (\ )$$
$$a \wedge b$$

Conjunction and adjunction are not only senses requiring completion but, what is more important, they are at the same time schemata, i.e. rules for the generation of tasks from other tasks. These rules are as follows:

(1) $A, B \Rightarrow A \wedge B$

(2) $A \Rightarrow A \vee B$
$B \Rightarrow A \vee B$

We will first discuss the rule of conjunction. In the reception of the tasks on our part, the horizon functions as a guiding principle, or as a categorical form, as Husserl writes. When we write '*a*, *b*,…, then 'et' is the categorical form according to which we receive the task.[33] We then carry out the rule or schema 'et' which nevertheless still remains unexpressed. The result of the operation is the first thing which is expressed. The mute comma ',' is first of all recognized for what it is and then expressed as the categorical form 'et'. The performance of a schematic operation is itself a mute operation on the part of the horizon and hence represents a transition from muteness to expression by means of language. We can illustrate what is meant here with the example of time-progression, i.e. the succession of the Nows and, more generally, of the instances of This. The Now is Now, Now, Now…, but *as* Now. The This is This and That, *as* This. An operation is a succession, and as a succession it is a synthesis. In this regard I would like to stress once more that the ostension of the This, which as such constitutes the task, cannot be separated from the This itself. The This is necessarily This and That.

Furthermore it must be noted that our choice of task according to the categorical forms of 'et' and 'vel' is arbitrary. For the content of the task is irrelevant. The same holds for the content of a temporal succession.

This formal nature of succession makes two operations possible.

The This as such is the dialectic of This and That, i.e. of a Something and an Other. The Other is itself Something. The This is This and This, *as* This. Here we have the structure of the conjunction.

However, Something is at the same time an Other and is indifferent to an Other. As a result it is possible to add an Other to Something or Something to an Other. This is the structure of the adjunction.

Using conjunction and adjunction we can now illustrate three aspects of the schematic operation:

(a) We perform an operation according to a schema. As such the operation is mute. According to Husserl, the act remains in this case unexpressed. The intention is absorbed by the intentum in such a way that the *intentum* is the invariant of the intention. The intention expresses itself as the *intentum*. The mute operation is expressed by means of language, and language is meaning – intention, i.e. the horizon which precedes the *intentum*, with the result that the intention, by acting, reaches forward into the horizon.

(b) According to Husserl, conjunction and adjunction are categorical forms. These categorical structures constitute a preformation-system in the sense that the *formed* content is different from the original content.

Husserl, however, hypostasized the categorical forms as pure ideas, as 'Ideen an sich'. But I maintain that the correlation of matter and form should be defined in terms of a schematic operation. The horizon, i.e. the form of the conjunction or the form of the adjunction, is formally independent of content. On the other hand, the horizon is only real in the consciousness of particular objects; that is to say, it is dependent on the content.

(c) What is the reason for this formal independence and material dependence? This is the same as asking: What is the link between the world-horizon and the consciousness of particular objects? Seen from the perspective of the consciousness of particular objects, the horizon, i.e. the categorical forms 'et' and 'vel' has the character of a proposition: it proposes (*proponiert*) something. The horizon is the freedom of action (*Vermöglichkeit*) of operation of subjectivity. Thus the act of proposing is itself an operation. Moreover, the horizon is dependent on the content. And the operation is governed by the horizon. The one capacity of the horizon engenders two modes of action: a proposing one and an operative one.

The horizon is temporal. Protention is the term which Husserl uses for the horizon in its function as a model for the mute operation. Protention must be fulfilled and is fulfilled in the act of presentation and in the act of receding into the past. The schematic operation is thus a *temporal succession*. Every succession has a beginning and has an arbitrary beginning and arbitrary end to the extent that it is indifferent to the content. The fulfilled intention is not only verbalized, but its verbalization is at the same time the retention of the past operations which are retained in the acts of presentation. As a result of the retention, the operation has the power to oppose the protention. Thus schematic operations can be defined as forms of dialogue by reason of their temporal constitution. Schematic operations are dialogue-definite. The proponent has to maintain his thesis against the opposition of the opponent. A thesis can either be maintained or it cannot be maintained in a dialogue.

We can thus define conjunction and adjunction in the following manner:

(1) If the proposition $A \wedge B$ is proposed, then the proponent can only maintain his thesis if, and only if, he can prove *A and B* to the opponent.

(2) If the proposition $A \vee B$ is proposed, then the proponent can only maintain his thesis if, and only if, he can prove either *A or B* to the opponent.

Negation is not itself a schematic process. It must be noted here that an operation has its own characteristic structure. An arbitrary beginning posits an arbitrary end, i.e. the termination of the operation. In a particular case, it is possible that the operation cannot be carried out effectively. The dialogue then reverts to its starting point. The horizon becomes empty. The horizon can become empty only because it is itself meaning-intention, that is, an empty Prius. The reversal of the progression constitutes a deficient mode of the emptiness. The horizon is freedom of action; it is the capacity of subjectivity. This freedom makes a return to the starting point and thus a new beginning possible. Insofar as the negation is the reversal of the progression it can be characterized as follows:

(a) It is formally independent, but it is dependent with regard to its content.

(b) It is defined by the form of the dialogue and is a characteristic type of dialogue. According to Lorenzen it is defined as follows:
If the proposition $\neg A$ is proposed, then the proponent can only maintain his thesis, if, and only if, he succeeds in refuting the proposition A. But if the opponent can prove the proposition A, the proponent loses.

Schematic operations are iterative acts to the extent that they are time-progressions; for the fulfillment of the intention is an act of reflection. This fulfillment constitutes the limit of reflection, which *limit* is, in turn, itself the beginning of a new iterative process. This iteration is then formally independent and materially dependent. The process of iteration makes up the fundamental structure of the dialogue. The logical subjunction is the manifestation of this iterative process. Subjunction can be defined as follows:

If the proposition $A \sqsubset B$ is proposed, then the proponent can only maintain his thesis in the following case: If the opponent proves the proposition A, then the proponent must be able to prove the proposition B. Otherwise he is defeated.

We can extend the system of propositional logic by introducing quantifiers. We can interpret an existential proposition as an infinite adjunction, and a general proposition as an infinite conjunction. The infinity which we are discussing here is not an actual infinity but rather the potential infinity of an iterative process.

The extension of propositional logic through elementary predicate-logic means an elevation to a higher level of reflection. I said above that the task is mainly characterized by ostension and reference and that the task itself is a schematic operation. Now indication (*Hinweis*) is the intention (empty intention) which is to be fulfilled and which is fulfilled. This means, in other words, that the This extends itself into something Other. Although the intention is expressed by means of its intentum. nevertheless the process of iteration *qua* iteration remains unexpressed, It is therefore a new kind of reflection which perceives iteration as such, i.e. temporality qua temporality.

The first one to give a full account of the dual modality of iterative reflection was Hegel in his work *The Phenomenology of Mind*, especially in the chapter entitled 'Sense Certainty'. He distinguishes between the initial act of meaning of the This (cf. "The Now is this day; The Here is this tree")[34] and the act of demonstration that develops into ostension. This distinction forms the basis of Frege's philosophy of language. Husserl's *Logical Investigations* presuppose the conclusions arrived at by Frege.[35]

Frege calls the meaning of a sentence 'thought'.[36] Thought as such is expressed in the form of a sentence-question. A sentence-question can be either affirmed or negated. We must distinguish between the affirmation or negation of a sentence, on the one hand, and the attribution or the non-attribution of a predicate, on the other hand. More importantly, affirmation is the determination of the sentence-question in the deictical act. Negation is the absence of determination. According to Frege the thought is thus the sense of the name of a truth-value (*der "Sinn des Names eines Wahrheitswerts"*).[37]

According to Husserl, the meaning of a sentence-question is meaning-intention. Meaning-intention is fulfilled in the intuitive act; the task as meaning is ostension, deictical act. Therefore, ostension is the fulfillment of a sentence-question. Hence we can characterize the horizon as a sentence-question.

In fulfilling its horizon, the intention constitutes itself as a synthesis of the many intentions. But it must be emphasized here that the invariant is only of thematical importance to the extent that its relation to the intentum is involved.

The fact that the horizon does constitute itself as an invariant meaning, which has varied objective correlations, attests to a new level of reflection. We arrive at a new and higher level of reflection in that the horizon constitutes itself as the very same meaning, which refers to objects in numerous and varied ways. What I mean is for instance: The Now is the one Now of many Nows. The Here is the one Here of many Heres. The sentence-question thus becomes a word-question (*Wortfrage*).[38]

The relationship between the one This and the many instances of This corresponds to the relationship between the horizon and matter. The task can therefore be divided into an invariant meaning on the one hand, a constant, the sense of which requires completion with regard to its relations to objects, and into an expression of the sense requiring completion, namely the variable, on the other hand. If we have only one variable, we can write:

$$P(x)$$

In as far as a variable can be bound in two different ways, the iteration of horizon and matter is a double iteration. First of all, a variable can be bound by a universal quantifier; this bond is defined by the form of the dialogue. So when the proponent asserts a general proposition, he is proposing a formally independent proposition which is dependent on the content. If he asserts $\wedge_x P(x)$, then he can only maintain his assertion against the opponent, if he can defend $P(y)$ for every y chosen by the opponent.

The variable can also be bound by means of an existential quantifier.

This proposition too is formally independent and is defined by the form of the dialogue.

If $\bigvee_x P(x)$ is proposed, then the proponent can only maintain his thesis if he can prove $P(y)$ for every arbitrary y chosen by himself.

III

In the natural attitude the act of presentation prevails over protention. We are conscious of the world-horizon, but this consciousness always remains in the background. Synopsis manifests the difference between the world-horizon and the world. As pure synopsis it is *affirmative*. The *life-world* itself is affirmative. And in the logical language-game, the difference between the horizon and the world is also an affirmative one. But then we have to ask: does a qualitative difference still exist between the language of the *life-world* and the logical language-game?

We must now direct our attention to a particular dialectic of the *life-world*. The task as such is expressed in the form of a sentence-question. The sentence-question calls for an affirmation or a negation. Its affirmation consists in the deictical act. We have already characterized the horizon as a sentence-question. For this reason we are able to question our consciousness: Is this table round? Consciousness answers this question in the affirmative, that is to say, by a deictical act. So, if one goes around to the other side of the table and thus changes one's perspective, then one can ask once more: 'looked at from this side, is the table round'? Consciousness affirms this question also by means of a new deictical act. In both cases the sentence proposed by consciousness must be verified by means of an affirmative deictical act. Now the second perspective is the blind side, i.e., the emptiness of the first perspective, so that the first perspective already refers to the second one. Thus the change of perspectives is the real act of fulfillment. However, the horizon as well as the act of changing perspectives, remains intuitive and therefore unexpressed. Consciousness is also affirmative here because of its intuitive character. But if it is affirmative in both cases, then it can legitimately be called a dogmatic consciousness because it does not express the motion which lies in the change of perspectives i.e. in the question-horizon. It *is* the question horizon which directs, or guides, the progression of the sentence-questions. The difference between the world-horizon and the consciousness of particular objects proves to be a methodical, iterative progression. In this progression, the intention continues to be absorbed by the intentum, but then as a result of a *method.* This consciousness is, according to Husserl, a technical one. The natural attitude is converted into a technical one. The consciousness of the *life-world* necessarily evolves into a dogmatic consciousness. If natural consciousness surmounts its dogmatic position, then it can advance from the natural viewpoint to higher levels of reflection.

But if this is true, then is not the natural standpoint a completely false consciousness? Is it consciousness which refuses to comprehend this progression?

The synopsis of the instances of This is indeed expressed in the technical progression as a synthesis. On the other hand, every synopsis as such implies a synthesis. Schematic operations can obviously only be learned by practice. Therefore they presuppose the everyday knowledge of the world, that is to say, the 'general-thesis' of the world. They already presuppose that the world-horizon is actualized in the consciousness of particular objects and further that this synthesis is a passive-temporal-synthesis.

The synthesis of the world-horizon and its actualisation must already be existing and be a habitual part of our everyday world. Only under these premisses is learning by practice possible and can we operate according to schematical rules. Therefore a methodical progression requires a special type of negation. Its characteristic structure is the affirmation of negativity, i.e. the elimination of oppositions. According to Husserl, this affirmation can be suspended; namely with regard to technical progression, we practise *epoché*. In the act of *epoché* subjectivity becomes conscious of the horizon which has already become habitual for us. Subjectivity becomes conscious of its own genesis. To the extent that the world-horizon is only real in the consciousness of particular objects, is this actualisation the commencement of a methodical progression. Subjectivity becomes conscious of the world-horizon as a particular object. Hence the world becomes contradictory. The natural attitude becomes a dogmatic consciousness to the extent that it tends to convert negation into affirmation. On the contrary, this contradiction is perpetuated through the progression. The horizon and the consciousness of particular objects permanently mediate one another. However the progression remains affirmative: The opposition between the world and the world-horizon is not resolved in a methodical progression. The progression perpetuates the opposition. In a methodical progression there is no reflection upon the opposition. This opposition establishes itself through a methodical progression as the very sphere of subjectivity, i.e. as society.

Hence the *life-world* has a threefold function:

(1) It establishes a sphere of habituality with regard to the horizon and the consciousness of particular objects. It is a prerequisite of language-competence and of learning by practice.

(2) It is the basis and driving force of the technical world.

(3) It contains the fundaments of society.

Technical progression is the totality of human capacities. By means of it this society is constituted as the society of free human beings.

IV

The task posed by the *life-world* merges into a schematic operation. Mankind has in the task the totality of human possibilities at its free disposal. The merging of the task into a schematic operation is accomplishes in the process of counting. This will now be illustrated.

First of all, the specific character of the task must be demonstrated, and iteration, *qua* iteration, must be defined.

A schematic operation is based on time-progression. Time-progression is an iterative bi-unity. The horizon, i.e. the protention, presents itself. The initial act of presentation fades away while fulfilling the horizon, and the act of presentation is renewed again, so that this new presentation is confronted with another empty horizon. The progression recommences.

Time-progression constitutes a dual sequence. The beginning is subdivided into two moments: namely the act of presentation and the act of fading away. This can be illustrated by means of a tree diagram.

Because the horizon precedes every act of presentation, the tree diagram continues to ramify.

What makes this ramification possible? The horizon is the *limit*. First of all, the presentation constitutes the outmost limit of the time-progression. The moments passing away are at the same time preserved; according to

Husserl, they are *retained.* How can the act of passing away be retained? Insofar as the horizon has, in perpetuating itself, become indifferent to its content, the starting point of the progression becomes arbitrary and its limit as well. The initial act of presentation fulfills the horizon, thus perpetuating itself as limit until it reaches an arbitrarily posited end. This end is again a new starting point. The perpetuation of the beginning is the perpetuation of the horizon itself. Therefore the limit is not static but, on the contrary, it is dynamic; it is a developing limit.

We defined the horizon as limit. The horizon makes it possible for us to reiterate freely the acts of the commencements.

The initial act has to fulfill the emptiness. A new commencement emerges from the act of fulfillment and from its passing away. The new commencement, that is, the limit of the preceding progression, retains the previous act. As the initial act, it is at the same time the first step of the progression and, as such, mute to the extent that the horizon is an empty-intention. The second step is the act of presentation. This fulfillment of the empty intention, which is as such a possible new starting point, also remains mute. The second step retains the beginning in its fading away. The arbitrary beginning plays in the schematic succession the role of a constantly recurring element, which is *One.*

If the progression continues, then the third step does not only retain the immediately preceding starting point, but the retention itself is also iterated; namely the result is the retention of the preceding retention. The immediately past Now is already anticipated by the original past Now so that every new fading away causes the original act to recede further back into the past.

In the retention of the preceding retention, the act of fading away as such is characterized by difference and identity. In the time-succession the fading away becomes the unit of iteration. It is in the second step of this succession that the limit constitutes the above mentioned unit, which, as such, is pure One; and so on for the third and following steps. Thus the progression proves to be iterative bi-unity. Hence we can schematically expand the tree in one direction.

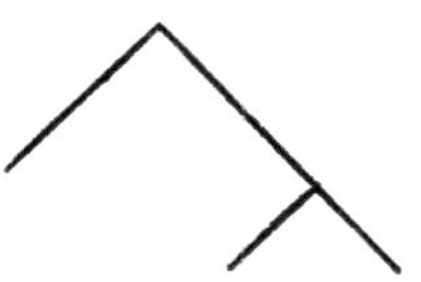

By reason of the new passing away, that is, because of the third, second and first retentions, this unit again must be characterised as bi-unity. Therefore we have to complete our sketch by adding the third retention:

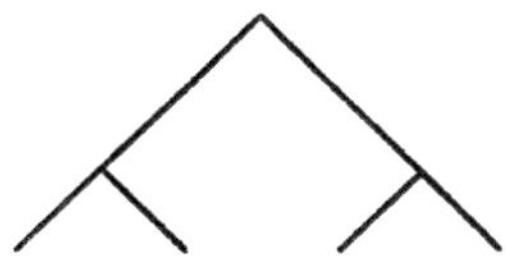

Two aspects of the dual sequence should be stressed. We have on the one hand the succession of Ones, of which the foremost element is the limit and on the other hand, the retention of the preceding retentions.

As we mentioned above, the commencements are subdivided into iterative bi-unities. These beginnings are themselves bi-unities of the preceding beginnings. It is the protention which makes it possible for the retention to retain the beginning.

The second step retains the beginning of the first step thus constituting the iterative unit. And so, the mute operation turns out to be the sequence of ordinal numbers (*Schrittzahl*, cf. Lorenzen). Limit means the moment in which the intention is fulfilled, and thus it is the foremost element of the progression. For this reason the limit makes a reflection upon the unity of the dual sequence possible. So, not only do we have the unit of iteration, but the iteration itself becomes an object of reflection.

The most elementary task therefore proves to be 'counting'.

The act of fulfilling the intention can only manifest itself as the horizon by means of the one beginning or the starting point of many beginnings. If we call the horizon 'concept', then number would be a statement about a concept, as Frege says. For only by means of the capacity of retention, can that which is retained be expressed. But it must be emphasized that the concept itself is generated in the mute operation. By means of retention the horizon expresses itself as number. This is possible because the task is terminated in the fulfillment of the horizon and because it reaches out again into the unlimited horizon. To the extent that the task posed by the life-world is itself a number, it becomes mathematical. Because of the fact that the task becomes mathematical, new operations are admissible.

Especially in his *Crisis* Husserl explained these difference between the horizon and the world as one of iteration.

This iteration has proved to be the essence of number. It is in the genesis of number that the language-game of the *life-world* becomes a mathematical language-game.

NOTES

(All references are to the German Editions)

[1] *Grundriss der Logistik*, 2nd ed., Paderborn, 1962.
[2] *Ibid.*, p. 13.
[3] *Die Krisis der europäischen Wissenschaften und die transzendentale Phänomenologie*, Martinus Nijhoff, The Hague, 1954, pp. 457–8.
[4] Ms. K III 6, p. 176a (1934–36).
[5] Cartesianische Meditationen, in: *Cartesianische Meditationen und Pariser Vorträge*, Martinus Nijhoff, The Hague, 1950, p. 183.
[6] For example, cf. *Formale Logik*, Berlin, 1967.
[7] *Logische Untersuchungen*, Vol. II, 5th edition, part 1, Tübingen, 1968, p. 1.
[8] *Die Krisis der europäischen Wissenschaften ...*, p. 60.
[9] *Philosophische Untersuchungen*, Frankfurt A.M., 1967, p. 20.
[10] *Ibid.*, p. 24.
[11] *Die Krisis der europäischen Wissenschaften ...*, p. 369.
[12] *Ibid.*, p. 370.
[13] *Ibid.*, p. 146
[14] *Ibid.*
[15] *Ibid.*
[16] *Ibid.*, p. 369.
[17] *Ibid.*, p. 146.
[18] Cf. *Logische Untersuchungen*, Vol. II, part 1, §9, pp. 37 ff.
[19] *Logische Untersuchungen*, Vol II, part 2, 5th edition Tübingen, 1968 p. 8 and pp. 32ff.
[20] Cf. L. Eley, *Metakritik der formalen Logik. Sinnliche Gewißheit als Horizont der Aussagenlogik und elementaren Prädikatenlogik*, Martinus Nijhoff, The Hague, 1969.
[21] Cf. *Logische Untersuchungen*, Vol. II, part 1, p. 40.
[22] *Logische Propädeutik oder Vorschule des vernünftigen Redens*, Mannheim, 1967, pp. 110ff.
[23] *Ideen zu einer reinen Phänomenologie und phänomenologischen Philosophie*, Vol. I, Martinus Nijhoff, The Hague, 1950, pp. 62–3.
[24] *Die Krisis der europäischen Wissenschaften ...*, p. 145.
[25] *Phänomenologie des Geistes, Philosophische Bibliothek Meiner*, Vol. 114, Hamburg, 1952, p. 82.
[26] Cf. L. Eley, *Metakritik der formalen Logik ...*, and: L. Eley, "*Die Krise des Apriori in der transzendentalen Phänomenologie Edmund Husserls*", Martinus Nijhoff, The Hague, 1962 (*Phaenomenologica*, Vol. 10).
[27] *Ideen zu einer reinen Phänomenologie ...*, Vol. I, p. 63.
[28] Cf. Beilage III, in: *Die Krisis der europäischen Wissenschaften*, pp. 365ff.
[29] Cf. *Logische Untersuchungen*, Vol. II, part 1, pp. 35–6.
[30a] Cf. *Die Krisis der europäischen Wissenschaften*, pp. 365ff.
[30] Cf. M. Theunissen, *Der Andere. Studien zur Sozialontologie der Gegenwart*, Berlin, 1965, pp. 19–20.
[31] Cf. *Logische Untersuchungen*, Vol. II, part 1, pp. 79ff.
[32] For the notion 'zulässig' cf. P. Lorenzen, *Formale Logik*, Berlin, 1970, pp. 61–2.

[33] Cf. *Logischen Untersuchungen*, Vol. II, part 2, §51, pp. 159ff.
[34] *Phänomenologie des Geistes*, p. 81 and pp. 85–6.
[35] Cf. L. Eley, *Metakritik der formalen Logik* ...
[36] Cf. *Logische Untersuchungen*, part 1: Der Gedanke. In: Gottlob Frege, *Kleine Schriften* (ed. by I. Angelelli), Darmstadt, 1967, pp. 342f.
[37] *Grundgezetze der Arithmetik*, Vol. I, 2nd edition, p. 7.
[38] Cf. *Logische Untersuchungen*, Vol. II, part 1, p. 50.

ANNEX

ROMAN INGARDEN'S LETTER TO EDMUND HUSSERL

Note from the Editor: Dallas Laskey's paper refers to Husserl's letters to Ingarden, which have been published recently in a seperate volume. In this correspondence all the letters of Ingarden to Husserl have been, however, destroyed during the war. The following letter is the only one which Ingarden by chance kept a copy of. It has appeared in the Polish translation in Ingarden's volume *Z badań nad filozofją współczesną*, PWN, Warszawa, 1963, p. 452–72. We have received the German original from the Polish Academy of Science in Warszawa with the permission to publish it in the *Analecta* in this particularly appropriate context.

This letter is considered by Ingarden as the first attempt of his criticism of Husserl's Idealism. The Polish translation contains several comments, which Ingarden made in 1963 before its publication. In view of the fact that it is the first publication of the original text in German we leave the comments for a subsequent publication of the English translation.

Roman Ingarden

Der Brief an Husserl über die VI. Untersuchung und den Idealism (Końskie, Ende Juli 1918)

Sehr geehrter Herr Professor!

Vor allem bitte ich um Entschuldigung, dass ich so lange nichts von mir hören liess. In Krakau war ich aber sehr beschäftigt, weil ich den fertigen Teil meiner Arbeit noch vor den Ferien abschicken woltte. Und es ist insofern gelungen, als das, was ich daran zu tun hatte, erledigt wurde. Wenn ich den Tekst trotzdem nicht abgeschickt habe, so lag es nur daran, dass ich in Krakau keinen Menschen finden konnte, der mir den Text ordentlich und für einen anständigen Preis abschreiben wollte. Ich habe freilich irgend jemanden die ersten 50 Seiten zur Probe gegeben, die aber kein günstiges Resultat ergeben hat. So habe ich mich entschlossen, den Text nach Freiburg zu schicken und dort bei dem Fräulein, das mir schon einmal Hilfe geleistet hat, abschreiben zu lassen. Indessen war dazu nötig,

Tymieniecka (ed.), Analecta Husserliana, Vol. II, 357–374.

dass ich zunächst die Änderungen, die ich gemacht habe, noch einmal abschreibe, weil sie in der vorhandenen Handschrift zu wenig lesbar sind. Ich habe es schon zum Teil getan. Da ich aber hier einige Stunden täglich mit meinen Schülern beschäftigt bin, ausserdem positiv so viel ich kann, arbeite, so konnte diese mechanische Arbeit noch nicht beendigt werden. Das alles machte mir aber unmöglich noch auf die ordentliche Erledigung der Korrespondenz Zeit zu finden.

Über meine äusseren Angelegenheiten wissen Sie wahrscheinlich schon von Frl. Stein. Ich gehe im Herbst nach Lublin. Wie sich aber die Situation in letzter Zeit darstellt, bestehen Möglichkeiten einer Übersiedlung nach Warschau, wahrscheindlich aber nicht früher als im nächsten Jahre. Ich will die Zeit benutzen um in Lublin in Stille eine Arbeit über die Methode und Sinn der Erkenntnistheorie zu schreiben. Meine Tätigkeit in Krakau hat einen gewissen Erfolg gehabt.

Frl. Stein hat mir vor einigen Wochen geschrieben, dass Sie wiederum an dem Problem des "Idealismus" arbeiten. Da mich die Sache ausserordentlich interessiert, habe ich mich sofort an die Arbeit gesetzt. Leider musste daran meine Ausarbeitung des letzten Teiles der Bergson-Arbeit leiden, aber die Arbeitstunden der letzten Wochen werden ihr im Grunde nur zu-gute-kommen. Ich habe noch einmal die V. und VI. Untersuchung ganz gründlich studiert, ausserdem alles wesentliche in dieser Hinsicht aus den "Ideen" durchdacht. Ich kann natürlich – und leider – nicht sagen, dass ich zum Abschluss gekommen bin. Aber wenigstens weiss ich, was ich nicht für haltbar halten kann.

Es wird Sie, Herr Professor, vieleicht interessieren, wenn ich einiges darüber schreibe.

Zunächst über die VI. Untersuchung. Ich glaube, sie muss neu herausgegeben werden in *derselben* Gestalt, wie sie vorliegt. Man müsste sie nur mit einer Reihe von Bemerkungen versehen, in welchen nur auf die schwachen Stellen, und auf die Gründe, weswegen sie schwach sind, hingewiesen sein würde. So dass man nur die an den entsprechenden Stellen sich eröffnenden tieferen Probleme andeute, die in eine *neue* Reihe von Betrachtungen münden und die im grossen Teil in den "Ideen" schon durchgeführt sind. So würde das Fragliche mit einem Fragezeichen des Verfassers versehen, die neuen Probleme angedeutet, was eine wertvolle Anregung für den Leser sein würde und die Wege und Zusammenhänge mit dem Standpunkt der "Ideen" angezeigt. – Eine solche Ausführung der neuen Auflage hätte die Vorzüge, dass sie verhältnismässig wenig Arbeit kosten würde, sachlich viel geben könnte und zugleich den Charakter des Werkes vollkommen unangetastet lassen würde. Auch terminologisch würde ich nichts oder fast nichts ändern.

Rein theoretisch gesprochen kann mich die VI. Untersuchung in folgenden Punkten nicht befriedigen (um das positiv Wertvolle und wirklich Schöne hier nicht mehr hervorzuheben): 1) Vor allem ist das Problem der Erkenntnis in einem Moment angegriffen, wo die eigentliche Erkenntnis (das "Kennenlernen") schon vollzogen ist und es sich im Grunde bloss um ein Wiedererkennen handelt. Wir *haben* schon "eine Idee" von dem Gegenstande und es handelt sich darum, ob der anschaulich auftretende Gegenstand mit der "Idee" (der signitiven Intention) zusammenstimmt (die Intention erfüllt) bzw. ob die Intention mit dem Gegebenen streitet. Es scheint mir, so müssten die Analysen der VI. Untersuchung durch eine Betrachtung der ursprünglichen Erkenntnis untergebaut werden, wo wir erst lernen, den Gegenstand "zu begreifen", wo wir noch keine Idee von ihm haben und erst die "Idee" bilden und eventl. weiterhin umbilden und an den Gegenstand "anpassen". Es müsste der Ursprung der signitiven Intention – oder besser – der Erfassung der Idee des (zunächst) individuellen Gegenstandes, und in der Folge der signitiven Begriffe aufgezeigt werden. (Dies müsste auch hinsichtlich der Wahrnehmungsanalyse in den "Ideen" durchgeführt werden.) Und diese Betrachtung mündete erst in die Untersuchung der Beziehungen zwischen "Anschauung" und "Denken", die im Wesentlichen schon geleistet ist. Es müsste der noetische Unterschied der anschaulichen (intuitiven) und signitiven Intentionen genauer herausgestellt werden, so wie eine rein noematisch (bzw. in gewissem Sinne "ontisch") gerichtete Betrachtung der "Bedeutung" durchgeführt werden.

(2) Die VI. Untersuchung krankt – und das ist das wichtigste Manko – an einer Untersuchung des Wesens des "Gegenstandes" bzw. des Seins (enger der Realität). Im Grunde ist das "Sein" dogmatisch behandelt. Hier liegt der grosse Fortschritt der "Ideen", dass sie dieses Problem angegriffen haben. (Über die dort vorliegende Lösung werde ich etwas später bei dem Problem des Idealismus schreiben.) Im Zusammenhang damit konnte einerseits das Problem der Ausweisung und – wo dies möglich ist – das der *letzten* Ausweisung nicht die erschöpfende Lösung finden. Eine derartige Lösung des Realitätsproblems, wie sie in den

"Ideen" vorliegt, forderte natürlich auch eine andere Stellung des Erkenntnisproblems selbst. Wobei freilich das Problem der Erkenntnis der idealen Gegenständlichkeiten für sich gestellt und von dem vorigen ganz anders gefasst werden müsste.

(3) Die Auffassung der kategorialen Akte als fundierter (Akte) scheint mir nicht ausreichend zu sein. Es gibt ja viele fundierte Akte, die gar nicht *"kategorial"* sind.

(4) Ich müsste die Scheidung zwischen gegenständlichen und logischen Kategorien stärker betonen, ausserdem das, was Sie "sinnliche Einheit" nennen, auch als *kategorialen* Bau betrachten und den letzten näher ausarbeiten. In dieser Richtung würde Aristoteles sehr hilfreich sein. Dabei müsste auch das Humé'sche Problem der Ausweisung der Kategorien gelösst werden.

(5) Mit der Ansicht, dass bei den kategorialen Akten (im Sinne der VI. Unters.) als intuitive Repräsentanten die Momente der Akte der kategorialen Funktionen dienen, kann ich mich nicht befreunden. Es scheint mir, dass hier manche Hume'sche Anklänge mitgewirkt haben. Speziell bei den prädikativen kategorialen Beziehungen würde diese Stellung eine *echte* Erfüllung ausschliessen und letzten Endes skeptisch sein. Wie es sich in dieser Hinsicht mit den echten logischen Kategorien ("und", "oder" und dergl.) sowie mit den kategorialen Formungen der Gegenständlichkeiten höherer Stufe verhält, weiss ich augenblicklich nicht. Jedenfalls kann mich die Lösung der VI. Untersuchung nicht befriedigen. Die kategoriale Anschauung – wie auch die Wesensanschauung im spezifischen Sinne muss einer tiefer gehenden Analyse unterworfen werden.

Um aber zu dem Problem des Idealismus zu kommen, über welches ich eigentlich schreiben wollte und das mich schon seit einigen Jahren quält: Es scheint mir, dass sich unter diesem Namen verschiedene und grundsätzlich verschiedene Probleme bergen. Gewöhnlich in der Literatur laufen diese Probleme ineinander und oft wird ein System "idealistisch" genannt, das im Grunde nicht so zu nennen wäre. Ich will hier nicht über die Aequivokationen bzw. über alle hier liegenden Probleme reden; aber es scheint mir, dass man in Bezug auf Sachen, die *uns* interessieren, 3 Problemgruppen abscheiden muss. (1) die ontologischen, (2) die metaphysischen (in etwas anderem Sinne als Sie dies Wort gebrauchen), (3) die erkenntnistheoretischen Probleme. Es bestehen natürlich zwischen allen Gruppen wesentliche Zusammenhänge.

(1) Das ontologische Hauptproblem des "Idealismus" scheint mir das Problem der Wesensgemeinschaft (bzw. Identität) zwischen "Sein" (bzw. enger: Realität) und *reinem* Bewusstsein zu sein. Die Identifizierung der beiden Wesen nenne ich "Idealismus". In diesem Sinne ist die Stellung der "Ideen" *zunächst* entschieden *nicht* idealistisch. Niemand hat vielleicht die Wesensheterogenität der Realität und des reinen Bewusstseins so stark betont (wie Sie in den "Ideen"). Indessen – soweit ich Sie verstehe – ändert sich im Laufe der Untersuchung diese ursprüngliche Stellung, so dass im Grunde die Wesensverschiedenheit eigentlich geleugnet wird, oder in die Scheidung zwischen Noesis und Noema umgedeutet. Dies geschieht einerseits dadurch, dass einerseits die Realität (um das Problem enger zu fassen und vom idealen Sein abzusehen) *nur* als Vermeintes "und darüber hinaus ein Nichts" gefasst wird, andererseits aber das Wesen des Bewusstseins auf das Noema ausgedehnt wird. Denn im Grunde ist – nach Ihrer Auffassung – z.B. das Ding als Korrelut einer unendlichen Wahrnehmungsmannigfaltigkeit nichts anderes als ein noematischer Sinn (und "darüber hinaus ist es ein Nichts"). stärker tritt diese Auffassung hervor beim Eintritt in die konstitutive Betrachtung. Ding bedeutet da nichts anderes als ein System von Noemata-Schichten bzw. wenn es sich um das anschauliche Ding handelt, als eine Schicht in diesem System. Wir kommen am Ende zu der Gleichung:

Ding = ein eigentümlich gebautes Noema-Bewusstsein. Sehr deutlich tritt dies noch in dem Satz der "Ideen", der einen etwas anderen Punkt berührt: "Die Realität... entbehrt wesenmässig... der Selbständigkeit. Sie ist nicht etwas in sich Absolutes und bindet sich nur sekundär an Anderes, sondern sie ist im absoluten Sinne gar nichts, es hat gar kein absolutes Wesen, es hat die Wesenheit von etwas, das prinzipiell *nur* Intentionales, nur Bewusstes ist (S. 94). (Hier wird die Selbständigkeit berührt, auf die ich gleich zurückkomme). Dass Sie zugleich sagen, die Realität sei etwas durch Abschattungen, Bekundungen Gegebenes, das Bewusstsein aber nicht, kann bei der Festhaltung der anfänglichen These der Wesensheterogenität als ein innerer Widerspruch im System betrachtet werden. Bei Leugnung der Wesensheterogenität kann diese Behauptung festgehalt werden, denn sie besagt, idealistisch gesprochen, nichts anderes dass es im Gebiet des Bewusstseins verschiedenartig gebaute Noemata sowie verschieden geartet Sinnzusammenhänge zwischen Noemata bzw. zwischen verschiedenen Schichten von Noematen, sowie verschieden geartete Zusammenhänge zwischen Akt (Noesis) und Noema gibt. Von etwas dem Bewusstsein wesensfremden braucht dabei nicht gesprochen zu werden. Die Noemata selbst sind immanent gegeben und sind absolut. Die noematischen Sinnzusammenhänge sind z.B. der Art, dass eine Mannigfaltigkeit von Ansichtnoematen (die relativ mehr erfüllt sind, als das sie durch *einen* Sinn umspannende Dingnoema) *einem* anderen Noema zugehört (entweder so, dass es die Mannigfaltigkeit sinngemäss umspannt oder so, dass es sich als eine neue Sinnschicht (die eventl. zu beseitigen ist) auf dem Ansichtsnoema als ein das Ganze durchgreifender, in der Beziehung des Abgeschatteten zum Abschattenden (vaag gesprochen) stehender oberer Sinn konstituiert, der die untere Schicht in Bezug auf die Erfüllungszusammenhänge – transzendiert. Dies gebe ich natürlich nur als *ein* Beispiel aus einer grossen Mannigfaltigkeit von solchen Sinnzusammenhängen. Die Konstitutionsprobleme bzw. ein wesentlicher Teil von ihnen (und zwar die noematische Konstitution) sind nichts anderes als das System von Fragen, die diese verschiedenartige Sinnzusammenhänge in Abhängigkeiten von einander betreffen.

Wäre der Idealismus in diesem Sinne haltbar, so wäre die Theorie der Konstitution mit der Metaphysik identisch, ein Teil von ihr mit der Metaphysik der Aussenwelt bzw. mit der Naturwissenschaft identisch.

Ich vermag mich diesem Idealismus nicht anschliessen. Die Wesensheterogenität zwischen Bewusstsein und Realität (bzw. allgemeiner: Sein) kann ich nicht aufgeben. Verhältnismässig klar sehe ich freilich den Wesensunterschied nur in der Gegenüberstellung: reale Aussenwelt-Bewusstsein. Wie es sich dagegen mit der Gegenüberstellung: psychisches Subjekt – reines Bewusstsein verhält, sehe ich nicht ganz klar. Bei der Einengung des Problems auf die reale Aussenwelt glaube ich den Satz behaupten zu können: "Alles Reale ist räumlich oder im Räumlichen fundiert". Das reine Bewusstsein dagegen ist wesensmässig raumlos, was *keine* positive Bestimmtheit, sondern eine vollkommene Privation bedeutet. Würde es bloss reines Bewusstsein geben, so würde man überhaupt nicht auf den Gedanken kommen, dem Bewusstsein Raumlosigkeit zuzusprechen. Man würde es ausschliesslich mit anderen positiven und dem Raum völlig heterogenen Bestimmtheiten charakterisieren. Das Wesenfaktum der Ausgedehntheit (Ausbreitung) der Empfindungsdaten bedeutet für mich keine Schwierigkeit, denn es ist eben die grosse Frage, ob Empfindungsdaten als "Bewusstsein" zu betrachten sind. Als eine

Grundheterogenität zwischen der realen Aussenwelt und dem Bewusstsein betrachte ich den Unterschied der *Seinweise*, den Unterschied in der Gegebenheitsweise beachte ich natürlich auch, aber das gehört zu einer anderen Problemgruppe. Als Folge des ersten Satzes ergeben sich Behauptungen der Art wie: Alles aussenweltlich Reale und Selbständige ist abstückbar und eventl. durch Trennung zerstörbar, oder – insofern es sich um Selbständigkeiten handelt, die nur im Räumlichen fundiert sind, durch Abstückung des Fundierenden zerstörbar (alle Prozesse sind unselbständig). Der Unterschied der Seinsweise besteht darin, dass das aussenweltlich Reale wesensmässig ein "stummes" (dunkles, stilles, "unbewusstes") Sein führt, während Bewusstsein eben im Selbst-bewusst-sein existiert (eine sich selbst durchglühende Glut, ein sichselbst (ohne Reflexion) *durchlebendes* Meinen ist. Sein des Bewusstseins liegt im *Vollzug* dieses *intuitiven* Meinens, das hinsichtlich seiner "Helligkeit" verschiedene Grade – wenn das Wort erlaubt ist – bis zum höchsten vollkommenen Bewusstsein (es handelt sich natürlich um lauter qualitativ heterogene Unterschiede) annehmen kann, aber nie völlig fehlen kann und bei *jedem* Erlebnis als Fundierungsschichte enthalten sein muss (vgl. Brentano und Log. Unters. V.).

Diesen fundamentalen Unterschied in der Seinsweise vermag ich nicht zu beseitigen. Ich gestehe gern, dass diese Auffassung – übrigens ist sie gar nicht neu – mit verschiedenen Schwierigkeiten zu kämpfen hat. Unter anderem ist zu fragen, was die verschiedenen Noemata sind, so weit sie anschauliche Fülle sind. Ich würde sie schwerlich dem "Bewusstsein" in echtem Sinne zurechnen können. Andererseits sind sie keine "Dinge"der realen Aussenwelt. Müssen sie aber notwendig *entweder* dies *oder* jenes sein? In dieser Hinsicht kann eine endgültige Antwort nur die noematische konstitutive Betrachtung geben. Diese aber kann den eben aufgestellten Satz nicht in Frage stellen.

Und nun einige Disskussionsbemerkungen: Gewiss kann jedes reale Ding ein "intentionaler Gegenstand" sein. Aber dies erschöpft sein Wesen nicht. Ausserdem gehört es überhaupt zu seinem Wesen nicht, dass es ein intentionaler Gegenstand *ist*. Ohne wahrgenommen oder sonstwie vermeint zu sein kann es existieren. Ein Bewusstsein aber, das in schlichter Weise nicht ein Selbst-durchleben wäre – *für sich* irgendwie *gegenwärtig* eben im Vollzug der Intuition, was nur ein anderes Wort für das Bewusstsein ist – ist ein Nonsens. So könnte es nicht sein. *Hier* liegt der Grund, weswegen wir von den "Durchlebnissen" überhaupt etwas "wissen" können (ohne auf sie reflektiert zu haben). Und das Wesensfaktum, dass das Erlebnis in der Reflektion immer auftritt als ein solches, das schon früher war, ist vielmehr nur ein Ausdruck, eine Folgeerscheinung, die ohne die wesentliche Seinsweise des Bewusstseins nicht zustandekommen könnte.

(Verzeihen Sie, Lieber Herr Professor, alle diese Ketzreien. Aber die müssen heraus. Ausserdem ist es eben diese innere Gärung, die mir meine Bergsonarbeit nicht zu beenden erlaubt.[1] Sie muss durchgemacht werden. Es muss viel Falsches in allem, was ich schreibe, stecken, aber vom Falschem kann man sich nur befreien, wenn man es sich zunächst als Wahrheit entgegenstellt.)

Die Realität ist nur, soweit sie etwas "in sich" ist. Das wofür sie vermeint werden kann, ist ihr eigentlich irrelevant. Sie ist das, was sie "in sich" ist und als solche ist, sie ist ein in sich jederzeit vollendetes, allseitig bestimmten Sein. In der Welt gibt es keine Unbestimmtheit, es sei denn als Unbestimmtheit einer Potenz, die selbst *voll* bestimmt ist. Schon dadurch (deswegen) kann ich das Ding nicht als (für?) die unendliche, sich einstimmig ausweisende und motivierende Wahrnehmungmannigfaltigkeit der Dingnoemata und auch nicht als den einheitlich diese Mannigfaltigkeit durchwaltenden Sinn (die Idee im Kantischen Sinn) betrachten. Die "Ideen" sagen an einigen Stellen identisch Dasselbe ("Transzendenz des Dinges") und doch – wenn sie das Ding reduzieren, dass es ein Vermeintes und *nur* ein Vermeintes ist und "darüber hinaus ein Nichts", so leugnen sie freilich die Transzendenz ausdrücklich nicht, aber deuten sie dermassen um, dass es eigentlich auf diese Leugnung hinauskommt. Denn "Transzendenz" bedeutet dann nur *entweder* einen bestimmten Bau des Wahrnehmungsnoemas bzw. ein bestimmtes Verhältnis zwischen dem in der Wahrnehmung intendierten Sinn und manchen als Unterbau diemenden anschaulichen Komponenten des Noemas, *oder* es bedeutet eine bestimmte Beziehung zwischen noematischen Sinnen die zu verschiedenen Schichten der Objektivation gehören. Ob das aber wirklich als Transzendenz betrachtet werden kann? Ich weiss, die Transzendenz hat, wie alles in der erkenntnistheoretischen Betrachtung, einen "Sinn" und dieser Sinn besteht eben in den angedeuteten Beziehungen bzw. weist sich in ihnen aus, aber *ist* die Transzendenz ein Sinnzusammenhang?

Der Standpunkt der "Ideen" in den besprochenen Fragen führt zu einem Problem, das unlösbar ist, soweit man nicht ein dem reinen individuellen Bewusstsein *fremdes wirkliches* Sein voraussetzt. Eigentlich streng genommen gibt es auf dem Standpunkt der "Ideen" nur ein einziges Bewusstsein, das *absolut* ist, das nicht einmal "mein" genannt werden. darf. Deswegen dürfte man die nähere Bezeichnung "individuelles" ganz streichen. Ich sehe gar nicht ein, weswegen das Sein anderes Bewusstseins etwas *mehr* als nur "Vermeintes" zu betrachten wäre, um so mehr als die "Einfühlung" in der äusseren Wahrnehmung fundiert ist. Wie die "Ideen" ganz richtig behaupten, gehört es zum Wesen des Bewusstseins nicht, dass sich in ihm eine Welt konstituieren müsste. Es muss also noch nach einem Prinzip der *faktisch* vorgefundenen Welt gefragt werden. Vom Standpunkt der "Ideen" aus kann es natürlich nicht in der Welt selbst gesucht werden. So kommt man auf den Gedanken der Teleologie mit der Setzung der Gottheit, die durch andere "intuitive" Bekundungen gegeben sein würde. Dagegen ist zunächst nichts einzuwenden. obwohl es ohne Hinzunahme irgend eines *ethischen* Prinzips nur eine Verschiebung des Problems ist. (Dann müssten aber die *ethischen* Werte transzendent und *absolut* sein.)

Aber ungeachtet dessen bestehen zwei Möglichkeiten: *Entweder* hat Descartes recht und Gott ist eine clara et distincta perceptio (idea), dann kann er in der Reduktion *nicht* ausgeschaltet werden, ist *immanent* und ist im Grunde nichts anderes als die "im Bewusstsein waltende Vernunft". Damit ist aber das Problem nicht gelöst, sondern nur in anderen Terminis gestellt (oder ist es eine Scheinlösung). Oder Descartes hat nicht recht, Gott ist transzendent und kann ausgeschaltet werden. Dann aber sehe ich nicht ein, weswegen Gott ein *absolutes* Sein, ein "An-sich-sein" und nicht *bloss* ein Vermeintes sein soll. Ich begreife es wohl, dass es verschiedenartige Transzendenzen geben kann, aber ist die Gottes-Transzendenz eine *Transzendenz* (und man behält den Standpunkt der "Ideen" bei), so ist Gott ebenso ein bloss Vermeintes und es besteht keine Notwendigkeit, dass Er ein "an sich", ein absolutes Sein sein müsste.

Hier kann mich ein Einwand treffen. Sie würden vielleicht sagen: Es liegt nicht an der Transzendenz als solchen, sondern an der bestimmten *Art* der Welt-Transzendenz, dass die Welt bzw. "Reales" nicht ein im absoluten Sinne "an sich" sein kann.

Ich würde darauf zunächst nichts antworten. Aber der Einwand ist mir sehr willkommen, weil er die Frage zum Bewusstsein bringt, weswegen eigentlich das "Reale" nicht ein im absoluten Sinne (was *nicht* "immanent" bedeuten soll, denn das wäre ganz selbstverständlich) "an sich" sein kann? Was bewirkt es?

Dem Sinne der Erfahrung nach ist das (eventl. vermeintlich) Gegebene nur dann "Realität", wenn es nicht *meine* "Einbildung" ist. Dabei "Einbildung" bedeutet mehr als eine Vermeintheit die ungenügend motiviert ist. Andererseits bedeutet "Realität" mehr eventl. etwas anderes als "im bisherigen Erfahrungsverlauf genügend motivierte Vermeintheit" Genügende Motivierung einer Vermeintheit in der objektiven Vernunft besagt nicht nur, dass in den Erfahrungssynthesen kein Widerstreit vorhanden ist, sondern vor allem, dass in jedem Glied der Erfahrungsynthesen sich etwas kundgibt, was "an sich" (in sich) ist. Dass ich in den betreffenden Erfahrungsakten auf etwas stosse, das "Nicht-Ich" und nicht Bewusstsein, sondern etwas dem Bewusstsein Fremdes und für sich Seiendes ist. "Objektiv gültig" meint einerseits "vernünftig motiviert", andererseits "dem Sein an sich entsprechend". Gewiss: die Wahrheit ist, erkenntnistheoretisch betrachtet, eine Adaequation zweier Sinne. Aber es gehört eben zum "Sinne" des Sinnes eines realen Gegenstandes, dass er Sinn dieses Gegenstandes ist, dass er "in" ihm *inkarniert* ist. Im Moment, wo es diesen Gegenstand nicht gibt, gibt es auch diesen Sinn nicht. Das bedeutet aber: sowohl dem Sinne der Erfahrung wie dem der Wahrheit nach ist die Realität ein Sein, das vom Bewusstsein, von jedwedem noematischen Sinn *verschieden* und eine "in sich" seiende Entität ist. Dieses Verschiedensein vom Bewusstsein bzw. vom noematischen Sinn bestimmt eben eine Bedeutung von "Transzendenz". In diesem Sinne schliesst die Transzendenz des realen Gegenstandes sein "an sich sein" nicht aus, sondern *fordert* es gerade, falls der Gegenstand überhaupt existieren soll. Rein ideal gesprochen besteht natürlich die leere Möglichkeit, dass alle Vermeintheiten eben nichts mehr als Vermeintheiten sind, die Möglichkeit irgendeines "transzendentalen Scheins" z.B. à la das Vaihingersche "Als-ob" (dessen Wirklichkeit übrigens ausgewiesen werden müsste); dann aber ändert sich der von mir hier festgehaltene *Sinn* der Realität nicht, er ist da vielmehr vorausgesetzt – gerade deswegen heisst es "transzendentaler *Schein*". Andererseits sind diese "Als-ob-Sinne" selbst *garnicht* transzendent, wenigstens in dem besherigen Sinne. Sie sind es natürlich als ideale Gegenständlichkeiten, was ein Problem für sich ist. Sollte es aber eine Realität geben, so müsste sie notwendig von diesen Als-ob-Sinne prinzipiell verschieden sein.

Es gibt aber einen anderen Sinn von der "Transzendenz", die sich in zwei Unterbedeutungen spaltet. (1) In Bezug auf eine *einzelne* Wahrnehmung: der die Wahrnehmung bzw. das Wahrgenommene als solches durchgreifende (in gewissem Sinne der oberste, das Ganze umfassende) Sinn geht (und auf verschiedene s.z.s. mehrschichtige Weise) über das wahrhaft Erfüllte hinaus (nicht bloss faktisch, sondern wesensmässig). (2) Dass der Sinn der Realität (des realen Gegenstandes) über den Erfüllungssinn jeden *endlichen* Mannigfaltigkeit von Wahrnehmungssynthesen notwendig *hinaus-geht*.

Schliesst vielleicht dieser Sinn von der "Transzendenz" das "an-sich-sein" der Realität aus? Es scheint mir, dass er entweder für die uns interessierende Frage ohne Bedeutung ist, oder eine Lösung des Problems in *meinem* Sinne fordert.

(I) Entweder nämlich betrifft diese Transzendenz nur eine Bestimmtheit des noematischen Baues, dann ist sie für unsere Frage irrelevant. Ebenso verhält es sich, wenn der "Sinn des realen Gegenstandes" im obigen Satze (2) als eine Bedeutung bzw. ein Begriff (oder ein Wesen) gefasst wird. In diesem Falle handelt es sich nämlich um eine Beziehung zwischen zwei verschiedenartigen *Sinnen*, und wie diese Beziehung auch aussehen mag, ist sie jedenfalls für die Frage, ob die Realität ein "in sich Seiendes" ist, ohne Bedeutung.

(II) Oder bedeutet der (Ausdruck) "Sinn des realen Gegenstandes" den in dem realen Gegenstande *inkarnierten* Sinn. Dann aber fordert diese Transzendenz, dass der reale Gegenstant "an sich" sei, denn dann nur kann der "Sinn des realen Gegenstandes" existieren und in der Transzendenzbeziehung zu Erfüllungsmannigfaltigkeiten stehen.[2]

Als prinzipiell andere Probleme betrachte ich die sich hier aufdrängenden Probleme der Ausweisung. Im konkreten Falle die Frage: was berechtigt uns den inkarnierten Sinn der Realität bzw. die Realität selbst zu setzen. Diese Frage führt zu einem neuen Sinn von Transzendenz und steht mit einer neuen Reihe von Problemen, Ausweisungsfragen, erkenntnistheoretischen Fragen – im Zusammenhang. Diese werden wir später besprechen. Hier sei nur bemerkt, dass man folgende zwei verschiedene Sachen nicht vermengen darf: Die eben berührte Transzendenz im neuen Sinne besteht darin, dass der Seinsmodus der Realität über den im *endlicher* Erfahrungsmannigfaltigkeit ausweisbaren (erreichbaren) *Set-*

zungswert hinausreicht. Ohne uns mit diesem wesentlich erkenntnistheoretischen Satze hier zu beschäftigen bemerke ich: man soll den Seinsmodus mit dem Setzungswert nicht vermengen. Meidet man diese Vermengung, so ist der von mir angestrebte Sinn von der Realität gesichert Vollzieht man aber diese Vermengung bzw. identifiziert man bewusst beides auf Grund der Voraussetzung Realität-noematischer Sinn-Bewusstsein, so muss man zugeben, dass dann eigentlich kein Grund zu der Aufstellung der Transzendenz III bestünde oder dass wenn der betreffende Satz über die Transzendenz III richtig wäre, man auf einen ganz unerhörten Widersinn innerhalb der Bewusstseinssinnzusammenhänge stossen würde. Dieser Satz birgt aber gar keinen prinzipiellen Widersinn, falls man die (angedeutete) Vermengung meidet.

So sehe ich – wenigstens der bisherigen Betrachtung nach – den Grund nicht, der das "an-sich-Sein" (das mehr als intentionalen Sein) der Realität ausschliessen würde. Es ist jetzt bloss die Frage, ob und was dieses Resultat erschüttern könne? Natürlich kann ich irren. Aber den Fall vorausgesetzt, dass die obigen Behauptungen eingesehen sind, ist zu fragen, ob es eine Möglichkeit gibt, die das Gesagte in Frage stellen könnte. Es war eine Sinn- bzw. eine Wesensuntersuchung des *konstituierten* Sinnes der Realität. Sie ist richtig, wenn es ein "Wesen" gibt, das den hier aufgestellten Behauptungen adaequat entspricht. Was verbürgt uns diese Existenz (vorausgesetzt die Gegebenheit)? Das Wesen selbst ist natürlich bewusstseinsfremd. In diesem Sinne ist es transzendent. Dem Satze über die apriorische Erkenntnis nach gibt zu diesem Wesen *einen* Akt, in dem es adaequat erschaubar ist. Die Transzendenz II und III besteht also nicht. (Gegebenheit durch Abschattungen gibt es hier natürlich nicht. Eine *ganz genaue* Analyse der diesbezüglichen Sachlagen wäre notwendig, sie ist bis jetzt nicht geleistet.) Nach dem Satz über das Apriori ist diese Erkenntnis unerschütterbar. Anders gesprochen: "es gibt" das zugehörige Wesen. Es ist also möglich, eine absolut setzende Erkenntnis von einem Transzendenten (I) Gegenstand zu haben. Dann kommt der Gedanke: bleibt alles bestehen, wenn man eine konstitutive Betrachtung durchführt?

Zunächst, was bedeutet eine "konstitutive Betrachtung"? Es scheint mir, dass es verschiedenartige Problemgruppen gibt, die unter diesem Namen zu führen wären.

Vor allem: (1) Konstitution als *statischer Aufbau* des Aktes und des

Noemas. Die Herausstellung verschiedener Momente des Noemas u.d. Aktes, die Erfassung der Funktion, die einzelne dieser Momente spielen, sowie der Sinnzusammenhänge und Abhängigkeiten, die zwischen den Momenten bestehen bzw. die zwischen Noema-Noesis (Akt) bestehe. Konkret: Wie ein *vorgegebener* Akt-Noema gebaut ist. (Eidetisch).

(2) Daran schliesst sich eine tiefer gehende Betrachtung mit dem leitenden Gedanken: Wie *muss* ein Akt-Noema gebaut sein, falls der oberste Sinn des Noemas ein bestimmter sein soll (damit er sein könnte). Das heisst: was in den die unteren Schichten bildenden Momenten enthalten sein muss, bzw. wie die Grenzen der Variabilität in den einzelnen Momenten laufen können, damit der oberste Sinn des Noemas als gegebener auftreten könnte (zugleich, was in den Aktmomenten enthalten sein muss).

Bei (1) stehen wir auf dem Boden der absoluten Faktizität, obwohl es eine eidetische Betrachtung ist. Der oberste Sinn (der volle Satz) des Noemas ist als gültig vorausgesetzt (als rechtmässig), ebenso die Existenz des Aktes, und die Frage geht nur nach dem wesensfaktischen Aufbau. Im zweiten Falle ist der volle Sinn des Noemas vorausgesetzt und es wird nach den in dem Aufbau des Noemas und des Aktes gründenden Bedingungen der Möglichkeit – den notwendigen und den hinreichenden – gefragt. Es wird eigentlich einem noematischen obersten vollen Sinn eine Sphäre von "möglichen" Noemata und Akte zugeordnet. Die Variabilitäten, die da auftreten, sind auch "statisch" zu fassen. Oder anders gesprochen: wir betrachten ein einzelnes – aus dem Bewusstseins- bzw. Noema-Strom herausgefasstes Noema-Akt. Wir verfolgen hier nicht die kontinuierlichen Synthesen, die zwischen mehreren Akten-Noematen zustandekommen können und die eventuellen Abhängigkeiten des Sinnes eines bestimmten Noemas von anderen Noematen (Akten) und den synthetischen Zusammenhängen.

(3) Folgt eine Betrachtung der kontinuierlichen Synthesen zwischen Noematen-Akten und weiterhin der hier obwaltenden Gesetzmässigkeiten in betreff der Bedingungen der Möglichkeit der anschaulichen Gegebenheit der Sinne (der Noemata).

(4) Konstitutive Rechtsbetrachtung. Hier liegt die Stätte der letzten erkenntnistheoretischen Entscheidungen. Der Gang der Betrachtung ist ein umgekehrter als in den vorigen Fällen. *Nicht* der Oberste, volle noematische Sinn (eines schlichten Aktes zunächst, dann der höher

gebauten, fundierten usw. Akte) ist als "gültig" vorausgesetzt, sondern die *letzten*, ursprünglichen, in sich nicht mehr "konstituierten" ("untersten") Elemente des Noemas müssen als *absolute Faktizitäten festgestellt* und *wesensmässig* ihrem Gehalt und Form nach untersucht werden. Und es muss weiterhin untersucht werden, ob der Gehalt (und die Form) dieser Elemente, sowie der weitere Bau des ganzen Noemas sowie des zugehörigen Aktes der Art sind, dass ein zu setzender Noematischer oberster Sinn bei dem Enthaltensein jener Elemente im Noema, *notwendig* gegeben sein muss. Als Richtschnurr dient hier ein ursprünglicher Gegensatz zwischen "subjektiv" (bewusstseinsartig) und "objektiv" (dem Bewusstsein fremdes "an sich-Sein"). Als Ausgang muss die *absolute Erfassung* des *Wesens* des *Bewusstseins* (des Aktes) dienen. Andererseits müssen, wie gesagt, die letzten (untersten) ursprünglichen Elemente des Noemas in ihrer absoluten Faktizität festgestellt und wesensmässig untersucht werden, endlich muss der Gehalt der anzusetzenden (gegenständlichen) noematischen Sinne klar erfasst, sowie auch der Sinn der "Objektivität" zunächst vorkonstitutiv untersucht werden, um dann als das Anzusetzende und zu Setzende (falls es sich in der konstitution als gültig *ausweist*) ingestellt zu werden. Der oberste Sinn des Noemas (Gegenstandsnoema) weist sich dann aus als "gültig", wenn – wie gesagt – der Gehalt und die Form der letzten ursprünglichen Elemente (Momente) der Art sind, dass bei ihrem Enthaltensein im Noema jede der aufeinander gebauten Schichten bis zum Gegenstandsnoema des betreffenden Erkenntnis-Erlebnisses (Akt plus Noema) *notwendig* vorhanden bzw. gegeben sein muss. Aber diese Notwendigkeit kann man noch verschieden verstehen. Man könnte meinen: der noematische Sinn sei "notwendig", soweit ein entsprechender Akt (bzw. Aktkomponente) existiert. Indessen nicht um diese Notwendigkeit handelt es sich. Gewiss, es besteht eine notwendige Korrelation zwischen Noesis und Noema, so dass einer bestimmten Intention ein ganz bestimmtes Intendiertes (als solches) notwendig entspricht. Die Feststellung dieses Zusammenhanges aber würde uns bei unserem Ziele der Herausstellung der Geltung der betreffenden Erkenntnis, sowie der Existenz des letztlich "Objektiven" nichts helfen. Vielmehr handelt es sich hier um eine Notwendigkeit, die nicht ausschliesslich in den Akten gründet. Genauer gesagt: Jeder Akt führt mit sich ein Vermeintes, aber auch umgekehrt die Akte (bzw. *manche* von den Akten) sind wesensmässig durch Elemente des Noemas motiviert.

Es handelt sich also um die Frage, ob die Akte, korrelativ die noematischen Sinne, die konstitutiv auszuweisen sind, eben durch die letzten ursprünglichen noematischen Elemente notwendig motiviert sind, oder ob ihr Vorhandensein auf andere aussernoematische Momente *relativ* ist, so dass auch der oberste noematische Sinn an dieser Relativität teilhat. Jeder Akt bedeutet Aktivität und diese Aktivität ist in vielen Fällen unentbehrlich zur Erfassung des Gegenstandes. Im Bereiche dieser Aktivität muss man aber zwischen einer konstruktiven und "willkürlichen", eventl. durch andere als rein erkenntnismässige Motive bedingten Aktivität und der Aktivität unterscheiden, die als *passive*, den noematischen letzten Elementen ihrem Sinne nach *folgende* Aktivität zu bezeichnen wäre. Jedem noematischen Element folgt eine "Auffassung" (Akt) und korrelativ ein Aufgefasstes als solches. Wenn dieses "folgen" ein derartiges ist, dass der "Akt" den noematischen Elementen passiv-aktiv folgt (wobei eine mehrfache Übereinanderbauung möglich ist: noematisches ursprüngliches Moment – Auffassung I, – Aufgefasstes I – Auffassung II, – Aufgefasstes II – bis zum obersten noematischen Sinn) und *ausschliesslich* durch letzte noematische Elemente (bzw. durch schon als "gültig" ausgewiesene noematische Sinne) motiviert ist, so ist der noematische oberste Sinn *notwendig* gegeben, der Gegenstand weist sich als existierend aus. Diese Notwendigkeit ist natürlich keine reine. Sie ist durch die absolute Faktizität der ursprünglichen noematischen Elemente verunreinigt, bedingt. Im entgegengesetzten Falle weist sich der Sinn konstitutiv nicht aus (obwohl er ein in der vorkonstitutiven Betrachtung vermeintlich gegebener ist) der Gegenstand existiert nicht "an sich", er ist in irgend einer Weise "relativ".

Die konstitutive Rechtsbetrachtung zerfällt in zwei Problemgruppen: (1) Die Ausweisung der materialen Sinne – materiale konstitutive Rechtsbetrachtung (– Metaphysik bzw. ihr Grundstück), und (2) die Ausweisung der Katogorien – formal konstitutive Rechtsbetrachtung (formale Rechtslehre der Kategorien).

(5) Konstitution als ideelle Genesis der Noemata. Im Hinblick darauf, dass in der Reihe der Noemata, die zu einem identischen Gegenstand gehören, manche Noemata der Art die Sinne der anderen bedingen, dass die "früher" erfasst werden müssen (als dieselben), kann man die Zusammenhänge, die zwischen einzelnen (und in verschiedenen Objektivierungsschichten liegenden) bestehen, im Sinne einer Aufeinanderfolge auffassen,

so dass dann die durch einen identischen Gegenstandssinn umspannte Mannigfaltigkeit der Noemata als eine Entwicklungsreihe (bzw. Continuum) des endgültig konstituierten noematischen Sinnes aufgefasst werden kann. Die Herausstellung der diesbezüglichen Gesetzmässigkeiten (die natürlich nicht nur die Sphäre der Noemata, sondern auch die der Akte umspannen würden) sowie der hier vorliegenden Möglichkeiten bildete eine ideelle Genesis der noematischen Sinne. Eine noetisch gerichtete, aber wesentlich auf die noematische Sphäre bezogene Untersuchung – mit Berücksichtigung individueller Unterschiede bzw. Typen – würde da eine ideelle Entwicklungstheorie möglicher Typen der Erkenntnissubjekte (Instinkt, Intellekt usw.) sein. Alles in der Sphäre der Immanenz. Hier wäre ein Teil der apriorisch-theoretischen Grundlagen der genetischen Psychologie zu suchen.

Ein genauer Entwurf der Problematik der oben angedeuteten Untersuchungssphären, sowie der Bedingungen der Möglichkeit einer jeden der Untersuchungen, ist ein der wichtigsten Aufgabenthemata, die sobald wie möglich zu erledigen wäre.

Jetzt können wir zu unserem eigentlichen Thema zurückkehren. Die Frage war: kann etwas unser Resultat (unter der Voraussetzung des Ausschlusses eines "Fehlers" in der Betrachtung) über das Wesen der Realität in Frage stellen? Die Antwort liegt jetzt nahe. Um sie aber vollständig zu machen ist noch Folgendes zu bemerken. Es ist zwischen absolut *reinen* Wesen, denen prinzipiell keine Realisierungen entsprechen können (z.B. die geometrischen Gegenständlichkeiten) und Wesen, die Realisierungen (Wesen in der Realisierung) zulassen. Die ersten existieren, soweit sie in einem einheitlichen Anschauungssinn erfassbar sind. Die Bedingung ihrer Existenz macht die Gesetzmässigkeit der entsprechenden Anschauung aus. Die anderen dagegen sind durch das Wesen des konkreten Seins als solchen bedingt und existieren als Wesen nur soweit sie die durch das Wesen des konkreten Seins vorgeschriebenen Bedingungen erfüllen. Andererseits gibt es Wesen, die existieren *müssen*, falls es andere ganz bestimmte Wesen gibt. Im ganz *radikalen* Sinne des Wortes gibt es überhaupt kein einziges solches Wesen, das *notwendig* existieren *müsste*. Irgendwo müssen wir notwendig auf ein *Faktum* der Existenz stossen. Nun gibt es innerhalb dieser Fakta solche, deren Seinsthesis – im Moment, wo sie einmal da ist, eine unaufhebbare (jeden Zweifel ausschliessende) "absolute" ist, und andere, wo dies nicht statthat. Zugleich gibt es unter den Wesen, die Inkarnierungen im konkreten Sein zulassen, solche, deren Inkarnierungen eine absolute Seinsthesis aufweisen, unter den letzten wiederum solche, die unabhängig andere, die – wegen der früher besprochenen konstitutiven Rechtszusammenhänge *notwendig* existieren müssen, *falls* die ersteren existieren. Durch die immanente bzw. absolute Erfassung der Inkarnierungen der ersteren ist zugleich die Existenz der anderen absolut gesichert.

Jetzt können wir sagen, was uns eine konstitutive Betrachtung nach der Erfassung des Sinnes der Realität geben kann. Eine Erfassung des Sinnes (bzw. des Wesens) ist als eine apriorische Erkenntnis unbezweifelbar, falls sie eben eine strenge Erfassung ist. Nicht alles aber, was "gegeben" ist, ist rechtmässig, "wirklich" gegeben. Die konstitutive Betrachtungen I–IV und vor allem die IV konstitutive Rechtsbetrachtung, können uns erstens sagen, ob die vermeintliche Erfassung eine wirkliche Erfassung ist. Zweitens aber kann uns die konstitutive Betrachtung sagen, ob der Sinn (das Wesen) der Realität *notwendig* existiert (scil. objektiv im oben angedeuteten Sinne), rechtmässig ist oder nicht, *falls* die ursprünglichen Elemente des Noemas absolut festgestellt sind. Sollte dies wirklich der Fall sein, dann ist unser Resultat betreffs des Wesens der Realität gesichert. Im anderen Falle sind zwei Möglichkeiten vorhanden:

(1) die Erfassung war keine "wirkliche". Dann muss es eben ausgewiesen werden, dass und weswegen es der Fall ist; zugleich muss – falls die Idee der Realität keine in sich absurde ist (was ebenso auszuweisen wäre, wenn ...) – ein anderer, und zwar der "richtige" Sinn der Realität herausgestellt werden.

(2) Die Erfassung war eine "wirkliche" Erfassung, der Sinn der Realität ist der von uns intendierte, aber er ist nicht "notwendig". Er ist auf irgend etwas relativ, er ist eine "Konstruktion" und erschöpft sich darin. Es wäre in diesem Falle rechtmässig nicht zu "begründen", dass die Realität ein An sich-sein *ist*. Sie praetendierte freilich das zu sein, diesen Sinn zu haben, aber es wäre nur eine relative Vermeinung. Dann aber müsste zugleich gezeigt werden, wie es kommt, dass es zu dieser Praetention kommt (zu einem transzendentalen "Als-ob").

So mündet das ontologische Ausgangsproblem in eine erkenntnistheoretische Betrachtung. Bevor wir aber dazu übergehen, wollen wir andere ontologische Probleme der "Idealismus-Frage" andeuten.

Vorausgesetzt zunächst mal, dass der Sinn der Realität wirklich so zu fassen ist, wie ich es getan habe, und dass er auch in der konstitutiven Betrachtung erhalten bleibt, dass also alles Reale ein An sich-sein ist und die reale Aussenwelt ein dem Bewusstsein Wesensfremdes ist, entsteht zunächst eine ontologische Frage: Ist das Wesen der Realität ein selbständiges Wesen, und speziell ein dem Wesen des Bewusstseins gegenüber selbständiges Wesen – als *Wesen* – oder nicht?

Prinzipiell gesprochen liegen 4 Möglichkeiten vor:

(1) Realität (als Wesen) unselbständig, das Wesen-Bewusstsein selbständig, d.h. dass es ohne das Wesen Realität existieren könnte.

(2) Wesen der Realität und Wesen des Bewusstsein – selbständig.

(3) Wesen der Realität selbständig. Bewusstsein unselbständig.

(4) Beides unselbständig und aufeinander angewiesen.

So viel ich sehe verhalten sich die Sachen folgendermassen:

(Hier bricht der Text des Brulions des Briefes vom Juli 1918 ab.)

NOTES

[1] Bemerkung 1951: Die "Bergsonarbeit" war natürlich längst fertig – im Sommer 1918, da sie im Herbst 1917 als Doktorarbeit vorgelegt wurde, und das Examen fand am 16.I.1918 statt. Es handelt sich lediglich um die endgültige zum Druck vorbereitete Redaktion dieser Doktordissertation. Sie wurde erst im Juni 1920 an Husserl abgeschickt.

[2] Die letzte Bedeutung des Wortes kann nicht in Betracht kommen da ein Wesen als Wesen zur Realität nicht in der oben angedeuteten Transzendenzbeziehung stehen kann.